Energy Transfers by Conduction

Energy Engineering Set

coordinated by
Abdelhanine Benallou

Volume 2

Energy Transfers by Conduction

Abdelhanine Benallou

iSTE

WILEY

First published 2018 in Great Britain and the United States by ISTE Ltd and John Wiley & Sons, Inc.

ISTE Ltd
27-37 St George's Road
London SW19 4EU
UK

www.iste.co.uk

John Wiley & Sons, Inc.
111 River Street
Hoboken, NJ 07030
USA

www.wiley.com

Library of Congress Control Number: 2018944776

British Library Cataloguing-in-Publication Data
A CIP record for this book is available from the British Library
ISBN 978-1-78630-275-5

MIX
Paper from responsible sources
FSC® C013604
FSC www.fsc.org

Contents

Preface

"Never regret the time that was needed for doing good."

Joseph Joubert,
"De la sagesse [Wisdom]",
Pensées, essais et maximes [Thoughts, Essays and Maxims],
Chapter IX, 1866

For several years, I have cherished the wish of devoting enough time to the writing of a series of books on energy engineering. The reason is simple: for having practiced for years teaching as well as consulting in different areas ranging from energy planning to rational use of energy and renewable energies, I have always noted the lack of formal documentation in these fields to constitute a complete and coherent source of reference, both as a tool for teaching to be used by engineering professors and as a source of information summarizing, for engineering students and practicing engineers, the basic principles and the founding mechanisms of energy and mass transfers leading to calculation methods and design techniques.

But between the teaching and research tasks (first as a teaching assistant at the University of California and later as a professor at the École des mines de Rabat, Morocco) and the consulting and management endeavors conducted in the private and in the public sectors, this wish remained for more than twenty years in my long list of priorities, without having the possibility to make its way up to the top. Only providence was able to unleash the constraints and provide enough time to achieve a lifetime objective.

This led to a series consisting of nine volumes:

– *Volume 1*: Energy and Mass Transfers;

– *Volume 2*: Energy Transfers by Conduction;

– *Volume 3*: Energy Transfers by Convection;

– *Volume 4*: Energy Transfers by Radiation;

– *Volume 5*: Mass Transfers and Physical Data Estimation;

– *Volume 6*: Design and Calculation of Heat Exchangers;

– *Volume 7*: Solar Thermal Engineering;

– *Volume 8*: Solar Photovoltaic Energy Engineering;

– *Volume 9*: Rational Energy Use Engineering.

The present book is the *second volume* of this series. It concerns the study of conduction heat transfer.

As we will see, the calculation methods established in this book present multiple applications in engineering: heat exchanger design and sizing, energy conservation, thermal insulation, metal tempering, etc.

A series of exercises is presented at the end of the book, aimed at enabling students to implement new concepts as rapidly as possible. These exercises are designed to correspond as closely as possible to real-life situations occurring in industrial practice or everyday life.

Abdelhanine BENALLOU
June 2018

Introduction

Conduction is one of the most important heat exchange mechanisms. It occurs by contact-continuity between systems. Energy is transferred, step by step through a medium, by transmission of the excitations of an atom to the neighboring atoms.

The present volume from the *Energy Engineering* set is devoted to studying heat conduction in detail. It highlights the equations that govern the phenomenon, and then applies them to a number of practical situations. In this context, special attention is paid to calculation methods used in analyses aiming at *energy conservation*. This involves, for example, calculating energy fluxes dissipated through the inner walls of industrial furnaces, financial assessment of heat losses and the choice of the economic thickness for thermal insulation.

Within the same context, great importance is placed in calculations allowing the implementation of *regulations of thermal installations in buildings* which define building energy efficiency requirements that are mandated by different countries. Implementation of these *regulations usually* involves determining the heat losses passing through the walls of heated or air-conditioned buildings.

Such applications are not only important to observe applicable laws and regulations, but they generally prove to be instrumental in researching energy savings in furnaces, steam pipes, reactors, freezing circuits, etc.

Chapter 1 of this volume is dedicated to establishing general equations translating energy balances in conduction, in different coordinate systems (plane, cylindrical and spherical). Applying a balance approach, conduction equations are put in place for the most general situations: systems in transient or stationary state, with or without internal energy generation.

Subsequently (Chapter 2), these equations are used to solve a number of physical problems of interest to the engineer: heat transfer through a simple, plane wall, loss through a multi-layer wall composed of several inner walls and transfer through cylindrical jackets (steam pipes, for example) or spherical containers (storage reservoirs). The choice of the different application examples presented in this chapter was made based on problems usually encountered in industrial practices or in sustainable-development approaches.

Chapter 3 of this volume is reserved for thermal insulation. Within this context, special attention has been devoted to heat lagging materials, given their importance in actions aiming at reducing energy loss. The appendix provided at the end of this document presents a set of data that offers guidance in choosing which insulation to use. Moreover, this data enables the reader to perform the calculations required in order to validate a lagging material for a given project. Indeed, the technical and economic analyses carried out in this chapter show that thermal insulation can lead to important reductions in overall production costs.

Chapter 4 is reserved for thermal analyses of buildings where the study of heat losses through the walls (which may or not be glazed) of a building leads to the estimation of the annual costs of these energy losses. These analyses are conducted bearing in mind that, in different countries, new thermal regulations impose rigorous constraints on the permitted heat losses of new constructions.

Next, Chapter 5 examines conduction problems in the presence of energy generation. They correspond to practical situations that are encountered in exothermic reactors or in fuel tubes of nuclear reactors, or in radiators used to dissipate surplus heat resulting from electronic circuit operation.

Chapter 6 details the different methods for solving conduction problems in transient state. They provide answers to common engineering problems, particularly in the field of metal tempering (Cengel, 1998).

The last chapter of this volume is devoted to presenting a number of exercises and their solutions. Note that when choosing these exercises, our aim was to stay as close as possible to the reality of industrial operations and to practical energy engineering problems.

1

Fundamental Equations of Conduction

1.1. Introduction

In this chapter we explore the general equations that reflect energy balances in conduction. These equations will then be used to solve a number of physical problems of interest to the engineer.

Equations governing conduction are extremely useful, both for quantifying energy losses in industrial installations (steam pipes, furnaces, reactors, etc.) and for the application of thermal building regulations. In each case, a comprehensive implementation of these equations makes it possible to calculate the fluxes of energy losses if no action was taken. The analysis methods presented in the following sections offer the possibility of quantifying energy flows, as well as indicating the actions to implement in order to limit energy loss.

For each of the situations that are explored, special attention is paid to the practical applications of the equations established. Thus, problems relating to the thermal insulation of industrial installations are approached both technically, by assessing the fluxes of energy losses, and with regard to the profitability of the investments to be made in order to achieve thermal insulation. For each of the situations explored, special attention is paid to the practical applications of the equations established. Thus, thermal insulation problems are concretely dealt with by means of technical and economic analyses, which demonstrate the benefit in terms of reducing production costs.

In this chapter, the problems relating to conduction in the presence of energy generation are also examined. They correspond to practical situations encountered in exothermic reactors or in nuclear-reactor fuel bars.

1.2. General equations of conduction

Consider a solid, of arbitrary shape, in which we assume that there is conduction heat transfer.

In order for the development to be general, we will assume that there is heat generation in the solid, and that P is the generation power per unit volume.

Let us establish an energy balance on a microscopic element of this solid with sides dx, dy and dz (see Figure 1.1).

The general balance equation is as follows:

Input – Output + Generation = Accumulation

which can be written simply as:

$$I - O + G = A$$

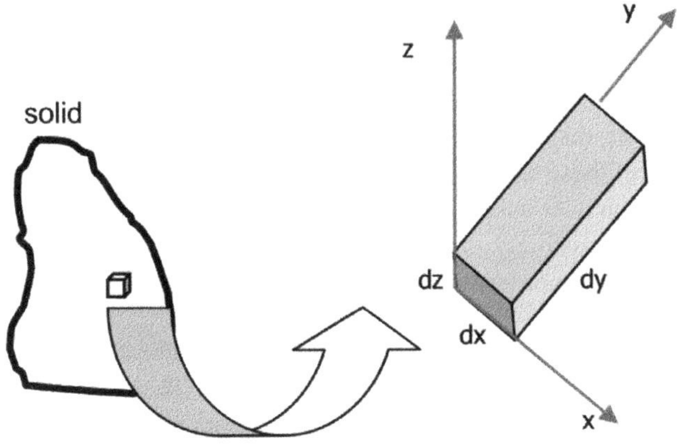

Figure 1.1. *Differential energy balance*

1.2.1. *Expressing the term* (I – O)

In the most general case, energy flows in the three directions, *ox*, *oy* and *oz*. Consequently, its flux entering or exiting the solid element considered will have three components: one component for each direction of propagation. These three components will be indicated as follows:

ϕ_x: flux in direction x

ϕ_y: flux in direction y

ϕ_z: flux in direction z

The flux vector is then written: $\mathbf{\Phi} = \begin{pmatrix} \phi_x \\ \phi_y \\ \phi_z \end{pmatrix}$.

1.2.1.1. *Expressing ϕ_x*

Conduction heat input is the energy flux in the direction x, expressed at abscissa x. It is given by: $\phi_{x_{\downarrow x}} = -\lambda_x (dydz)\left(\frac{\partial T}{\partial x}\right)_{\downarrow x}$.

Similarly, the output corresponds to the energy flux in the direction x, expressed at abscissa $x + dx$: $\phi_{x_{\downarrow x+dx}} = \phi_{x_{\downarrow x}} + \left(\frac{\partial \phi_x}{\partial x}\right)_{\downarrow x} dx$.

Hence: $(I-O)_x = \phi_{x_{\downarrow x}} - \phi_{x_{\downarrow x+dx}} = -\left(\frac{\partial \phi_x}{\partial x}\right)_{\downarrow x} dx.$

Therefore: $(I-O)_x = \frac{\partial}{\partial x}\left[\lambda_x \frac{\partial T}{\partial x}\right] dxdydz.$

1.2.1.2. *Expressing ϕ_y*

Using a similar development we obtain: $(I-O)_y = \frac{\partial}{\partial y}\left[\lambda_y \frac{\partial T}{\partial y}\right] dxdydz.$

1.2.1.3. *In direction z*

We have: $(I-O)_z = \frac{\partial}{\partial z}\left[\lambda_z \frac{\partial T}{\partial z}\right] dxdydz.$

1.2.1.4. *For the three directions*

When we consider the flow of energy in the three propagation directions, the term *(I – O)* is then given by: $I-O=(I-O)_x+(I-O)_y+(I-O)_z$.

Therefore: $$I-O=\left[\frac{\partial}{\partial x}\left(\lambda_x\frac{\partial T}{\partial x}\right)+\frac{\partial}{\partial y}\left(\lambda_y\frac{\partial T}{\partial y}\right)+\frac{\partial}{\partial z}\left(\lambda_z\frac{\partial T}{\partial z}\right)\right]dxdydz.$$

NOTE.– The equation giving the term (I – O), established above, is valid in the general case where the conduction behavior of the solid is different along the three directions of propagation; i.e., the heat conductivity is different for each direction of the space, $\lambda x \neq \lambda y \neq \lambda z$. This is the case for orthotropic materials.

In practice, we often encounter materials where conduction is identical in the three directions of propagation: the heat conductivity λ is the same in the three directions. This is the case for *isotropic* materials. Thus, the equation giving the term *I – O* for isotropic conduction is given by:

$$I-O=\left[\frac{\partial}{\partial x}\left(\lambda\frac{\partial T}{\partial x}\right)+\frac{\partial}{\partial y}\left(\lambda\frac{\partial T}{\partial y}\right)+\frac{\partial}{\partial z}\left(\lambda\frac{\partial T}{\partial z}\right)\right]dxdydz$$

For the sake of clarity, we will continue developing the balance equation for the isotropy case; the development is identical in the general case.

1.2.2. *The term "generation"*

P being the volumetric generation power, the energy generated in the volume element dxdydz is given by: $\phi_G = P\,dx\,dy\,dz$.

1.2.3. *The term "accumulation"*

If *m designates the mass* of the volume element considered and if C_p is the sensible heat of the material constituting it, the energy accumulated in this element is given by: $\phi_A = mC_p\frac{\partial T}{\partial t}$

The mass, *m*, can be expressed as follows: $m = \rho\,dx\,dy\,dz$.

Therefore: $\phi_A = \rho C_p\frac{\partial T}{\partial t}dxdydz.$

1.2.4. *Energy balance equation*

The balance equation can then be written as follows for isotropic materials:

$$\rho C_p \frac{\partial T}{\partial t} = \frac{\partial}{\partial x}\left(\lambda \frac{\partial T}{\partial x}\right) + \frac{\partial}{\partial y}\left(\lambda \frac{\partial T}{\partial y}\right) + \frac{\partial}{\partial z}\left(\lambda \frac{\partial T}{\partial z}\right) + P$$

This energy balance equation is often referred to as the *general conduction equation*.

NOTE.–

i) In the case of isotropy: λ is a scalar value.

We then have: $$\varphi = \begin{pmatrix} \varphi_x \\ \varphi_y \\ \varphi_z \end{pmatrix} = \begin{pmatrix} -\lambda \frac{\partial T}{\partial x} \\ -\lambda \frac{\partial T}{\partial y} \\ -\lambda \frac{\partial T}{\partial z} \end{pmatrix} = -\lambda \nabla T$$

Thus, for isotropic materials with constant λ, the conduction equation becomes:

$$\left(\frac{\rho C_p}{\lambda}\right)\frac{\partial T}{\partial t} = \frac{\partial^2 T}{\partial x^2} + \frac{\partial^2 T}{\partial y^2} + \frac{\partial^2 T}{\partial z^2} + \frac{P}{\lambda}$$

This equation is usually written as follows, using the Laplace notation:

$$\frac{1}{\alpha}\frac{\partial T}{\partial t} = \nabla^2 T + \frac{P}{\lambda} \quad \text{where} \quad \alpha = \frac{\lambda}{\rho C_p}$$

α is called the thermal diffusivity of the solid considered. It is a physical property of the material considered. It is easily verified that the dimensions of α correspond to a surface per unit time:

$[\alpha] = [L]^2\,[T]^{-1}$

The thermal diffusivity could then be interpreted as a surface propagation speed of heat.

The thermal diffusivity values for the usual materials are presented in the Appendix. Some orders of magnitude of α:

α Cu = 0.4 m^2/hr

α Steel = 0.05 m^2/hr

α Glass = 1.96 10^{-3} m^2/hr

α Wood = 3.5 10^{-4} m^2/hr

It should be noted that the thermal diffusivities of copper and steel are two orders of magnitude greater than those of glass or cork.

ii) In the case of anisotropic (non-isotropic) materials, directional thermal conductivities are usually grouped into a matrix (Λ). The conduction equations can be obtained from the general equation presented in section 1.2.4, taking into consideration the variability of λ or, where applicable, using the thermal conductivity matrix corresponding to the case considered (McAdams, 1954; Sherwood and Reed, 1957).

iii) In the case of orthotropy

Orthotropic materials have different conductivities for each of the different directions. However, the conductivity is the same in a given direction with a rotational symmetry around each axis.

Mathematically, the thermal conductivity of an orthotropic material is a diagonal matrix Λ. The energy flux density is then as follows:

$$\varphi = \begin{pmatrix} \varphi_x \\ \varphi_y \\ \varphi_z \end{pmatrix} = -\begin{pmatrix} \lambda_x & 0 & 0 \\ 0 & \lambda_y & 0 \\ 0 & 0 & \lambda_z \end{pmatrix} \begin{pmatrix} \frac{\partial T}{\partial x} \\ \frac{\partial T}{\partial y} \\ \frac{\partial T}{\partial z} \end{pmatrix} = -\Lambda \bullet \nabla T$$

Consequently, the conduction equation is given by:

$$\rho C_p \frac{\partial T}{\partial t} = \frac{\partial}{\partial x}\left(\lambda_x \frac{\partial T}{\partial x}\right) + \frac{\partial}{\partial y}\left(\lambda_y \frac{\partial T}{\partial y}\right) + \frac{\partial}{\partial z}\left(\lambda_z \frac{\partial T}{\partial z}\right) + P$$

iv) In the most general case of anisotropic materials, thermal conductivity is an arbitrary matrix (Λ). We then have:

$$\varphi = \begin{pmatrix} \varphi_x \\ \varphi_y \\ \varphi_z \end{pmatrix} = -\begin{pmatrix} \lambda_{xx} & \lambda_{xy} & \lambda_{xz} \\ \lambda_{yx} & \lambda_{yy} & \lambda_{yz} \\ \lambda_{zx} & \lambda_{zy} & \lambda_{zz} \end{pmatrix} \bullet \nabla T$$

In this context:

$$\rho C_p \frac{\partial T}{\partial t} = \frac{\partial}{\partial x}\left(\lambda_{xx}\frac{\partial T}{\partial x}\right) + \frac{\partial}{\partial y}\left(\lambda_{xy}\frac{\partial T}{\partial y}\right) + \frac{\partial}{\partial z}\left(\lambda_{xz}\frac{\partial T}{\partial z}\right) + \frac{\partial}{\partial y}\left(\lambda_{yx}\frac{\partial T}{\partial x}\right)$$

$$+\frac{\partial}{\partial y}\left(\lambda_{yy}\frac{\partial T}{\partial y}\right) + \frac{\partial}{\partial z}\left(\lambda_{yz}\frac{\partial T}{\partial z}\right) + \frac{\partial}{\partial z}\left(\lambda_{zx}\frac{\partial T}{\partial x}\right) + \frac{\partial}{\partial y}\left(\lambda_{zy}\frac{\partial T}{\partial y}\right) + \frac{\partial}{\partial z}\left(\lambda_{zz}\frac{\partial T}{\partial z}\right) + P$$

This case is rarely encountered in practice.

However, in most cases of interest to the engineer, the materials implemented can be considered isotropic, with thermal conductivities, λ, *variable or constant* (Brown and Marco, 1958). This is why the following section is dedicated to presenting the equation of conduction in different coordinate systems, for the isotropic case.

1.3. Equations of conduction in different coordinate systems

Anisotropic or orthotropic materials (variable thermal conductivity λ depending on the direction) are sometimes encountered in engineering calculations. This is particularly true when energy is called upon to propagate in wood or in salts of non-homogeneous crystalline structures. For such situations, the forms of the conduction equation that take account of this variability in thermal conductivity are to be used.

Yet in practice, we mainly encounter isotropic materials, that is, materials whose thermal conductivity, λ, is the same in all directions. Indeed, this is often the case for metals and alloys usually encountered in industrial equipment and devices. However, even in the isotropic case, λ is not necessarily constant. It is identical in all directions, but can vary with the temperature, for example. For this reason, this section is dedicated to presenting equations of conduction for the isotropic case, with either variable or constant thermal conductivities.

Moreover, the geometry of the solid determines which coordinate system to use. Thus, when the solid studied is a parallelepiped shape, for example, it makes sense to choose the Cartesian coordinates for analysis.

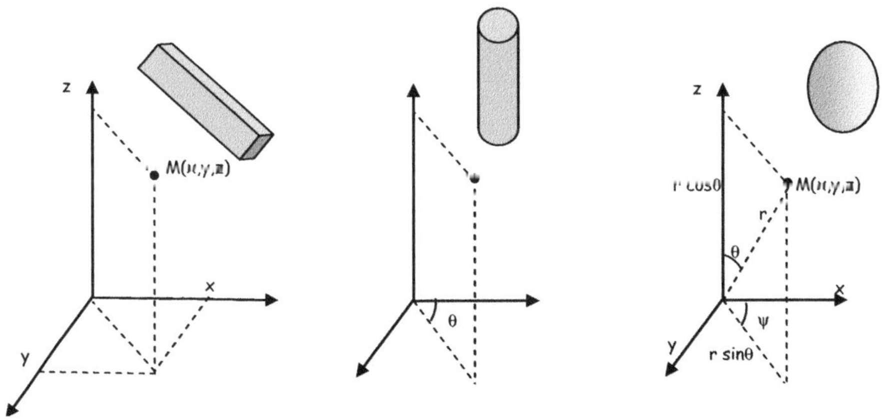

Figure 1.2. *The coordinate system depends on the shape of the solid considered*

Likewise, when the system studied is a pipe or a ball, we will opt for cylindrical or spherical coordinates, respectively.

The following sections present the equations of conduction in different coordinate systems for isotropic cases, but whose thermal conductivities can be either variable or constant.

1.3.1. *When λ is not constant*

– Rectangular coordinates:

$$\rho C_p \frac{\partial T}{\partial t} = \frac{\partial}{\partial x}\left(\lambda_x \frac{\partial T}{\partial x}\right) + \frac{\partial}{\partial y}\left(\lambda_y \frac{\partial T}{\partial y}\right) + \frac{\partial}{\partial z}\left(\lambda_z \frac{\partial T}{\partial z}\right) + P$$

– Cylindrical coordinates (r, θ, z):

$$\rho C_p \frac{\partial T}{\partial t} = \frac{1}{r}\frac{\partial}{\partial r}\left(\lambda r \frac{\partial T}{\partial r}\right) + \frac{1}{r^2}\frac{\partial}{\partial \theta}\left(\lambda \frac{\partial T}{\partial \theta}\right) + \frac{\partial}{\partial z}\left(\lambda \frac{\partial T}{\partial z}\right) + P$$

– Spherical coordinates (r, θ, ψ):

$$\rho C_p \frac{\partial T}{\partial t} = \frac{1}{r^2}\frac{\partial}{r}\left(\lambda r^2 \frac{\partial T}{\partial r}\right) + \frac{1}{r^2 \sin^2\theta}\frac{\partial}{\partial\psi}\left(\lambda \frac{\partial T}{\partial\psi}\right) + \frac{1}{r^2 \sin\theta}\frac{\partial}{\partial\theta}\left(\lambda \sin\theta \frac{\partial T}{\partial\theta}\right) + P$$

1.3.2. *When λ is constant*

– Rectangular coordinates:

$$\frac{1}{\alpha}\frac{\partial T}{\partial t} = \frac{\partial^2 T}{\partial x^2} + \frac{\partial^2 T}{\partial y^2} + \frac{\partial^2 T}{\partial z^2} + \frac{P}{\lambda}$$

– Cylindrical coordinates (r, θ, z):

$$\frac{1}{\alpha}\frac{\partial T}{\partial t} = \frac{\partial^2 T}{\partial r^2} + \frac{1}{r}\frac{\partial T}{\partial r} + \frac{1}{r^2}\frac{\partial^2 T}{\partial\theta^2} + \frac{\partial^2 T}{\partial z^2} + \frac{P}{\lambda}$$

– Spherical coordinates (r, θ, ψ):

$$\frac{1}{\alpha}\frac{\partial T}{\partial t} = \frac{\partial^2 T}{\partial r^2} + \frac{2}{r}\frac{\partial T}{\partial r} + \frac{1}{r^2 \sin^2\theta}\frac{\partial^2 T}{\partial\psi^2} + \frac{1}{r^2}\frac{\partial^2 T}{\partial\theta^2} + \frac{1}{r^2 \mathrm{tg}\theta}\frac{\partial T}{\partial\theta} + \frac{P}{\lambda}$$

1.3.3. *Simplified cylindrical and spherical coordinates*

Very often, temperature variations are confined to only two directions, and sometimes even only one. The latter cases are frequently encountered when the heat propagates from the center of a pipe to the outside under the action of a large temperature gradient in the r direction, while only a small gradient exists in the longitudinal direction. This also happens in the case of a sphere that is in a medium or bath of homogeneous temperature: it is obvious that, in this context, the existence of a temperature gradient between the center of the sphere and the bath will result in heat propagation along r only; due to the symmetry of the problem and the homogeneity of the bath, the other directions will not be involved in the transfer.

In these situations, simplified coordinates r and z are used for cylindrical systems, and r and θ, for spherical systems.

– Simplified cylindrical coordinates (r, z):

$$\frac{1}{\alpha}\frac{\partial T}{\partial t}=\frac{\partial^2 T}{\partial r^2}+\frac{1}{r}\frac{\partial T}{\partial r}+\frac{\partial^2 T}{\partial z^2}+\frac{P}{\lambda}$$

– Simplified spherical coordinates (r, θ):

$$\frac{1}{\alpha}\frac{\partial T}{\partial t}=\frac{\partial^2 T}{\partial r^2}+\frac{2}{r}\frac{\partial T}{\partial r}+\frac{1}{r^2\sin\theta}\frac{\partial}{\partial\theta}\left(\sin\theta\frac{\partial T}{\partial\theta}\right)+\frac{P}{\lambda}$$

1.3.4. *One-dimensional conduction*

In most problems of interest to the engineer, using a single heat-propagation direction is sufficient to be able to describe fairly accurately what happens in practice (Jakob and Hawkins, 1957).

Thus, the equations of conduction presented previously in Cartesian, cylindrical and spherical coordinates can be further simplified when a single propagation direction is retained. This direction is generally x for Cartesian coordinates and r for cylindrical and spherical coordinates, in which case we will speak of one-dimensional or one-directional problems. This type of problem is commonly encountered as a simplification in engineering calculations. In this case, the equations are:

– In rectangular coordinates: $\frac{1}{\alpha}\frac{\partial T}{\partial t}=\frac{d^2 T}{dx^2}+\frac{P}{\lambda}$

– In simplified cylindrical coordinates: $\frac{1}{\alpha}\frac{\partial T}{\partial t}=\frac{d^2 T}{dr^2}+\frac{1}{r}\frac{dT}{dr}+\frac{P}{\lambda}$

– In simplified spherical coordinates: $\frac{1}{\alpha}\frac{\partial T}{\partial t}=\frac{d^2 T}{dr^2}+\frac{2}{r}\frac{dT}{dr}+\frac{P}{\lambda}$

1.4. Reading: metal tempering

Tempering is a heat treatment used in metallurgy to improve the mechanical properties of metals. It enables the hardness or the overall resistance of a given metal to be increased. This treatment generally consists of heating the metal to raise its temperature, then cooling it at a given speed. The thermal stresses thus imposed lead to changes in the metal's crystalline structure.

As a result, with each metal having a crystalline form within which atoms of a so-called "interstitial" compound can be inserted, the solubility of these atoms in the mesh depends on the structure of the latter and its size. If a metal's temperature variations are conducted so that the phase change point is approached, they can then lead to changes in the crystalline structure, in particular with respect to its interstitial distances. We will thus pass from a mesh having important sites to a tighter mesh imprisoning interstitial atoms in the new crystalline structure.

Bearing in mind that the mechanical properties are closely related to the geometries of the crystalline meshes, it follows that the different temperature variations induce a stress in these meshes, and thus a change in the mechanical properties of the metal considered.

The origin of metal tempering dates back to the early 20th Century, when blacksmiths subjected the sharpened parts of a tool to this type of heat treatment, in order to increase their mechanical strength. The metal was heated to red before being immersed in water.

The temperature to which the metal is heated before cooling is significant. It is called the "phase change temperature". It must be chosen in such a way that it allows the "suspension" of the interstitial chemical compounds dissolved in the metal's crystalline structure.

The heating time, or the time during which the metal is subjected to high temperatures, is also significant. It must be long enough to ensure the transformation of the entire heated mass, but it must not be too long either.

The cooling time, for its part, is also a decisive parameter. Indeed, the latter must allow cooling of the entire mass considered at a great enough speed to lead to the imprisonment of interstitial chemical elements having diffused into the crystalline structure of the solid during the heating phase. This cooling operation is generally performed by soaking the metal part in a cooling fluid. The fluids used are either water or salt water to generate faster cooling.

As we will see in Chapter 6, the cooling speed will depend on the following three elements:

– The heat transfer in the metal considered, and therefore its thermal conductivity.

– The nature of the fluid used.

– The heat transfer between the tempering fluid and the solid, and therefore the convection heat transfer coefficient, h.

In Chapter 6 of this volume we will see how the conduction equations are used to accurately calculate the time required for a certain degree of heat penetration into a solid.

Knowledge of this time period will permit heating the material to a sufficiently high temperature for just long enough and, next, cooling it at an appropriate speed.

In this manner, chemical elements will be imprisoned in the crystalline structure creating, after cooling, tensions in the meshes which will contribute to improving the mechanical resistance of the tempered part.

Although mechanical resistance of metals can be augmented through tempering, it should be noted, however, that this thermal treatment often causes metals to lose their elasticity. Indeed, the modifications induced by tempering in the meshes of the crystalline structure constrain movements of the crystalline irregularities (vacancy defects or dislocations) which are important for elasticity. Thus, tempering results in a loss of the plastic deformation mechanism that allows a metal to return to its resting state following a deformation. Actually, tempering brings a metal's elastic resistance closer to its failure resistance.

2

Conduction in Steady State and Applications

2.1. Introduction

In this chapter, we focus on the study of conduction problems in steady or stationary state. In these cases, temperature will not be a function of time, but it will nevertheless depend on the coordinates that define the position of a given point of the solid considered.

Studying steady-state conduction is of particular importance for the engineer. Indeed, as we will see later on in this chapter, several practical problems come under this category.

But before going on to solve practical problems, we first suggest going over the simplifications applied to the equations of conduction when steady state is taken into consideration.

2.2. Equations of conduction in steady state

In steady state, the term $\frac{\partial T}{\partial t}$ is nil.

The general equation of conduction thus becomes: $\nabla^2 T + \frac{P}{\lambda} = 0$.

2.2.1. *Expressions in the different coordinate systems*

In different coordinate systems, the conduction equation is obtained by setting $\frac{\partial T}{\partial t} = 0$, in the expressions of the general equation of conduction that are presented in section 1.3. This gives the following equations:

– In rectangular coordinates:

$$\frac{\partial^2 T}{\partial x^2} + \frac{\partial^2 T}{\partial y^2} + \frac{\partial^2 T}{\partial z^2} + \frac{P}{\lambda} = 0$$

– In cylindrical coordinates:

$$\frac{\partial^2 T}{\partial r^2} + \frac{1}{r}\frac{\partial T}{\partial r} + \frac{1}{r^2}\frac{\partial^2 T}{\partial \theta^2} + \frac{\partial^2 T}{\partial z^2} + \frac{P}{\lambda} = 0$$

– In spherical coordinates:

$$\frac{\partial^2 T}{\partial r^2} + \frac{2}{r}\frac{\partial T}{\partial r} + \frac{1}{r^2 \sin\theta}\frac{\partial}{\partial \theta}\left(\sin\theta \frac{\partial T}{\partial \theta}\right) + \frac{1}{r^2 \sin^2\theta}\frac{\partial^2 T}{\partial \psi^2} + \frac{P}{\lambda} = 0$$

2.2.2. *Simplifications in the case of one-dimensional conduction*

As we have already pointed out, in most situations of interest to the engineer, it is sufficient to use a single heat propagation direction.

Thus, in the case where a single direction of heat propagation is considered, the steady-state conduction equations are given by:

– In rectangular coordinates:

$$\frac{d^2 T}{dx^2} + \frac{P}{\lambda} = 0$$

– In simplified cylindrical coordinates:

$$\frac{d^2 T}{dr^2} + \frac{1}{r}\frac{dT}{dr} + \frac{P}{\lambda} = 0$$

– In simplified spherical coordinates:

$$\frac{d^2T}{dr^2}+\frac{2}{r}\frac{dT}{dr}+\frac{P}{\lambda}=0$$

As indicated, it is possible to solve several practical problems in steady state by applying the conduction equation. In particular, such an application makes it possible to calculate heat loss fluxes through the walls of industrial equipment (furnaces, boilers, reactors, etc.) or building walls. In the following sections we will study several cases of one-dimensional conduction, corresponding to several practical situations.

2.3. Applying to single-layer walls

Consider a wall of thickness, *e*, and thermal conductivity, λ, (see Figure 2.1) the inner and outer faces of which are at temperatures T_1 and T_2, respectively. We wish to assess the flux of energy transferred by conduction through this wall.

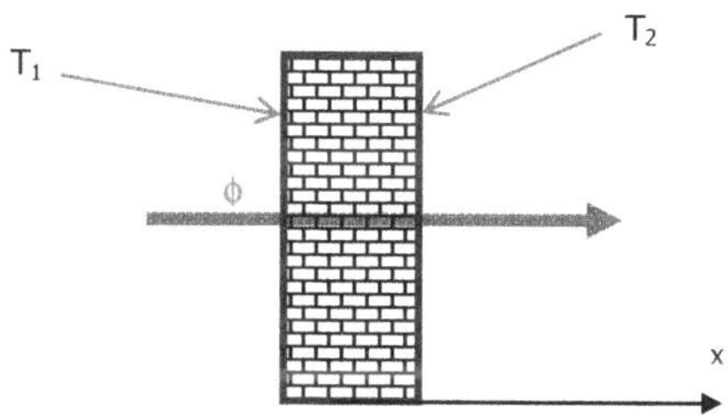

Figure 2.1. *Conduction through a single-layer wall*

It will be assumed that heat propagates in the *x* direction only.

As we have already pointed out, the study of conduction through a single-layer wall corresponds to several situations of practical interest for the engineer. This is the case, for example, when we examine heat losses through the wall of an industrial machine or across a building's concrete envelope.

Thus, with a single propagation direction, the equation of conduction for a single-layer wall is written:

$$\frac{d^2T}{dx^2}+\frac{P}{\lambda}=0$$

Assuming there is no generation ($P = 0$), the equation becomes:

$$\frac{d^2T}{dx^2} = 0$$

After a double integration, we obtain the temperature profile inside the wall:

$$T = c_1 x + c_2$$

c_1 and c_2 are two integration constants determined from the boundary conditions:

CL_1:	at $x = 0$:	T: T_0
CL_2:	at $x = e$:	T: T_1,

hence: $c_2 = T_0$, and $c_1 = \frac{T_1 - T_0}{e}$.

The temperature profile inside the wall is then fully determined as a function of x:

$$T = \frac{T_1 - T_0}{e} x + T_0$$

Knowledge of the temperature profile makes it possible to identify the flow of energy through the wall as follows:

$$\phi = -\lambda S \frac{dT}{dx}$$

i.e., the flux expression for a single-layer wall: $\phi = \frac{\lambda S}{e}(T_0 - T_1)$

or the flux density, φ: $\varphi = \lambda \frac{T_0 - T_1}{e}$.

2.4. Concept of thermal resistance

Let us rewrite the flux expression in the case of a single-layer wall:

$$\phi = \frac{\lambda S}{e}(T_0 - T_1)$$

Remember that the term $T_o - T_l$ corresponds to a thermal potential difference. By analogy with electrical systems (see Figure 2.2), the thermal resistance, R_{th}, is defined by: $R_{th} = \frac{e}{\lambda S}$.

Subsequently, the flux can be written as a potential difference divided by the thermal resistance:

$$\phi = \frac{T_0 - T_1}{R_{th}}$$

This expression is similar to that which gives the electric current, i, as a function of the potential difference, $V_0 - V_1$:

$$i = \frac{V_0 - V_1}{R}$$

Figure 2.2. *Electrical analogy of a single-layer wall*

2.5. Applying to composite or multi-layer walls

While the single-layer wall represents an interesting practical case for the engineer, study of conduction through walls composed of several layers of different materials presents even greater importance. Indeed, in most practical cases, a wall consists of a series or a succession of layers, each representing, in theory, a single-layer wall (see Figure 2.3). As a matter of fact, a furnace wall is usually constituted of several envelopes including refractory bricks, metal protection, thermal insulation, etc. The same remark stands for building walls: very often the building envelope is constituted by a couple of walls separated by insulation. Moreover, external stone coverings and internal paint or decoration layers constitute additional thermal resistances.

Figure 2.3 shows a typical situation of a multi layer wall.

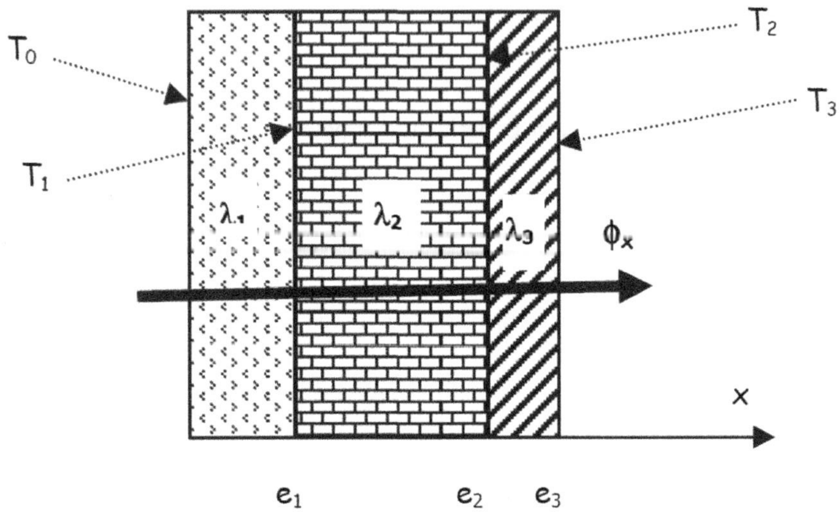

Figure 2.3. *Conduction through a composite wall*

The composite wall shown in Figure 2.3 consists of three layers of different materials, of thermal conductivities λ_1, λ_2 and λ_3 respectively. Each of these layers will constitute a thermal resistance.

Let us assume that temperatures T_0, T_1, T_2 and T_3 are known.

As in the previous case, our objective is to determine the expression of the flux, ϕ_x, of the energy passing through the wall.

Assuming that heat propagation is one-dimensional (preferentially in the x direction), the flux, ϕ_x, passing through the composite wall can be expressed in different ways, as a function of one of the thermal resistances:

– as a function of R_1: $\phi_x = \dfrac{T_0 - T_1}{R_1}$ with $R_1 = \dfrac{e_1}{\lambda_1 S}$;

– as a function of R_2: $\phi_x = \dfrac{T_1 - T_2}{R_2}$ with $R_2 = \dfrac{e_2}{\lambda_2 S}$;

– as a function of R_3: $\phi_x = \dfrac{T_2 - T_3}{R_3}$ with $R_3 = \dfrac{e_3}{\lambda_3 S}$.

In a similar way to that used for electrical resistances in series, we show here that *thermal resistances in series add up to give an equivalent resistance, R_{eq}*. Indeed, equations expressing the fluxes as functions of thermal resistances may be rewritten as follows:

$$T_0 - T_1 = R_1 \phi_x$$

$$T_1 - T_2 = R_2 \phi_x$$

$$T_2 - T_3 = R_3 \phi_x$$

It should be noted that the same flux ϕ_x passes through the different resistances.

By adding member to member we obtain: $T_0 - T_3 = (R_1 + R_2 + R_3)\ \phi_x$.

i.e.: $\phi_x = \dfrac{T_0 - T_3}{R_1 + R_2 + R_3}$, or: $\phi_x = \dfrac{T_0 - T_3}{\dfrac{e_1}{\lambda_1 S} + \dfrac{e_2}{\lambda_2 S} + \dfrac{e_3}{\lambda_3 S}}$.

Thus, we can define equivalent resistance R_{eq} such that flux ϕ_x is written:

$$\phi_x = \frac{T_0 - T_3}{R_{eq}}$$

As a result, the equivalent resistance is defined in this case by:

$$R_{eq} = \sum_{i=1}^{3} R_i = \frac{e_1}{\lambda_1 S} + \frac{e_2}{\lambda_2 S} + \frac{e_3}{\lambda_3 S}$$

In general, the flux will be given, for *n* walls in series, by:

$$\phi_x = \frac{T_0 - T_n}{\sum_{i=1}^{n} R_i}$$

We deduce the following rule therefrom:

RULE.– For walls in series, *the thermal resistances are added up.*

2.6. Applying to cylindrical walls

Cylindrical walls are frequently used in industry in hot-fluid pipes, steam pipes, fuel rods in nuclear reactors, fuel storage tanks, etc.

The study of conduction in cylindrical walls is of great importance, particularly since it allows the heat dissipations through these walls to be calculated.

Consider the cylindrical conductor of thermal conductivity, λ, represented in Figure 2.4. Assume that the temperatures, T_1, of the inner wall ($r = r_1$) and T_2 of the outer wall ($r = r_2$) are known.

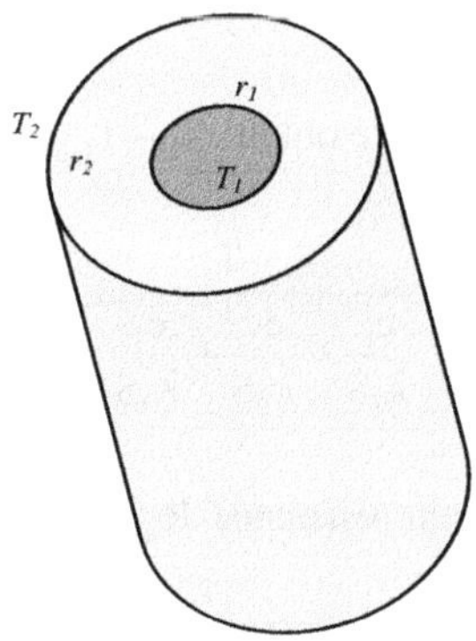

Figure 2.4. *Cylindrical wall*

Given that in such situations heat propagates essentially in direction r, the steady-state conduction equation is given by:

$$\frac{d^2T}{dr^2} + \frac{1}{r}\frac{dT}{dr} + \frac{P}{\lambda} = 0$$

As we are not considering any generation in the wall, we have: $P = 0$.

The equation of conduction is then simplified to: $\frac{d^2T}{dr^2} + \frac{1}{r}\frac{dT}{dr} = 0$,

with the general boundary conditions:

CL_1: at $r = r_1$: $T = T_1$

CL_2: at $r = r_2$: $T = T_2$

For integration purposes we proceed with a function change: consider function $u(r)$, defined by:

$$u(r) = \frac{dT}{dr}$$

Consequently: $\frac{du(r)}{dr} = \frac{d^2T}{dr^2}$.

The differential equation becomes: $\frac{du(r)}{u(r)} = -\frac{dr}{r}$.

Integration of this equation then yields: $Ln(u) = -Ln(r) + K = Ln\left(\frac{1}{r}\right) + K$, where *K* is an integration constant.

Taking the exponential of the two members of the equation, we obtain: $u = e^K\left(\frac{1}{r}\right)$.

Or alternatively: $\frac{dT}{dr} = \frac{A}{r}$, where: $A = e^K$.

A second integration gives: $T(r) = A\ Ln(r) + B$.

Integration constants, *A* and *B*, determined using the boundary conditions, are as follows:

$$A = \frac{T_2 - T_1}{Ln\left(\frac{r_2}{r_1}\right)} \quad \text{and} \quad B = T_1 - (T_2 - T_1)\frac{Ln(r_1)}{Ln\left(\frac{r_2}{r_1}\right)}$$

Thus, the temperature in the tube wall satisfies the following distribution:

$$T(r) = T_1 + \frac{(T_2 - T_1)}{Ln\left(\frac{r_2}{r_1}\right)}Ln(r) - (T_2 - T_1)\frac{Ln(r_1)}{Ln\left(\frac{r_2}{r_1}\right)}$$

Or: $T(r) = T_1 + (T_2 - T_1)\left[\dfrac{\text{Ln}\left(\dfrac{r}{r_1}\right)}{\text{Ln}\left(\dfrac{r_2}{r_1}\right)}\right]$.

The flux expression is then determined by derivation of the temperature:

$$\phi(r) = -\lambda S(r)\left(\frac{dT}{dr}\right)$$

with $S(r) = 2\pi rL$, where L is the length of the pipe considered.

Yet: $\dfrac{dT}{dr} = \dfrac{T_2 - T_1}{r\ \text{Ln}\left(\dfrac{r_2}{r_1}\right)}$.

Hence: $\phi(r) = -\lambda\ (2\pi rL)\dfrac{T_2 - T_1}{r\ \text{Ln}\left(\dfrac{r_2}{r_1}\right)}$.

i.e.: $\phi = 2\pi L\lambda\ \dfrac{T_1 - T_2}{\text{Ln}\left(\dfrac{r_2}{r_1}\right)}$.

Or the flux per unit length of the pipe: $\dfrac{\phi}{L} = 2\pi\lambda\ \dfrac{T_1 - T_2}{\text{Ln}\left(\dfrac{r_2}{r_1}\right)}$.

NOTES.–

i) Neither ϕ nor $\dfrac{\phi}{L}$ depend on r.

ii) The flux density, φ, on the other hand, does depend on r.

Indeed: $\varphi(r) = \frac{\phi(r)}{S(r)} = -\lambda \ \frac{dT}{dr} = \lambda \frac{T_1 - T_2}{r \ Ln\left(\frac{r_2}{r_1}\right)}$.

iii) The flux may be rewritten by introducing the thermal resistance: $\phi = \frac{\Delta T}{R_{th}}$.

We deduce therefrom the definition of thermal resistance for cylindrical walls: $R_{th} = \frac{1}{2\pi L\lambda} Ln\left(\frac{r_2}{r_1}\right)$

2.7. Applying to composite cylindrical walls

In practice, composite cylindrical walls are more often encountered than single-layer cylindrical walls. Indeed, pipes transporting hot fluids or steam are usually covered with a layer of thermal insulation and, in most cases, other layers of materials providing mechanical protection. Thus, the cylindrical wall is composed both of the material constituting the pipe, and of the different insulating or protective layers.

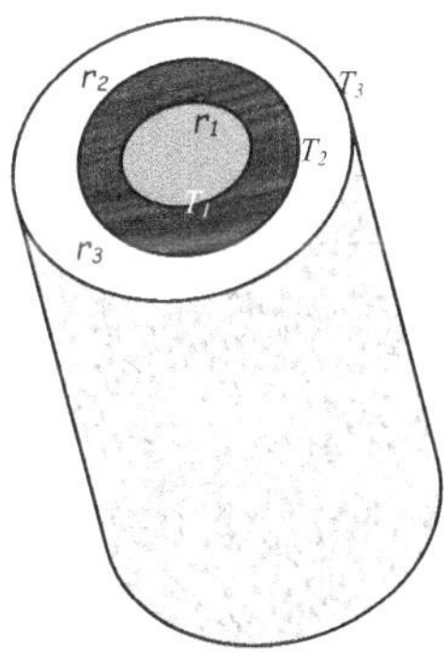

Figure 2.5. *Composite cylindrical wall*

Each of these cases represent situations involving composite cylindrical pipes, similar to that represented in Figure 2.5.

If we consider a cylindrical wall composed of two thicknesses defined by radii r_1, r_2 and r_3, each layer will correspond to a thermal resistance. We will therefore have a series of two layers of materials.

If we apply similar reasoning to that used in section 2.5, it will be readily demonstrated that the resistances of cylindrical walls in series are added together. The thermal flux, ϕ, will then be expressed as follows, as a function of the driving potential difference, T_1-T_3, and the equivalent thermal resistance, R_{eq}:

$$\phi = \frac{T_1 - T_3}{R_{eq}} = \frac{T_1 - T_3}{\sum_{i=1}^{2} R_{th_i}},$$

with: $R_{th_i} = \frac{1}{2\pi L \lambda_i} \mathrm{Ln}\left(\frac{r_{i+1}}{r_i}\right)$.

i.e.: $\phi = \frac{2\pi\, L\, (T_1 - T_3)}{\frac{1}{\lambda_1}\mathrm{Ln}\left(\frac{r_2}{r_1}\right) + \frac{1}{\lambda_2}\mathrm{Ln}\left(\frac{r_3}{r_2}\right)}$.

2.8. Applying to spherical walls

In industry, we often encounter conduction through spherical walls, notably in storage reservoirs.

The study of conduction in spherical walls is of particularly great importance since it allows for calculation of the heat dissipations through these walls.

Consider the spherical conductor of thermal conductivity, λ, represented in Figure 2.6. Whilst not reducing the general scope of the problem studied, let us assume that the temperatures, T_1, of the inner wall ($r = r_1$) and T_2 of the outer wall ($r = r_2$) are known.

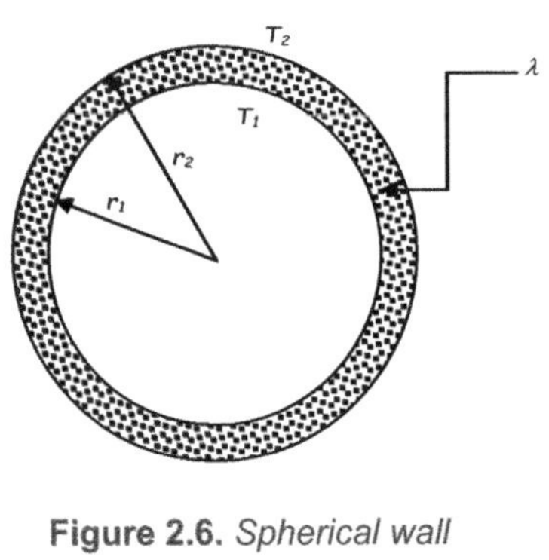

Figure 2.6. *Spherical wall*

In such situations, with the heat propagating essentially in direction r, the steady-state conduction equation will be given by:

$$\frac{d^2T}{dr^2}+\frac{2}{r}\frac{dT}{dr}+\frac{P}{\lambda}=0$$

Hence, after integration: $T=-\frac{K_1}{r}\frac{dT}{dr}+K_2$.

With the boundary conditions:

CL_1: $r = r_1$; $T = T_1$
CL_2: $r = r_2$; $T = T_2$

Hence: $K_1=r_1r_2\frac{T_1-T_2}{r_1-r_2}$ and $K_2=T_2+r_1\frac{T_1-T_2}{r_1-r_2}$.

i.e.: $$\frac{T-T_2}{T_1-T_2}=\frac{1-\frac{r_2}{r}}{1-\frac{r_2}{r_1}}.$$

The flux density is given by: $\varphi(r)=-\lambda\frac{dT}{dr}$.

By differentiating the temperature profile, we obtain: $\frac{dT}{dr}=\frac{K_1}{r^2}=\frac{r_1r_2}{r^2}\frac{T_1-T_2}{r_1-r_2}$.

Hence: $\varphi(r)=-\lambda\left(\frac{r_1r_2}{r^2}\right)\left(\frac{T_1-T_2}{r_1-r_2}\right)$.

The flux density is therefore in $\frac{1}{r^2}$.

We note that the flux, ϕ, on the other hand, is independent of r. Indeed:

$$\phi(r)=4\pi r^2\varphi(r)=-4\pi\lambda r^2\left(\frac{r_1r_2}{r^2}\right)\left(\frac{T_1-T_2}{r_1-r_2}\right)$$

i.e.: $\phi(r)=4\pi\lambda r_1r_2\left(\frac{T_1-T_2}{r_2-r_1}\right)$.

NOTE.–

The flux may be rewritten as follows, by introducing the thermal resistance:

$$\phi = \frac{\Delta T}{R_{th}} = (T_1 - T_2)\frac{4\pi\lambda\ r_1 r_2}{r_2 - r_1}$$

We deduce therefrom the definition of thermal resistance for spherical walls:

$$R_{th} = \frac{r_2 - r_1}{4\pi\lambda\ r_1 r_2}$$

2.9. Case of composite spherical walls

In practice, composite spherical walls are more often encountered than single-layer spherical walls. Indeed, spherical storage reservoirs are usually covered with a layer of thermal insulation and, in most cases, other layers of materials providing mechanical protection. We thus find ourselves in situations similar to those represented in Figure 2.7.

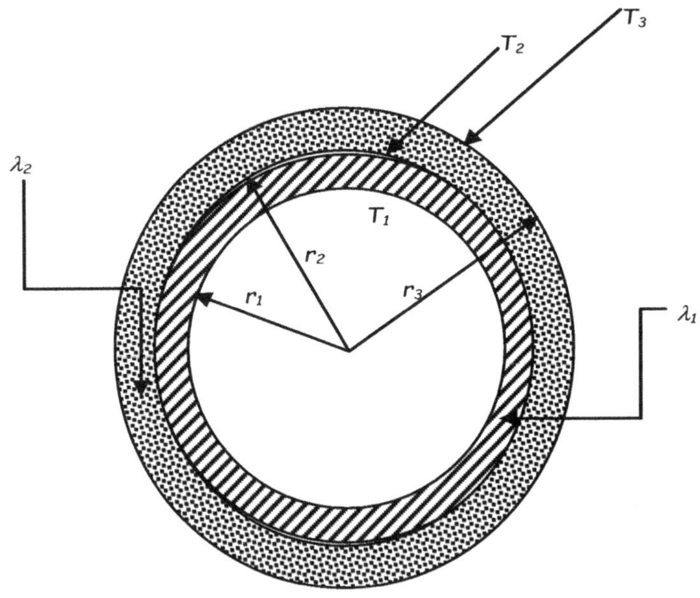

Figure 2.7. *Spherical storage reservoir*

Thus, where we consider a spherical wall composed of two thicknesses defined by radii r_1, r_2 and r_3, each layer will have a corresponding thermal resistance, defined by:

$$R_{thi} = \frac{r_{i+1} - r_i}{4\pi\lambda\ r_i r_{i+1}}$$

In the situation being considered, we have two layers of materials in series. Using the same kind of reasoning as that applied in sections 2.5 and 2.7, we can readily demonstrate that the resistances of spherical walls in series are added together. The thermal flux, ϕ, is then expressed as follows, as a function of the driving potential difference, T_1-T_3, and the equivalent thermal resistance, R_{eq}:

$$\phi = \frac{T_1 - T_3}{R_{eq}} = \frac{T_1 - T_3}{\sum_{i=1}^{2} R_{th_i}}$$

$$\text{i.e.: } \phi = \frac{4\pi\ \lambda\ (T_1 - T_3)}{\frac{1}{\lambda_1}\frac{r_2 - r_1}{r_1 r_1} + \frac{1}{\lambda_2}\frac{r_3 - r_2}{r_3 r_2}}.$$

NOTE.–

i) This expression will make it possible to calculate the flux density, φ, of the heat losses towards the outside, provided that the temperatures are known, that is, temperatures T_1 of the inner wall and T_3 of the outer surface. In general, T_1 and T_3 are not directly known. The temperatures usually given are those of the fluid located inside the storage tank, T_i, and the ambient air, T_e.

ii) Furthermore, temperatures T_1 and T_3 are linked to temperatures T_i and T_e by the equations of convection, which determine thermal exchanges between the internal fluid and the inner surface of the reservoir on the one hand, and between the outer surface and the surrounding environment on the other. These conditions, known as "convective boundary conditions" are of the type:

$\varphi = h_e S(T_3 - T_e)$: for transfer between the outer wall, which is at temperature T_3, and the surrounding environment, which is at T_e;

$\varphi = h_i S(T_i - T_1)$: for transfer between the fluid, which is at T_i, and the inner wall, at temperature T_1

The temperatures of the inner and outer faces are thus determined using the convective boundary conditions.

iii) This remark is applicable to spherical walls, cylindrical pipes and plane walls.

The following sections present a study of the problems with convective-type boundary conditions, for single-layer walls, cylindrical pipes and spherical walls, respectively.

2.10. Convective-type boundary conditions: case of a single-layer wall

Let us consider the case of a single-layer wall of thermal conductivity λ, subjected to internal and external convections (see Figure 2.8). We will assume that the wall has a thickness Δx, and that the external and internal fluids are at temperatures T_e and T_i, respectively.

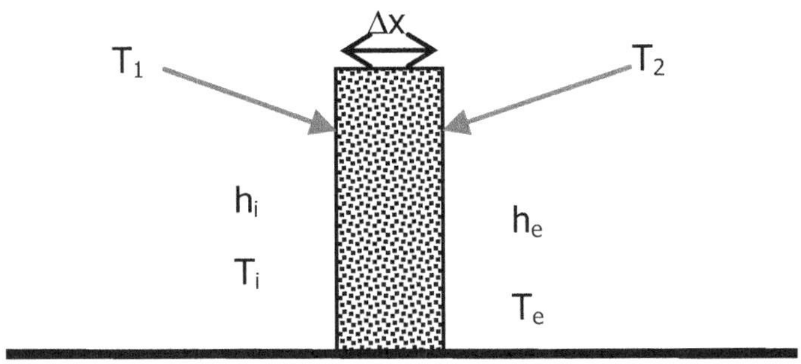

Figure 2.8. *Single-layer wall with internal and external convections*

In a situation such as this, the heat flux encounters *three thermal resistances*:

– an *internal convection* resistance;

– a resistance to *conduction in the wall*;

– an *external convection* resistance.

These three resistances will need to be taken into account in order to determine the flux as a function of T_i and T_e.

2.10.1. *Internal convection resistance*

Using the convection equation, the flux, ϕ, may be expressed as a function of the temperature difference $(T_i - T_1)$ by the relation:

$$\phi = h_i A (T_i - T_1),$$

A being the transfer area.

Let us define the *internal convection resistance* as follows: $R_i = \dfrac{1}{h_i A}$.

We therefore obtain: $\phi = \dfrac{T_i - T_1}{R_i}$.

2.10.2. *External convection resistance*

We can also express the flux, ϕ, as a function of temperatures T_2 and T_e. In this case, we will introduce the convection resistance on the outside of the wall, R_e:

$$\phi = \frac{T_2 - T_e}{R_e} \text{ with } R_e = \frac{1}{h_e A}$$

2.10.3. *Conduction resistance*

ϕ can also be expressed as a function of temperatures T_1 and T_2 alone. The resistance to be taken into consideration is that which exists between these two points, that is: the resistance to conduction in the wall: R_c.

$$\text{i.e.: } \phi = \frac{T_1 - T_2}{R_c} \text{ with } R_c = \frac{\Delta x}{\lambda A}.$$

2.10.4. *Expressing the flux as a function of T_i and T_e*

The above expressions of the flux can be rewritten as a function of the equivalent resistance (resistances in series):

$$\phi = \frac{T_i - T_e}{R_i + R_c + R_e} \quad \text{i.e.: } \phi = \frac{T_i - T_e}{\dfrac{1}{h_i A} + \dfrac{\Delta x}{\lambda A} + \dfrac{1}{h_e A}}$$

Figure 2.9 presents the equivalent electrical diagram in this case:

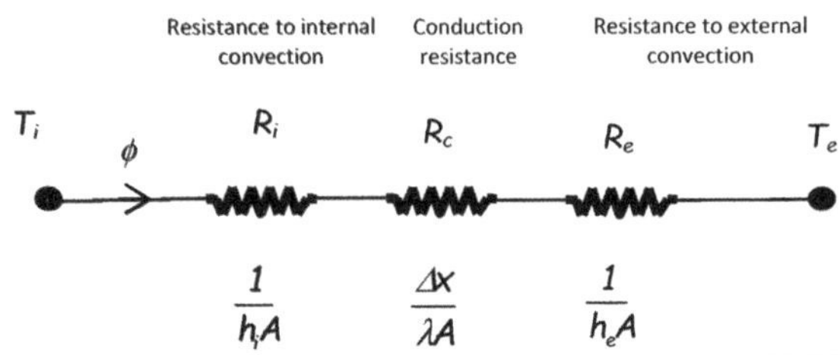

Figure 2.9. *Electrical analogy of a wall with convections*

2.10.4.1. *Illustration: calculating heat losses through a building wall*

An architect in charge of a new project explains to you that, for esthetical reasons, he envisages building the wall which separates the rooms from the outside as a single shield composed of special bricks of thermal conductivity λ. The air temperature inside the rooms is T_i, whereas the temperature of the external surrounding environment is equal to T_e.

We will assume $T_e < T_i$.

Questions

Before going further in his project, the architect requires an order of magnitude of the heat losses towards the outside. He asks you to:

1) calculate the amount of heat that would be lost to the outside per m^2 of wall and per year;

2) calculate the temperatures of the inner and outer surfaces of the wall: T_1 and T_2.

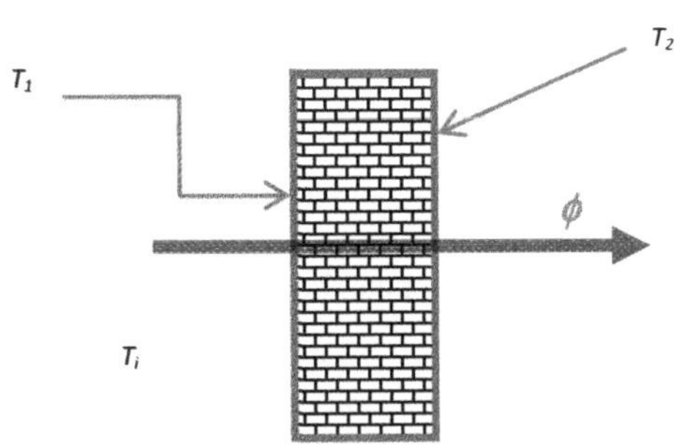

Figure 2.10. *Heat losses through a building wall*

The architect provides you with the following data:

$T_i = 22$ °C $\quad$ $T_e = 8$ °C

$h_i = 65$ W/m^2 °C $\quad$ $h_e = 85$ W/m^2 °C

thermal conductivity of the wall: $\lambda = 7$ W/m °C

wall thickness: $e = 30$ cm

Solutions

<u>*1) Heat lost per m^2 of wall and per year*</u>

Using A to designate the transfer area between the wall and the outside, the heat flux is given by:

$$\phi = \frac{T_i - T_e}{\frac{1}{h_i A} + \frac{e}{\lambda A} + \frac{1}{h_e A}}$$

Or: $\phi = \dfrac{(T_i - T_e)A}{\frac{1}{h_i} + \frac{e}{\lambda} + \frac{1}{h_e}}$.

i.e. the flux per m^2 of wall: $\varphi = \dfrac{\phi}{A} = \dfrac{(T_i - T_e)}{\frac{1}{h_i} + \frac{e}{\lambda} + \frac{1}{h_e}}$.

NUMERICAL APPLICATION.–

$T_i = 22$ °C $\quad$ $T_e = 8$ °C

$\lambda = 7$ W/m °C $\quad$ $e = 30$ cm

$h_i = 65$ W/m^2 °C $\quad$ $h_e = 85$ W/m^2 °C

Hence: $\dfrac{\phi}{A} = \dfrac{(22-8)}{\frac{1}{65} + \frac{0.3}{7} + \frac{1}{85}}$

$\varphi \approx 200$ W/m^2 $\varphi = 1{,}752$ kWh/m^2/year

2) Calculating T_1 and T_2

To calculate T_1, we use the expression of flux density as a function of $T_i - T_1$:

$$\phi = h_i (T_i - T_1)$$

Hence: $T_1 = T_i - \frac{\varphi}{h_i}$

> NUMERICAL APPLICATION.–
>
> $T_i = 22$ °C
>
> $h_i = 65$ W/m^2 °C
>
> Hence: $T_1 = 22 - \frac{200}{65}$
>
> **$T_1 = 18.9$ °C**

Likewise, to calculate T_2, we use the expression of flux density as a function of $T_2 - T_e$: $\varphi = h_e(T_2 - T_e)$

Hence: $T_2 = T_e + \frac{\varphi}{h_e}$

> NUMERICAL APPLICATION.–
>
> $T_e = 8$ ° $Ch_e = 85$ W/m^2 °C
>
> Hence: $T_2 = \frac{200}{85} + 8$ **$T_2 = 10.3$ °C**

2.11. Composite walls with convective boundary conditions

Figure 2.11 presents a composite wall with internal and external convections.

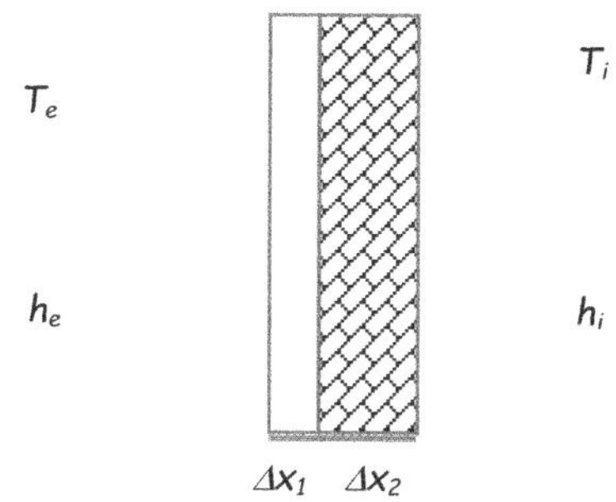

Figure 2.11. *Composite wall with internal and external convections*

If we use A to identify the transfer area, the flux is obtained by dividing the temperature difference, T_i - T_e, by the equivalent thermal resistance, i.e.:

$$\phi = \frac{T_i - T_e}{\frac{1}{h_i A} + \frac{\Delta x_1}{\lambda_1 A} + \frac{\Delta x_2}{\lambda_2 A} + \frac{1}{h_e A}}$$

This relation may be generalized for a wall composed of n layers of thickness, e_j ($1 \leq j \leq n$), and thermal conductivities, λ_j ($1 \leq j \leq n$), respectively, subject to convective boundary conditions:

$$\phi = \frac{T_i - T_e}{\frac{1}{h_i A} + \sum_{j=1}^{n} \frac{e_j}{\lambda_j A} + \frac{1}{h_e A}},$$

where:

h_i and h_e are the convection transfer coefficients, on the inside and on the outside of the wall, respectively.

A is the transfer area between the wall and the outside.

2.11.1. *Illustration: calculating heat losses through the walls of an industrial furnace*

The wall of an industrial furnace may be considered to be composed of the following four layers:

– a first layer having a thickness, e_1, is composed of refractory bricks of thermal conductivity, λ_1;

– a second (thickness, e_2) is constituted of clay bricks of thermal conductivity, λ_2;

– a third: glass wool (thickness, e_3, thermal conductivity, λ_3);

– a mechanical protective sheath of thickness, e_4, composed of steel of thermal conductivity, λ_4.

The temperature inside the furnace is assumed to be constant and equal to T_i.

Questions

The plant manager wishes to get an idea of the extent of the energy flux passing from the furnace to the outside.

You know the average temperature of the surrounding environment, T_a. Your supervisor therefore asks you to calculate:

1) the flux density of heat losses to the outside, were the furnace jacket to be composed of thicknesses e_1 and e_4 only;

2) the flux density of heat losses to the outside, were the jacket to be composed of all of the inner walls, e_1, e_2, e_3 and e_4;

3) the monetary value of these losses, in Dh per year, in both cases.

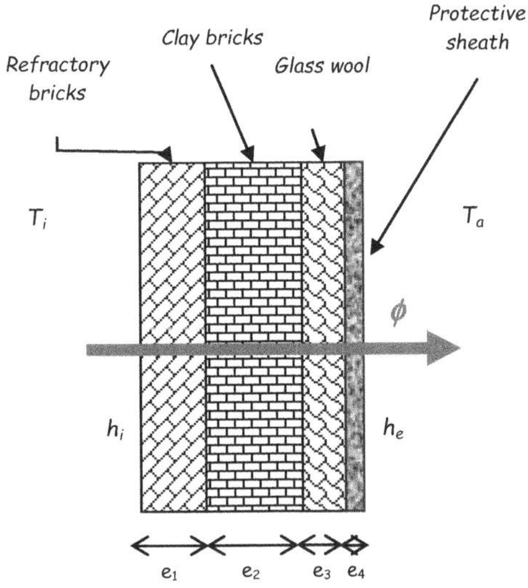

Figure 2.12. *Energy losses through the walls of an industrial furnace*

The technician in charge of running the furnace provides you with the following data:

$T_i = 650$ °C $T_a = 32$ °C

$h_i = 90$ W/m^2 °C $h_e = 48$ W/m^2 °C

$e_1 = 30$ cm $\lambda_1 = 27$ W/m °C

$e_2 = 35$ cm $\lambda_2 = 5.2$ W/m °C

$e_3 = 15$ cm $\lambda_3 = 0.04$ W/m °C

$e_4 = 0.3$ cm $\lambda_4 = 45$ W/m °C

Annual operating time: $\tau = 8{,}700$ hrs/year

Energy cost: $C_E = 3.2$ Dh/kWh

Transfer area between the furnace and the outside: $A = 60$ m^2

Solutions

1) Flux density when the furnace jacket is composed of e_1 and e_4 only

In this case, there are four thermal resistances:

– internal convection resistance: $R_i = \dfrac{1}{h_i A}$;

– two conductive resistances: $R_1 = \dfrac{e_1}{\lambda_1 A}$ and $R_4 = \dfrac{e_4}{\lambda_4 A}$; and

– an external convection resistance: $R_e = \dfrac{1}{h_e A}$.

The flux, ϕ_1, is then given by: $\phi_1 = \dfrac{T_i - T_a}{\dfrac{1}{h_i A} + \dfrac{e_1}{\lambda_1 A} + \dfrac{e_4}{\lambda_4 A} + \dfrac{1}{h_e A}}$.

Hence the flux density: $\varphi_1 = \dfrac{T_i - T_a}{\dfrac{1}{h_i} + \dfrac{e_1}{\lambda_1} + \dfrac{e_4}{\lambda_4} + \dfrac{1}{h_e}}$.

NUMERICAL APPLICATION.–

$T_i = 650$ °C $T_a = 32$ °C

$h_i = 90$ W/m^2 °C $h_e = 48$ W/m^2 °C

$e_1 = 30$ cm $\lambda_1 = 27$ W/m °C

$e_4 = 0.3$ cm $\lambda_4 = 45$ W/m °C

$$\varphi_1 = \frac{650-32}{\frac{1}{90}+\frac{0.3}{27}+\frac{0.003}{45}+\frac{1}{48}}$$ $\boldsymbol{\varphi_1 = 14.33}$ **kW/m²**

2) Flux density of the thermal losses towards the outside

In this case, there are six thermal resistances:

– the internal convection resistance: $R_i = \dfrac{1}{h_i A}$;

– four conductive resistances: $R_j = \dfrac{e_j}{\lambda_j A}$ with $1 \le j \le 4$; and

– the external convection resistance: $R_e = \dfrac{1}{h_e A}$.

The flux, ϕ_2, is then given by: $\phi_2 = \dfrac{T_i - T_a}{\dfrac{1}{h_i A} + \sum_{j=1}^{4} \dfrac{e_j}{\lambda_j A} + \dfrac{1}{h_e A}}$.

Hence the flux density: $\varphi_2 = \dfrac{T_i - T_a}{\dfrac{1}{h_i} + \dfrac{e_1}{\lambda_1} + \dfrac{e_2}{\lambda_2} + \dfrac{e_3}{\lambda_3} + \dfrac{e_4}{\lambda_4} + \dfrac{1}{h_e}}$.

NUMERICAL APPLICATION.–

$T_i = 650$ °C $T_a = 32$ °C

$h_i = 90$ W/m² °C $h_e = 48$ W/m² °C

$e_1 = 30$ cm $\lambda_1 = 27$ W/m °C

$e_2 = 35$ cm $\lambda_2 = 5.2$ W/m °C

$e_3 = 15$ cm $\lambda_3 = 0.04$ W/m °C

$e_4 = 0.3$ cm $\lambda_4 = 45$ W/m °C

$$\varphi_2 = \frac{650-32}{\frac{1}{90}+\frac{0.3}{27}+\frac{0,35}{5.2}+\frac{0.15}{0.04}+\frac{0.003}{45}+\frac{1}{48}}$$ **φ_2 = 160 W/m²**

3) Monetary value of the losses in both cases

a) Wall composed of e_1 and e_4 only

In this context:

– the flux of heat losses is: $\phi_1 = A\ \varphi_1$;

– the amount of energy lost per year is: $Q_1 = \tau\ \phi_1 = \tau\ A\ \varphi_1$;

– the cost of these losses is therefore obtained by multiplying Q_1 by the energy unit cost, C_E, i.e.:

$$C_1 = C_E\ \tau\ A\ \varphi_1$$

NUMERICAL APPLICATION.–

φ_1 = 14.33 kW/m²

Annual operating time: τ = 8,700 hrs/year

Energy cost: C_E = 3.2 Dh/kWh

Transfer area between the furnace and the outside: A = 60 m²

Thus: C_1 = (3.2) 8,700 (60) 14.33 **C_1 = 23,94 Million Dh/year**

b) Wall composed of four thicknesses: e_1, e_2, e_3 and e_4

Within this context, the flux density is noted φ_2.

As in the previous case, the cost of the losses takes the following form:

$$C_2 = C_E\ \tau\ A\ \varphi_2$$

NUMERICAL APPLICATION.–

φ_2 = 0.16 kW/m²

Annual operating time: τ = 8,700 hrs/year

Energy cost: C_E = 3.2 Dh/kWh

Transfer area between the furnace and the outside: $A = 60$ m^2

Thus: $C_2 = (3.2)$ 8,700 (60) 0.16 **C_2 = 267.64 Dh/year**

2.12. *Parallel resistances with convective boundary conditions*

Figure 2.13 presents another type of composite wall that, whilst uncommon, could be encountered in certain building walls or in some rare cases of industrial equipment walls.

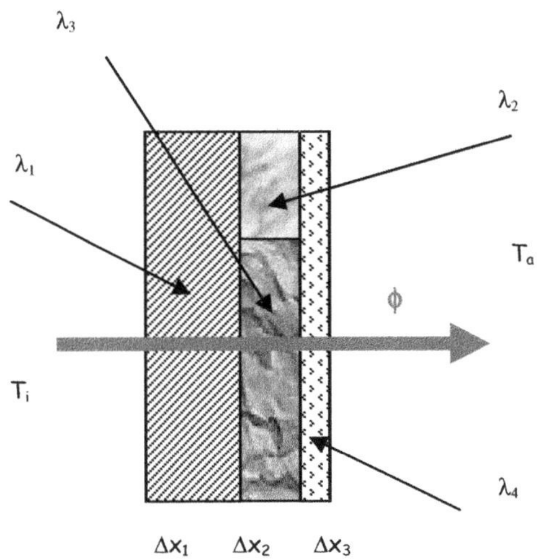

Figure 2.13. *Complex composite wall with internal and external convections*

The transfer area associated with the walls of thermal conductivity λ_1 and λ_4 is noted A.

The walls of conductivities λ_2 and λ_3, on the other hand, have transfer areas that differ from A. We will note them A_2 and A_3, respectively, with $A_2 + A_3 = A$.

In relation to conduction heat transfer, the walls of thermal conductivities λ_2 and λ_3 can be considered as thermal resistances arranged in parallel. For this configuration, the electrical analog is set out in Figure 2.14.

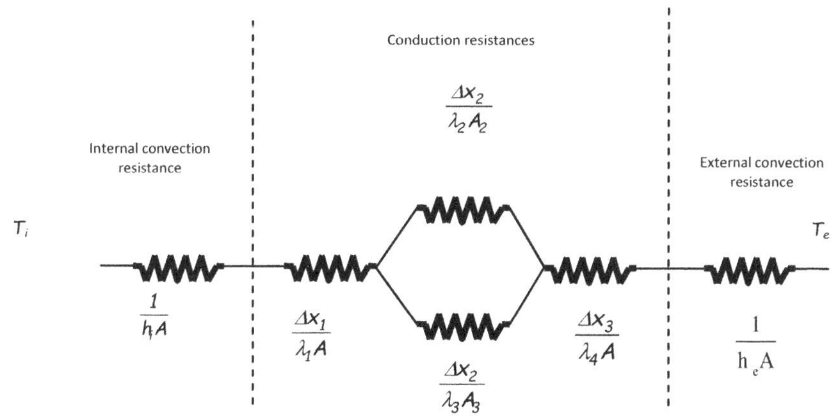

Figure 2.14. *Electrical analogy*

In this case, the equivalent resistance of the two walls in parallel, A_2 and A_3, is obtained by calculating the total of the inverses of the individual resistances, i.e.:

$$\frac{1}{R_{\lambda_2+\lambda_3}} = \frac{\lambda_2 A_2}{\Delta x_2} + \frac{\lambda_3 A_3}{\Delta x_2}$$

or: $R_{\lambda_2+\lambda_3} = \frac{\Delta x_2}{\lambda_2 A_2 + \lambda_3 A_3}$.

The flux is then given by: $\phi = \dfrac{T_i - T_e}{\dfrac{1}{h_i A} + \dfrac{\Delta x_1}{\lambda_1 A} + \dfrac{\Delta x_2}{\lambda_2 A_2 + \lambda_3 A_3} + \dfrac{\Delta x_3}{\lambda_4 A} + \dfrac{1}{h_e A}}$.

2.12.1. *Illustration: composite wall with parallel thermal resistances*

The wall of an industrial induction furnace (see Figure 2.15) is composed of a first layer of steel (thickness Δx_1, thermal conductivity, λ_a). The second thickness, Δx_2 (see Figure 2.15) is composed of two materials of thermal conductivities λ_1 (transfer area S_1) and λ_2 (transfer area S_2), respectively. The third layer, of thickness Δx_3, is a galvanized sheet, of thermal conductivity λ_3, having to assure mechanical protection.

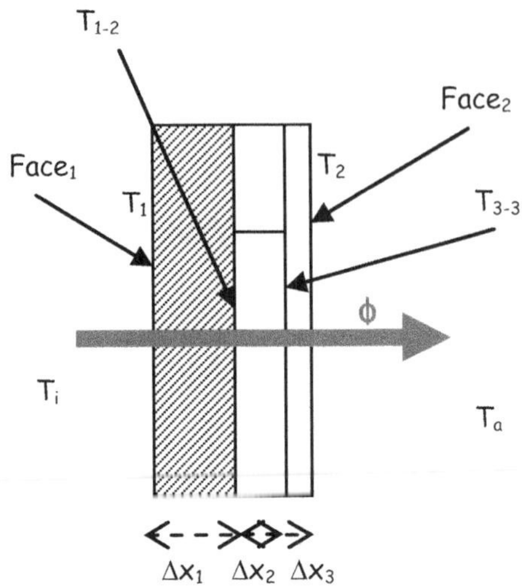

Figure 2.15. *Composite wall of an induction furnace*

The temperature of Face 1 of this wall, T_1, is assumed to be known. Temperature T_2 of Face 2, those of the interfaces and that prevailing inside the furnace, T_i , are unknown, however. Only the ambient temperature in the workshop, T_a, is known.

Questions

1) Give the electrical analog that represents the heat exchange between the inside of the machine and the surrounding environment. Be sure to represent T_i, T_a, and the different resistances.

2) Place on the electrical analog temperatures T_1 (of Face 1), T_2 (of Face 2) and those of interfaces: T_{1-2}, for the interface between the steel and the composite thickness and T_{2-3} for the interface between the composite thickness and the galvanized sheath.

3) Calculate the value of the different resistances to heat transfer.

4) Determine the equivalent resistance of the analog circuit, based on the parameters of the problem.

5) Deduce therefrom the expression of the flux density, φ, as a function of T_i, T_a, λ_1, λ_2, λ_3, Δx_1, Δx_2, Δx_3 and of the ratio, $\sigma = \frac{S_1}{S}$.

6) Calculate the value of φ.

7) Calculate the unknown temperatures, including $T_{1\text{-}2}$ and $T_{2\text{-}3}$.

The plant's design office provides you with the following data:

$\lambda_a = 65$ W/m °C $\quad\lambda_1 = 10$ W/m °C

$\lambda_2 = 7$ W/m °C $\quad\lambda_3 = 80$ W/m °C

$\Delta x_1 = 15$ cm $\quad\Delta x_2 = 8$ cm $\quad\Delta x_3 = 35$ mm

$h_i = 4.65$ kW/m^2 °C $\quad h_e = 0.55$ kW/m^2 °C

transfer areas: $S_1 = 1.25$ m^2; $S_2 = 0.5$ m^2; $S = S_1 + S_2 = 1.75$ m^2

known temperatures: $T_a = 28$ °C; $T_1 = 465$ °C

Solutions

1) Electrical analog circuit

The different resistances present are:

– the internal convection resistance, noted R_i;

– the conduction resistances in the walls, noted R_a, R_1, R_2 and R_3;

– the external convection resistance, R_e.

The electrical analog is then:

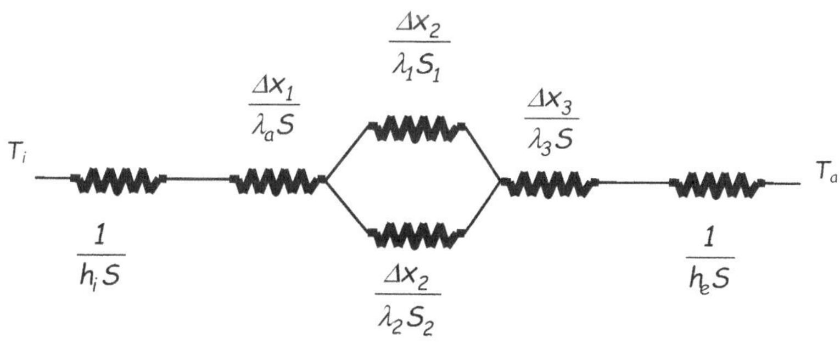

Figure 2.16. *Electrical analogy of the wall of an induction furnace*

2) Placing temperatures T_1, T_2, $T_{3\text{-}2}$ and $T_{2\text{-}1}$ on the electrical analog

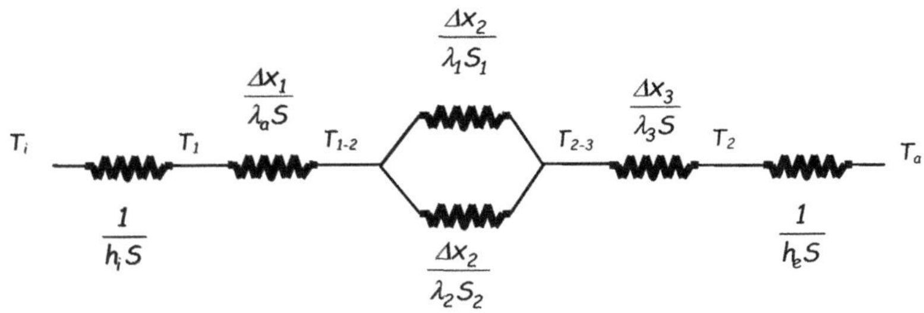

Figure 2.17. *Locations of the temperatures on the electrical analogy*

3) Calculating the thermal resistances

The internal convection resistance: $R_i = \frac{1}{h_i S}$.

NUMERICAL APPLICATION.–

$h_i = 4.65$ kW/m^2 °C$S = 1.75$ m^2

$R_i = \frac{1}{4.65(1.75)}$ $\quad$ $\boldsymbol{R_i = 0.123}$ **°C/kW**

The resistance to conduction in the steel: $R_a = \frac{\Delta x_1}{\lambda_a S}$.

NUMERICAL APPLICATION.–

$\lambda_a = 65$ W/m °C $\quad$ $\Delta x_1 = 15$ cm $\quad$ $S = 1.75$ m^2

$R_a = \frac{0.15}{65\ (1.75)}$ $\quad$ $\boldsymbol{R_a = 1.3\ 10^{-3}}$ **°C/kW**

Conduction resistance in the thermal conductivity material, λ_1:

$$R_1 = \frac{\Delta x_2}{\lambda_1 S_1}$$

NUMERICAL APPLICATION.–

$\lambda_1 = 10$ W/m °C $\Delta x_2 = 8$ cm $S_1 = 1.25$ m^2

$$R_1 = \frac{0.08}{10\ (1.25)}$$ $\boldsymbol{R_1 = 6.4\ 10^{-3}}$ **°C/kW**

Conduction resistance in the thermal conductivity material, λ_2:

$$R_2 = \frac{\Delta x_2}{\lambda_2 S_2}$$

NUMERICAL APPLICATION.–

$\lambda_2 = 7$ W/m °C $\Delta x_2 = 8$ cm $S_2 = 0.5$ m^2

$$R_2 = \frac{0.08}{7\ (0.5)}$$ $\boldsymbol{R_2 = 2.3\ 10^{-2}}$ **°C/kW**

Conduction resistance in the thermal conductivity material, λ_3:

$$R_3 = \frac{\Delta x_3}{\lambda_3 S}$$

NUMERICAL APPLICATION.–

$\lambda_3 = 80$ W/m °C $\Delta x_3 = 0.035$ m $S = 1.75$ m^2

$$R_3 = \frac{0.035}{80\ (1.75)}$$ $\boldsymbol{R_3 = 2.5\ 10^{-4}}$ **°C/kW**

External convection resistance:

$$R_e = \frac{1}{h_e S}.$$

NUMERICAL APPLICATION.–

$h_e = 0.55$ kW/m^2 °C $S = 1.75$ m^2

$$R_e = \frac{1}{0.55\ (1.75)}$$

$\boldsymbol{R_e = 1.04}$ **°C/kW**

4) Equivalent resistance of the analog circuit

$$R_{eq} = R_i + R_a + R_{1-2} + R_3 + R_e,$$

with $\dfrac{1}{R_{1-2}} = \dfrac{1}{R_1} + \dfrac{1}{R_2}$, or by replacing R_1 and R_2 with their expressions:

$$\frac{1}{R_{1-2}} = \frac{\lambda_1 S_1}{\Delta x_2} + \frac{\lambda_2 S_2}{\Delta x_2} = \frac{\lambda_1 S_1 + \lambda_2 S_2}{\Delta x_2}$$

i.e.: $R_{eq} = \dfrac{1}{h_i S} + \dfrac{\Delta x_1}{\lambda_a S} + \dfrac{\Delta x_2}{\lambda_1 S_1 + \lambda_2 S_2} + \dfrac{\Delta x_3}{\lambda_3 S} + \dfrac{1}{h_e S}$.

5) Expression of flux density

The transferred flux is given by:

$$\phi = \frac{T_i - T_a}{R_{eq}}$$

$$\phi = \frac{T_i - T_a}{\dfrac{1}{h_i S} + \dfrac{\Delta x_1}{\lambda_a S} + \dfrac{\Delta x_2}{\lambda_1 S_1 + \lambda_2 S_2} + \dfrac{\Delta x_3}{\lambda_3 S} + \dfrac{1}{h_e S}}$$

Hence: $\varphi = \dfrac{\phi}{S} = \dfrac{T_i - T_a}{\dfrac{1}{h_i} + \dfrac{\Delta x_1}{\lambda_a} + \dfrac{\Delta x_2}{\sigma(\lambda_1 - \lambda_2) + \lambda_2} + \dfrac{\Delta x_3}{\lambda_3} + \dfrac{1}{h_e}}$.

6) Calculating the value of φ.

As temperature T_i is unknown, we use the expression of the flux as a function of T_1 and T_a, which are known temperatures:

$$\varphi = \frac{\phi}{S} = \frac{T_1 - T_a}{S\, R_{éq_1}}$$

$$R_{eq_1} = R_a + R_{1-2} + R_3 + R_e$$

with $\dfrac{1}{R_{1-2}} = \dfrac{1}{R_1} + \dfrac{1}{R_2} = \dfrac{R_1 + R_2}{R_1 R_2}$.

i.e.: $R_{eq_1} = R_a + \dfrac{R_1 R_2}{R_1 + R_2} + R_3 + R_e$.

NUMERICAL APPLICATION.–

$T_1 = 465$ °C $\qquad T_2 = 28$ °C

$R_a = 1.3\ 10^{-3}$ °C/kW $\qquad R_1 = 6.4\ 10^{-3}$ °C/kW

$R_2 = 2.3\ 10^{-2}$ °C/kW $\qquad R_3 = 2.5\ 10^{-4}$ °C/kW $\qquad R_e = 1.04$ °C/kW

Hence: $R_{eq_1} = 1.3\ 10^{-3} + 0.023 + 2.5\ 10^{-4} + 1{,}04$ $\qquad R_{eq1} = 1.06$ °C/kW.

i.e., ultimately: $\varphi = \dfrac{T_1 - T_a}{S\ R_{eq_1}} = \dfrac{456 - 28}{1.75\ (1.06)}$. $\qquad$ **φ = 230.73 kW/m²**

7) Calculating the unknown temperatures

The only known temperatures are T_1 and T_a. Temperatures T_i, $T_{1\text{-}2}$, $T_{2\text{-}3}$ and T_2 will therefore need to be calculated.

These temperatures are obtained from the corresponding expressions of the flux, namely:

$$\phi = \frac{T_1 - T_a}{\dfrac{1}{h_i S} + \dfrac{\Delta x_1}{\lambda_a S} + \dfrac{\Delta x_2}{\lambda_1 S_1 + \lambda_2 S_2} + \dfrac{\Delta x_3}{\lambda_3 S} + \dfrac{1}{h_e S}} \quad \text{and} \quad \phi = h_i\, S\, (T_i - T_s)$$

To determine T_i, we use: $\phi = h_i S (T_i - T_1)$.

Hence: $T_i = T_1 + \dfrac{\varphi}{h_i}$

NUMERICAL APPLICATION.–

$T_1 = 465$ °C $\varphi = 230.73$ kW/m^2 $h_i = 4.65$ kW/m^2 °C

$T_i = 465 + \frac{230.73}{4.65}$ $T_i = 514.6$ °C

To determine $T_{1\text{-}2}$, we use: $\phi = \frac{T_1 - T_{1-2}}{R_a}$

Hence: $T_{1-2} = T_1 - S\,\varphi R_a$

NUMERICAL APPLICATION.–

$T_1 = 465$ °C $\varphi = 230.73$ kW/m^2 $S = 1.75$ m^2 $R_a = 1.3\ 10^{-3}$ °C/kW

$T_{1-2} = 514.6 - 230.73\ (1.75)\ (1.3\ 10^{-3})$ $T_{1\text{-}2} = 514.08$ °C

To determine $T_{2\text{-}3}$, we use: $\phi = \frac{T_{1-2} - T_{2-3}}{R_{1-2}} = \frac{(T_{1-2} - T_{2-3})(R_1 + R_2)}{R_1 R_2}$.

Hence: $T_{2-3} = T_{1-2} - \varphi S\ \frac{R_1 R_2}{R_1 + R_2}$

NUMERICAL APPLICATION.–

$T_{1\text{-}2} = 514.08$ °C $\varphi = 230.73$ kW/m^2 $S = 1.75$ m^2

$R_1 = 6.4\ 10^{-3}$ °C/kW $R_2 = 2.3\ 10^{-2}$ °C/kW

$T_{2-3} = 514.08 - 230.73\ (1.75)\ \frac{(6.4\ 10^{-3})(2.3\ 10^{-2})}{6.4\ 10^{-3} + 2.3\ 10^{-2}}$ $T_{2\text{-}3} = 512.06$ °C

To determine T_2, we use: $\phi = h_e S(T_2 - T_a)$.

Hence: $T_2 = T_a + \frac{\varphi}{h_e}$

NUMERICAL APPLICATION.–

$T_a = 28$ °C $\varphi = 230.73$ kW/m^2 $h_e = 0.55$ kW/m^2 °C

$$T_2 = 28 + \frac{230.73}{0.55} \qquad T_2 = 447.51 \text{ °C}$$

2.13. Composite cylindrical pipes with convective boundary conditions

Figure 2.18 shows a composite cylindrical pipe with internal and external convections. This configuration corresponds to a cylindrical pipe of thermal conductivity, λ_1, of internal radius, r_1, and external radius, r_2, transporting a fluid of average temperature, T_i. The pipe is covered with a protective layer of radius, r_3. The outer face of this layer is in contact with the surrounding environment, the temperature of which is T_e.

The convection transfer coefficients inside the pipe, h_i, and between the outer face and the surrounding environment, h_e, are assumed to be known.

When $T_i > T_e$, an energy flux, ϕ, will pass from the inside to the outside of the tube.

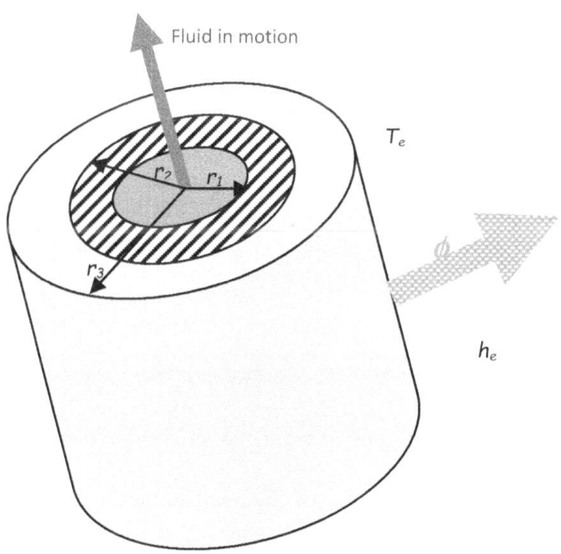

Figure 2.18. *Pipe with internal and external convections. For a color version of this figure, see www.iste.co.uk/benallou/energy2.zip*

Under these conditions, we have four resistances in series:

– the internal convection resistance, with a convective coefficient, h_i;

– the two conduction resistances in the two solid layers of thermal conductivities, λ_1 and λ_2, respectively;

– the external convection resistance, with a convective coefficient, h_e.

As the resistances are in series, the flux is obtained by dividing the extreme temperature difference by the sum of the resistances:

$$\phi = \frac{T_i - T_e}{R_i + R_{\lambda_1} + R_{\lambda_2} + R_e}$$

If L is the length of the pipe being considered, the internal and external convective resistances, respectively, are given by:

$$R_i = \frac{1}{2\pi r_1 L h_i} \quad \text{and} \quad R_e = \frac{1}{2\pi r_3 L h_e}$$

The conduction resistances, for their part, correspond to:

$$R_{\lambda_1} = \frac{\mathrm{Ln}\left(\frac{r_2}{r_1}\right)}{2\pi L \lambda_1} \quad \text{and} \quad R_{\lambda_2} = \frac{\mathrm{Ln}\left(\frac{r_3}{r_2}\right)}{2\pi L \lambda_2}$$

The flux is then given by: $\phi = \dfrac{2\,\pi\,L\,(T_i - T_e)}{\dfrac{1}{r_1 h_i} + \dfrac{1}{\lambda_1}\mathrm{Ln}\left(\dfrac{r_2}{r_1}\right) + \dfrac{1}{\lambda_2}\mathrm{Ln}\left(\dfrac{r_3}{r_2}\right) + \dfrac{1}{r_3 h_e}}$

2.13.1. *Illustration: transfer through a composite cylindrical wall*

Let us consider a cylindrical pipe of inner radius, R_1, and outer radius, R_3.

The material constituting the part of the wall comprised between $r = R_1$ and $r = R_2$ admits a thermal conductibility, λ_1, whereas the part of the wall comprised between $r = R_2$ and $r = R_3$ is composed of a material of thermal conductibility, λ_2.

A fluid at uniform temperature, T_i, circulates inside this pipe. The outside surrounding environment is at temperature T_e ($T_i > T_e$).

We will use T_{pi} and T_{pe} to refer to the internal and external wall temperatures, and h_i and h_e to refer to the internal and external convection coefficients.

Questions

1) Calculate the flux, ϕ_{L1}, of energy lost per unit length of pipe.

2) Calculate T_{pi} and T_{pe}.

3) Between $r = R_3$ and $r = R_4$, this pipe is wrapped in a third material of thermal conductivity, λ_3. The flux per unit length is then ϕ_{L2}. What value must λ_3 have in order to obtain $\phi_{L2} = 50\% \ \phi_{L1}$?

<u>*Data:*</u>

$R_1 = 10$ cm $\quad R_2 = 12$ cm
$R_3 = 13$ cm $\quad R_4 = 15.5$ cm
$T_i = 160$ °C $\quad T_e = 20$ °C
$\lambda_1 = 50$ W/m °C $\quad \lambda_2 = 40$ W/m °C
$h_i = 20$ W/m^2 °C $\quad h_e = 10$ W/m^2 °C

Solutions

<u>*1) Calculating the flux, ϕ_{L1}*</u>

$$\phi_1 = \frac{2\ \pi\ L\ (T_i - T_e)}{\dfrac{1}{R_1 h_i} + \dfrac{1}{\lambda_1} Ln\left(\dfrac{R_2}{R_1}\right) + \dfrac{1}{\lambda_2} Ln\left(\dfrac{R_3}{R_2}\right) + \dfrac{1}{R_3 h_e}}$$

$$\text{Hence: } \phi_{L1} = \frac{\phi_1}{L} = \frac{2\pi(T_i - T_e)}{\dfrac{1}{R_1 h_i} + \dfrac{1}{\lambda_1} Ln\left(\dfrac{R_2}{R_1}\right) + \dfrac{1}{\lambda_2} Ln\left(\dfrac{R_3}{R_2}\right) + \dfrac{1}{R_3 h_e}}$$

NUMERICAL APPLICATION.–

$R_1 = 10$ cm	$R_2 = 12$ cm	$R_3 = 13$ cm	$R_4 = 15.5$ cm
$T_i = 160$ °C	$T_e = 20$ °C	$\lambda_1 = 50$ W/m °C	$\lambda_2 = 40$ W/m °C
$h_i = 20$ W/m^2 °C		$h_e = 10$ W/m^2 °C	

$$\phi_{L1} = \frac{2\pi(160-20)}{\frac{1}{(0.1)20}+\frac{1}{50}\text{Ln}\left(\frac{0.12}{0.10}\right)+\frac{1}{40}\text{Ln}\left(\frac{0.13}{0.12}\right)+\frac{1}{(0.13)10}}$$

$\boldsymbol{\phi_{L1} = 690}$ **W/m**

2) Calculating T_{pi} and T_{pe}.

$$\phi_l = 2\pi R_l\, L\, h_i\,(T_i - T_{pi}) \quad \text{and} \quad \phi_l = 2\pi R_3\, L\, h_e\,(T_{pe} - T_e)$$

Or: $\phi_{Ll} = 2\pi R_1\, h_i\,(T_i - T_{pi})$ and $\phi_{Ll} = 2\pi R_3\, h_e\,(T_{pe} - T_e)$.

Hence: $T_{pi} = T_i - \frac{\phi_{L1}}{2\pi R_1 h_i}$ and $T_{pe} = \frac{\phi_{L1}}{2\pi R_3 h_e} + T_e$.

NUMERICAL APPLICATION.–

$$T_{pi} = 160 - \frac{689.63}{2\pi(0.1)(20)} \quad \text{and} \quad T_{pe} = \frac{689.63}{2\pi(0.13)(10)} + 20$$

$\boldsymbol{T_{pi} = 105}$ **°C** and $\boldsymbol{T_{pe} = 104}$ **°C**

3) λ_3 so that ϕ_{L2} is half ϕ_{L1}

We have: $$\phi_{L1} = \frac{2\pi\ (T_i - T_e)}{\frac{1}{R_1 h_i}+\frac{1}{\lambda_1}\text{Ln}\left(\frac{R_2}{R_1}\right)+\frac{1}{\lambda_2}\text{Ln}\left(\frac{R_3}{R_2}\right)+\frac{1}{R_3 h_e}}$$

Likewise, the flux, ϕ_{L2}, is obtained by taking into consideration the additional resistance between R_3 and R_4:

$$\phi_{L2} = \frac{2\pi\ (T_i - T_e)}{\frac{1}{R_1 h_i}+\frac{1}{\lambda_1}\text{Ln}\left(\frac{R_2}{R_1}\right)+\frac{1}{\lambda_2}\text{Ln}\left(\frac{R_3}{R_2}\right)+\frac{1}{\lambda_3}\text{Ln}\left(\frac{R_4}{R_3}\right)\frac{1}{R_4 h_e}}$$

Hence, by imposing the condition, $\frac{\phi_{L1}}{\phi_{L2}} = 2$:

$$\frac{\left[\frac{1}{R_1h_i}+\frac{1}{\lambda_1}Ln\left(\frac{R_2}{R_1}\right)+\frac{1}{\lambda_2}Ln\left(\frac{R_3}{R_2}\right)+\frac{1}{\lambda_3}Ln\left(\frac{R_4}{R_3}\right)+\frac{1}{R_4h_e}\right]}{\left[\frac{1}{R_1h_i}+\frac{1}{\lambda_1}Ln\left(\frac{R_2}{R_1}\right)+\frac{1}{\lambda_2}Ln\left(\frac{R_3}{R_2}\right)+\frac{1}{R_3h_e}\right]}=2$$

λ_3 must then satisfy the following equation:

$$\left[\frac{1}{R_1h_i}+\frac{1}{\lambda_1}Ln\left(\frac{R_2}{R_1}\right)+\frac{1}{\lambda_2}Ln\left(\frac{R_3}{R_2}\right)+\frac{1}{\lambda_3}Ln\left(\frac{R_4}{R_3}\right)+\frac{1}{R_4h_e}\right]$$
$$=2\left[\frac{1}{R_1h_i}+\frac{1}{\lambda_1}Ln\left(\frac{R_2}{R_1}\right)+\frac{1}{\lambda_2}Ln\left(\frac{R_3}{R_2}\right)+\frac{1}{R_3h_e}\right]$$

i.e.: $$\lambda_3=\frac{Ln\left(\frac{R_4}{R_3}\right)}{\frac{1}{R_1h_i}+\frac{1}{\lambda_1}Ln\left(\frac{R_2}{R_1}\right)+\frac{1}{\lambda_2}Ln\left(\frac{R_3}{R_2}\right)+\frac{1}{h_e}\left(\frac{2}{R_3}-\frac{1}{R_4}\right)}.$$

NUMERICAL APPLICATION.–

$$\lambda_3=\frac{Ln\left(\frac{15.5}{13}\right)}{\frac{1}{0.1\,(20)}+\frac{1}{50}Ln\left(\frac{12}{10}\right)+\frac{1}{40}Ln\left(\frac{13}{12}\right)+\frac{1}{10}\left(\frac{2}{0.13}-\frac{1}{0.155}\right)}$$

$\boldsymbol{\lambda_3 = 0.12}$ **W/m °C**

2.14. Composite spherical installations with convective boundary conditions

Figure 2.19 shows a composite spherical chamber with internal and external convections. In practice, this situation corresponds to a spherical storage reservoir composed of a material of thermal conductivity, λ_1, of internal radius, r_1, and external radius, r_2, containing a fluid of average temperature, T_i. The reservoir is covered by a protective layer of radius, r_3. The outer face of this layer is in contact with the surrounding environment, the temperature of which is T_e.

The convection transfer coefficients inside the pipe, h_i, and between the outer face and the surrounding environment, h_e, are assumed to be known.

We will also assume that $T_i > T_e$. An energy flux, ϕ, will therefore flow from the inside to the outside of the tube.

The conduction is considered to be one-dimensional.

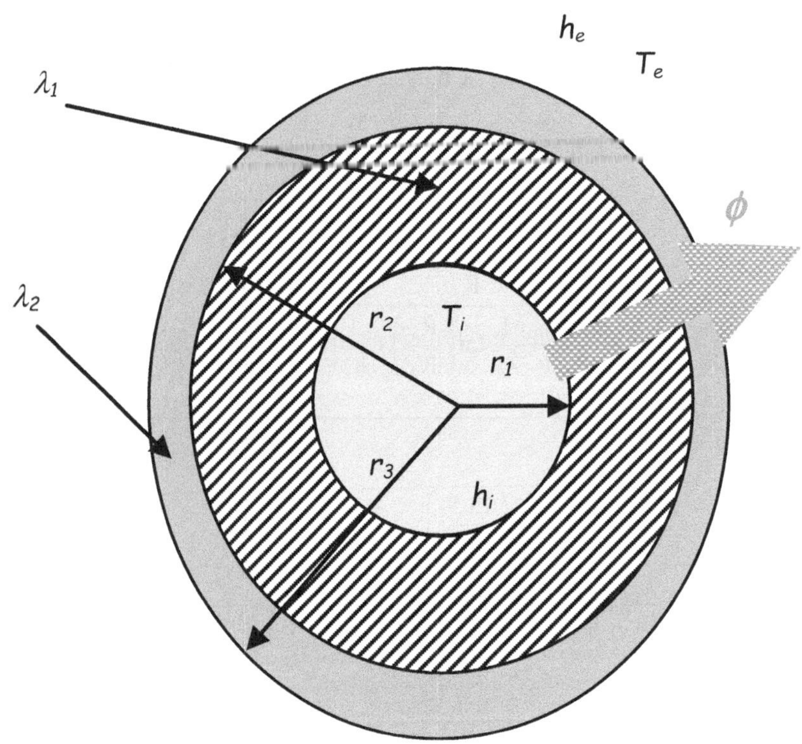

Figure 2.19 *Spherical reservoir with internal and external convections. For a color version of this figure, see www.iste.co.uk/benallou/energy2.zip*

Under these conditions, we have four resistances in series:

– the internal convection resistance, with a convective coefficient, h_i;

– the two conduction resistances in the two solid layers of thermal conductivities, λ_1 and λ_2, respectively;

– the external convection resistance, with a convective coefficient, h_e.

As the resistances are in series, the flux is equal to the extreme temperature difference divided by the sum of the resistances:

$$\phi = \frac{T_i - T_e}{R_i + R_{\lambda_1} + R_{\lambda_2} + R_e}$$

The internal and external convective resistances are given, respectively, by:

$$R_i = \frac{1}{4\pi r_1^2 h_i} \quad \text{and} \quad R_e = \frac{1}{4\pi r_3^2 h_e}$$

The conduction resistances, for their part, correspond to:

$$R_1 = \frac{r_2 - r_1}{4\pi\lambda_1 r_1 r_2} \quad \text{and} \quad R_2 = \frac{r_3 - r_2}{4\pi\lambda_2 r_3 r_2}$$

The equivalent resistance is therefore: $R_{eq} = R_i + R_{\lambda_1} + R_{\lambda_2} + R_e$

i.e.: $R_{eq} = \frac{1}{r_1^2 h_i} + \frac{1}{\lambda_1}\frac{r_2 - r_1}{r_2 r_1} + \frac{1}{\lambda_2}\frac{r_3 - r_2}{r_3 r_2} + \frac{1}{r_3^2 h_e}$.

The flux is then given by: $\phi = \dfrac{4\ \pi\ (T_i - T_e)}{\frac{1}{r_1^2 h_i} + \frac{1}{\lambda_1}\frac{r_2 - r_1}{r_2 r_1} + \frac{1}{\lambda_2}\frac{r_3 - r_2}{r_3 r_2} + \frac{1}{r_3^2 h_e}}$

3

Conduction Applications in Thermal Insulation

3.1. Introduction

Heat-lagging a wall means protecting it or insulating it thermally. This protection is generally obtained by placing a layer of *thermal insulation* around this wall.

A *thermal insulation* is a material that has a *low thermal conductivity*: approximately 0.3 to 0.4 W/cm^2 °C. To a certain extent, it acts as a barrier to conduction heat transfer.

When thermal insulation is affixed to a given wall, the objective is to *minimize energy losses* through this wall by protecting it from its surrounding environment; that is, by avoiding heat transfer with the outside.

The wall to be insulated may be hot or cold. In the case of a hot wall, we heat-lag it to reduce energy losses (calories) to the outside. In the case of a cold wall, the role of thermal insulation is to minimize heat access from the outside to the inside; in this case, we will speak of "minimizing frigorie losses".

Such situations are frequently encountered in industry and we usually have to make a number of choices during a heat-lagging project. Indeed, there are often several configurations possible, defined by the type of insulation to use, the thickness of this insulation, etc. But for each of the possible configurations, two essential questions will need to be answered:

– How can we choose the adapted thermal insulation?

– What procedure should be used to determine the thickness of this insulation?

The answer to these questions is closely linked to three significant parameters:

– the cost of electricity or energy in general;

– the cost of one m^2 of heat lagging for different thicknesses;

– the lifetime of the heat lagging considered.

In this chapter, we start by presenting the main materials usually used as heat laggings in industry and in construction. We then go on to study the two most typical heat-lagging situations, namely:

– insulation of plane walls;

– heat lagging of cylindrical jackets or spherical containers.

3.2. The main insulation materials

Numerous materials are used as heat laggings or thermal insulators owing to their *low thermal conductivity*. Two categories exist:

– *Natural materials* or derivatives of natural products, such as cork, rock wool, glass wool, sawdust, flax, hemp, sheep wool, etc.;

– *Synthetic materials* (polyurethane, expanded and extruded polystyrene and polyester), which are generally effective in terms of insulation, but they often present a high cost.

3.2.1. *Cork*

This is the natural product *par excellence*. It is extracted from the bark of the oak tree.

It exists commercially in the form of fairly spherical chips, which are practical for filling volumes intended to constitute the insulating barrier. This is the case, for example, for water heaters where the space left empty between the reservoir and the jacket is filled with cork chips.

Cork also exists in the form of plates, which are more suitable for insulating plane walls.

With a thermal conductivity around 0.43 W/m °C, cork is a very good thermal insulator of natural origin. However, it has poor fire resistance. In addition, its water resistance and resistance to rodents and chemicals is low.

Cork is nevertheless an insulator that offers *very good value for money*, although its lifetime is shortened when it is in contact with moisture; this is often the case in the event of leaks from insulated reservoirs or pipes connected to them.

3.2.2. *Sawdust and wood wool*

Sawdust is the byproduct obtained in lumber mills. With a thermal conductivity lower than that of wood, it is a good and inexpensive thermal insulator. However, when used directly to ensure thermal insulation, this product becomes more compact over time, and needs to be kept dry. It should be noted that this compaction tendency as well as the capacity to absorb moisture greatly reduces the insulating properties of sawdust.

To avoid these problems, different types of sawdust can be mixed to produce what is known as wood wool.

Wood wool is available commercially in the form of panels that are developed from wood fibers produced by different sawdust. These fibers are then bonded together with wood lignin.

Wood wool has very good heat insulating qualities:

– a low thermal conductivity: 0.059 W/m °C;

– a reasonable price;

– an acceptable lifetime.

It is one of the best insulators available on the market for building walls.

3.2.3. *Hemp*

Hemp is a plant that can be grown in different climates and its development does not require fertilizer or large amounts of rainfall.

This plant consists of fibers that have good abrasion resistance and, more generally, good mechanical strength characteristics.

In addition, the plant fibers have a low thermal conductivity, making it a good-quality heat lagging.

Figure 3.1. *Hemp wool*[1]

Moreover, hemp fibers present other natural qualities: they have fungicidal and antibacterial properties. These characteristics are significant as they increase hemp resistance to damage in humid environments.

In order to be used as thermal insulators, hemp fibers are processed to form a kind of wool presented loose or in the form of rolls.

Due to their low thermal conductivity (0.055 W/m °C), hemp fibers are often used in construction thermal insulation. They are sometimes mixed with insulating concretes or plaster between joists.

3.2.4. *Cellulose*

Cellulose is composed of natural fibers that are generally extracted from wood produced by eucalyptus trees. It is mainly used in the manufacture of paper pulp.

Different grades of paper are produced from this pulp. But for thermal insulation applications, fibers are mainly obtained from the *recycling of paper and newspapers*.

These fibers are used to make cellulose flakes which are available on the market, loose or in panel form, with thermal conductivities of between 0.04 and 0.05 W/m °C.

---

1 Source: https://encrypted-tbn0.gstatic.com/images?q=tbn:ANd9GcSAJ4MpwQrjMu4Kix_5LVD2pFcGmJSR64nqzjUCUMA1W7RwJjvDDQ.

Figure 3.2. *Cellulose wool*[2]

When used loose to fill the spaces to be insulated, provisions must be taken (wearing of masks) to protect against the inhalation of particles, as the fibers can cause pulmonary inflammation during their set up.

This technique of producing thermal insulation materials from the recycling of cellulose fibers is fairly widespread in the United States and Scandinavia, where thermal insulation made from recycled paper and newspapers accounts for almost 30% of the thermal insulation market.

3.2.5. *Glass and rock wools*

Figure 3.3. *Glass-wool roll*[3]

2 Source: https://encrypted-tbn0.gstatic.com/images?q=tbn:ANd9GcQ-yHkuJzQHYCI1bQaOb8zLSpLmJikYykRpN-pZnlrsBNY7mDWQ7g.

3 Source: https://encrypted-tbn0.gstatic.com/images?q=tbn:ANd9GcRPztx-MkKA-yMnL_8sPPPR17NH6Q3VRJzeWUDgcf_xCWKTJ-Sl.

Glass wool and rock wool are among the most widely-used thermal insulation materials in industry and in construction. Both insulators are mineral-fiber based. Glass wool is obtained by recycling glass. The recovered glass is subject to melting followed by centrifugation and blowing, in order to obtain extremely fine fibers of glass, which form a wool. Rock wool is generally derived from basalt by a process comparable to that of the manufacture of glass wool.

Glass wool and rock wool can be obtained off the shelf or in bulk, or in the form of rolls in a variety of thicknesses.

To insulate spaces of specified sizes, the dimensions suitable for the space or unit to be insulated are cut out from the roll. Glass-wool rolls are suitable both for the insulation of plane industrial walls and for curvilinear walls. This is possible thanks to the flexibility and workability of the panels obtained.

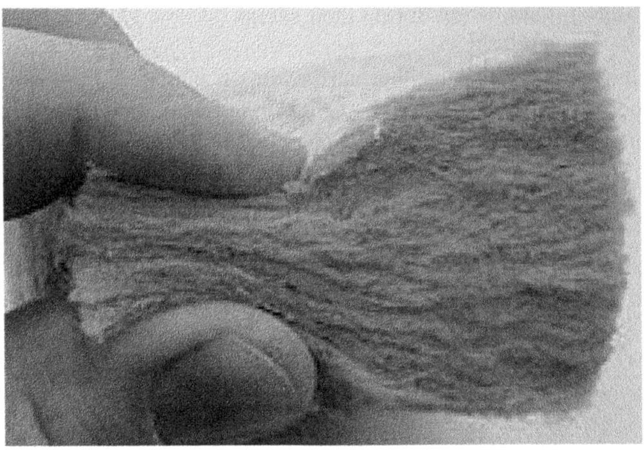

Figure 3.4. *Rock wool*[4]

For the same reasons, the rolls and panels of this material are also used to insulate building walls, even though for this type of use, expanded polystyrene remains significantly less expensive and less sensitive to moisture.

Glass wool and rock wool generally provide good heat lagging due to their low thermal conductivity (0.04 W/m °C), which is almost equivalent to that of polyurethane foam (see below). Yet, their good value for money makes them sought-after heat laggings.

4 Source: https://encrypted-tbn0.gstatic.com/images?q=tbn:ANd9GcS1S7K9j29tzB1DmQn6-zqOli309rxjzZNPzamfgPggXAklWdNy.

Nevertheless, they present the disadvantage of allowing the fixing of water molecules by condensation in moisture-loaded atmospheres. As a result, they are not recommended in environments where air is humid. Indeed, the fixing of water on these materials reduces their insulation efficiency.

3.2.6. *Polyurethane foam*

Polyurethane is a saturated hydrocarbon that is obtained by polymerizing ethane. It is an excellent thermal insulator thanks to its low conductivity: 0.027–0.029 W/m °C.

Polyurethane is used either in the form of a foam to be injected or as expanded panels to be attached to a flat surface.

The injectable foam is available commercially in spray form. Polyurethane foam is very practical for insulation of non-uniform spaces as the injection releases the pressurized polyurethane in foaming liquid state. This enables the injected foam to undergo expansion as the pressure is decreased, thus enabling all of the recesses of the space to be completely filled and insulated.

Figure 3.5. *Polyurethane foam*[5]

Given its excellent thermal insulation properties and its ability to heat lag even the most complex surfaces and volumes, polyurethane foam is very commonly used in industry.

5 Sources: https://encrypted-tbn0.gstatic.com/images?q=tbn:ANd9GcSAtwfbVO2pnbxJPtQvVY up1xGgrDifsV0c99ZUHnVQ3e3dlH8h and https://encrypted-tbn0.gstatic.com/images?q=tbn: ANd9GcT9MWD 4KMWbKbOctCZOuVe-oJxHqPTb0gCS1N8jdZsdsoIC9f.

However, it presents risks that should not be overlooked since polyurethane releases amines. Therefore, precautions must be taken when injecting polyurethane: airing of workshops, ventilation, etc.

Yet, despite these constraints and its relatively high price, polyurethane remains one of the most widely-used materials for thermal insulation, which can be explained by:

– its low thermal conductivity;

– the flexibility it offers for the heat, lagging of volumes and surfaces with complex geometry;

– its excellent moisture resistance.

3.2.7. *Expanded polystyrene*

Figure 3.6. *Polystyrene*[6]

Polystyrene is also a hydrocarbon. It is obtained by polymerizing styrene. When this polymerization is conducted under an air flow, *expanded polystyrene* is then obtained.

6 Source: https://encrypted-tbn0.gstatic.com/images?q=tbn:ANd9GcS6-aimFbyJoS3NSea3J0aDGw2Jq2iLINGOSijK0A5Dk1TvLwzFA.

Nevertheless, they present the disadvantage of allowing the fixing of water molecules by condensation in moisture-loaded atmospheres. As a result, they are not recommended in environments where air is humid. Indeed, the fixing of water on these materials reduces their insulation efficiency.

3.2.6. *Polyurethane foam*

Polyurethane is a saturated hydrocarbon that is obtained by polymerizing ethane. It is an excellent thermal insulator thanks to its low conductivity: 0.027–0.029 W/m °C.

Polyurethane is used either in the form of a foam to be injected or as expanded panels to be attached to a flat surface.

The injectable foam is available commercially in spray form. Polyurethane foam is very practical for insulation of non-uniform spaces as the injection releases the pressurized polyurethane in foaming liquid state. This enables the injected foam to undergo expansion as the pressure is decreased, thus enabling all of the recesses of the space to be completely filled and insulated.

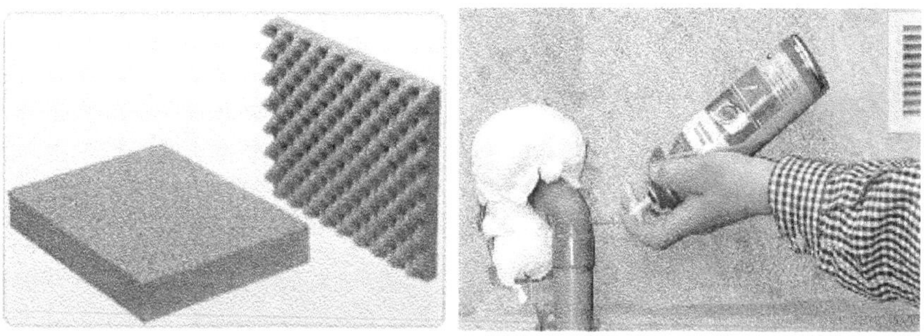

Figure 3.5. *Polyurethane foam*[5]

Given its excellent thermal insulation properties and its ability to heat lag even the most complex surfaces and volumes, polyurethane foam is very commonly used in industry.

5 Sources: https://encrypted-tbn0.gstatic.com/images?q=tbn:ANd9GcSAtwfbVO2pnbxJPtQvVY up1xGgrDifsV0c99ZUHnVQ3e3dlH8h and https://encrypted-tbn0.gstatic.com/images?q=tbn: ANd9GcT9MWD 4KMWbKbOctCZOuVe-oJxHqPTb0gCS1N8jdZsdsoIC9f.

However, it presents risks that should not be overlooked since polyurethane releases amines. Therefore, precautions must be taken when injecting polyurethane: airing of workshops, ventilation, etc.

Yet, despite these constraints and its relatively high price, polyurethane remains one of the most widely-used materials for thermal insulation, which can be explained by:

– its low thermal conductivity;

– the flexibility it offers for the heat, lagging of volumes and surfaces with complex geometry;

– its excellent moisture resistance.

3.2.7. *Expanded polystyrene*

Figure 3.6. *Polystyrene*[6]

Polystyrene is also a hydrocarbon. It is obtained by polymerizing styrene. When this polymerization is conducted under an air flow, *expanded polystyrene* is then obtained.

6 Source: https://encrypted-tbn0.gstatic.com/images?q=tbn:ANd9GcS6-aimFbyJoS3NSea3J0aDGw2Jq2iLINGOSijK0A5Dk1TvLwzFA.

The expanded polystyrene available commercially is in the form of fairly spherical particles. These particles are able to fill volumes of different geometries with great flexibility. This advantage is important in the packaging industry.

Expanded polystyrene also comes in the form of panels, which are particularly suitable for the building industry and that of isothermal packaging (used for ice cream, fish, meats, frozen cakes, etc.).

In this form, expanded polystyrene presents the great advantage of being waterproof. This creates a watertight barrier in walls and isothermal packaging and thus prevents moisture from spreading.

Figure 3.7. *Expanded-polystyrene panels*[7]

As such, expanded polystyrene is an excellent thermal insulator for the construction industry and it satisfies all of the cold-chain insulation requirements. This is thanks to:

– its relative solidity;

– its moisture resistance;

– its low weight;

– its reasonable cost;

– above all, its low thermal conductivity (0.036–0.042 W/m °C).

7 Source: https://encrypted-tbn0.gstatic.com/images?q=tbn:ANd9GcTm6W6godHV8_qTYIFPC55gcreX-yGKlrbeKDwvDQVeVHVZGX.

3.3. Choosing a suitable thermal insulator

Obviously, the choice of thermal insulation to select for a given use depends, first and foremost, on the products available on the market and their price. There is no point in choosing an excellent thermal insulation that is out of stock for a long period of time or that is not available in a given market.

Likewise, the cost of the heat lagging to be used must be taken into consideration when making the final decision. This cost will naturally depend on the area of transfer with the outside, as well as on the *thickness* retained. In the following subsections, we will see how the optimum thickness of the heat lagging can be determined according to the parameters of the problems considered.

The first task, however, is to define the nature of the heat lagging to be employed. This is obtained based on the physical parameters related to usage, such as the maximum operating temperature, the heat lagging's resistances (to moisture, fire, rodents and common chemicals) and its lifetime (that is, the time during which this insulation will resist whilst retaining its physical and thermal properties). Table 3.1 presents these parameters for the most common thermal insulation materials. A more detailed set of data has been compiled in the Appendix.

Insulating material	P ($kg\ m^{-3}$)	C_p ($J\ kg^{-1}\ °C^{-1}$)	λ ($W\ m^{-1}°C^{-1}$)	T_{Max} °C	Durability[(1),(2)] D	F	R	C	Lifetime
Cellulose (flakes)	200	890	0.045	< 100	μ	μ	μ	A	2 years
Hemp	180	900	0.055	< 100	A	μ	G	A	4 years
Kapok	50	950	0.035	< 100	μ	μ	μ	A	4 years
Rock wool	20	880	0.047	> 150	A[(4)]	G	G	G	5 years
	55	880	0.038	> 150	A[(4)]	G	G	G	
	135	880	0.041	> 150	A[(4)]	G	G	G	
Glass wool	8	875	0.051	> 150	A[(4)]	G	G	G	5 years
	15	880	0.041	> 150	A[(4)]	G	G	G	
	40	880	0.035	> 150	A[(4)]	G	G	G	
Expanded cork	120	2,100	0.044	> 100	A[(4)]	μ	G	A	2 years
Carpet	200	1,300	0.06	< 60	μ	μ	A	μ	
Polyurethane (foam)	32	1,300	0.03	≈ 100	G	μ[(3)]	A	A	5 years
	50	1,360	0.035	≈ 100	G	μ[(3)]	A	A	
	85	1,300	0.045	≈ 100	G	μ[(3)]	A	A	

Thermocompressed expanded polystyrene	18	1,300	0.041	≈ 85	A[4]	μ	μ	A[5]	4 years
	30	1,300	0.041	≈ 85	A[4]	μ	μ	A[5]	
Sawdust or wood wool	160	190	0.059	< 100	μ	μ	G	A	2 years

[1] Durability– D: against dampness, F: against fire, R: against rodents and C: against chemical products.
[2] Durability scale– G: good durability, A: average durability, μ: poor durability.
[3] Nevertheless does not spread flames.
[4] Requires a steam barrier (the insulating panels or felts are bonded onto bituminous kraft, grids or aluminum foil).
[5] Except in the presence of aggressive solvents.

Table 3.1. *Physical data of the main heat insulators*

NOTE.– It should be noted that the lifetime data presented in Table 3.1 correspond to average values. The actual lifetimes will depend on the usage conditions. Cork, for example, could have longer lifetimes than those indicated in the table if used in a moisture-free environment.

In practice, the most significant parameter is the *maximum temperature*, T_{max}, *supported by the heat lagging*. Consequently, we begin by checking that this constraint is satisfied.

Next comes the information available concerning the *operating conditions* (humidity in the workshop, fire risk, possible presence of rodents and potential exposure to chemicals), which will help limit the selection of eligible heat laggings, based on their *resistance* to moisture, fire, rodents and common chemicals.

This analysis will facilitate the final selection. This selection is usually operated based on thermal conductivity: *the heat lagging that presents the lowest conductivity is chosen.*

Thus, once the heat lagging type has been selected, it will be necessary to assess how much heat lagging will be needed in order to reach the sought level of heat-loss reduction. This comes down to determining the necessary heat-lagging thickness, which will enable its cost to be determined.

The following subsections describe the approaches that enable determination of the optimum heat-lagging thickness to be used in each practical situation (plane walls, cylindrical jackets or spherical containers).

3.3.1. *Optimum heat-lagging thickness for plane walls*

This is the case for the walls of industrial furnaces and some thermal machines. The classic problem is posed as follows: given that we know the temperatures on the inside and on the outside of the plane wall to be heat-lagged, what would be the optimum heat-lagging thickness? In other words: what thermal insulation thickness would correspond to the minimum operating cost?

To answer this question, we need to know:

– The expression of the residual heat losses, in other words those remaining once the heat lagging is put in place. These losses will depend on the thickness, "*e*", of the heat lagging used, and on the transfer area.

The cost of the heat lagging chosen, which depends on its thickness and surface area.

In general, we reason in terms of *annual costs*. We then define the function "total annual cost", denoted $C_T(e)$, taking into consideration both the cost that the residual heat losses would represent over the course of a year, and the annual cost of the heat lagging:

Total annual cost = cost of losses + annual cost of the heat lagging

$$C_T(e) = P(e) + C(e),$$

where the *annual cost of the heat lagging*, $C(e)$, is defined as its total cost (including installation costs) divided by its lifetime:

$$C(e) = \frac{P_u(e)\ S}{D_v},$$

with:

$P_u(e)$: the unit price (per m^2) of the heat lagging, expressed in Dh/m^2

S: the transfer area

D_v: the lifetime of the heat lagging

The *annual cost of the heat losses*, $P(e)$, is obtained by multiplying the annual flux of the losses by the energy price:

$$P(e) = \varphi(e)\ S\ D_F P_E,$$

with:

P(e): the annual cost of the losses, expressed in Dh/year

φ(e): the flux density of the heat losses, expressed in kcal/hm^2 or in kW/m^2

S: the total transfer area

D_F: the annual operating time of the unit to be heat-lagged

P_E: the energy unit price, expressed in Dh/kWh or in Dh/kcal

Thus, the "total annual cost" function, $C_T(e)$, is given by:

$$C_T(e) = \varphi(e)\ S\ D_F P_E + \frac{P_u(e)\ S}{D_v}.$$

$C_T(e)$ is therefore the sum of:

– a *decreasing function* of *e*: $P(e) = \varphi(e)\ S\ D_F\ P_E$; given that in principle the flux density of the losses, $\varphi(e)$, decreases as the thickness of the heat lagging increases;

– an *increasing function* of *e*: $C(e) = \frac{P_u(e)\ S}{D_v}$; given that the greater the thickness, the higher the unit price of the heat lagging, $P_u(e)$, will be.

Figure 3.8 shows, with respect to the heat-lagging thickness, *e*, the general shapes of the functions "annual cost of heat losses", *P(e)*, and "annual cost of heat lagging", *C(e)*.

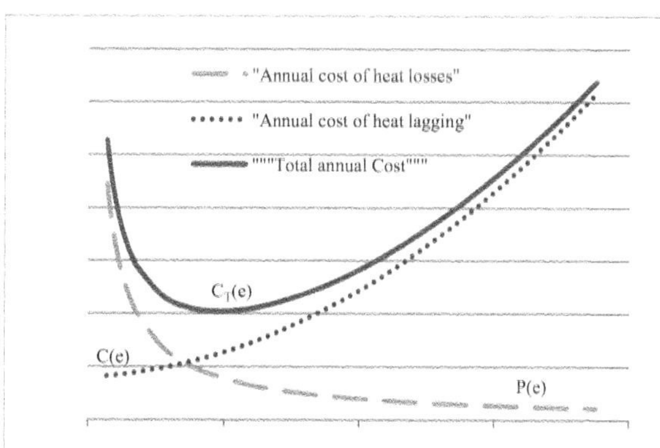

Figure 3.8. *Determining the optimum heat-lagging thickness. For a color version of this figure, see www.iste.co.uk/benallou/energy2.zip*

Thus, the function total annual cost, which is the sum of these two functions, one *increasing* and the other *decreasing*, will be a function of "e", which *admits a minimum*. This minimum value is the thickness of the heat lagging to be used, e^*, with e^* corresponding to the minimum total cost.

Analytically, the optimum thickness of the heat lagging, e^*, is obtained by canceling out the differential of the function total cost, $C_T(e)$:

$$\left.\frac{dC_T(e_c)}{de_c}\right|_{e_c=e_c^*}=0$$

3.3.1.1. *Illustration: cost of heat losses through the walls of an industrial furnace*

In this illustration, we wish to minimize the heat losses between the side walls of an industrial furnace and the surrounding environment. The four side walls of the furnace are composed of refractory bricks.

The outer face of the walls is in contact with the surrounding environment. The thickness, e, and the thermal conductibility, λ, are constant for all four walls.

The internal face of the walls is at temperature T_i, the surrounding environment is at temperature T_a, and the convective heat transfer coefficient between the outer face of the walls (at T_e) and the atmosphere is denoted as h.

Questions

A) In this part, we wish to assess the annual cost of the heat losses in the event that no heat lagging is applied to the lateral surface of the furnace.

1) Calculate the temperature, T_e.

2) Calculate the heat losses.

3) Assuming that the furnace is supplied by electricity and that its annual operating time (24 hours a day, without interruption) is D_F, what would these losses represent in monetary terms?

B) We now wish to reduce the heat losses using glass wool as insulation.

We apply a thickness, e_c, of this heat lagging onto the outer lateral faces of the wall. The heat conductibility of the heat lagging will be denoted as λ_c and its price per square meter depends on the thickness according to the function:

$$P_u(e_c) = \alpha e_c \text{ with } \alpha = 9{,}600 \text{ Dh/m}^3$$

1) Give, as a function of e_c, the expression of the flux of heat losses with heat lagging.

2) Give, as a function of e_c, the expression of the overall annual operating cost.

3) What is the optimum thickness of the heat lagging?

4) What, in monetary terms, are the savings realized (in million Dh/year)?

We will represent the four walls of the furnace using a single wall of the same thickness, *e*, presenting an equivalent lateral area, *S*.

Data:

$T_i = 750°C$ $T_a = 25°C$ $S = 20$ m^2
$\lambda = 27$ W/m°C $h = 40$ W/m^2°C $e = 35$ cm

Price of electricity: $P_E = 2.2$ Dh/kWh

Annual operating time: $D_F = 7{,}900$ hrs/year

Lifetime of the heat lagging: $D_v = 5$ years

Solutions

A) Without heat insulation

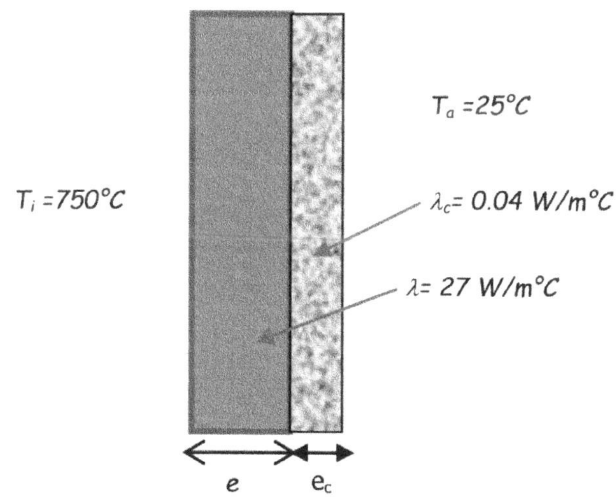

Figure 3.9. *Wall with heat lagging*

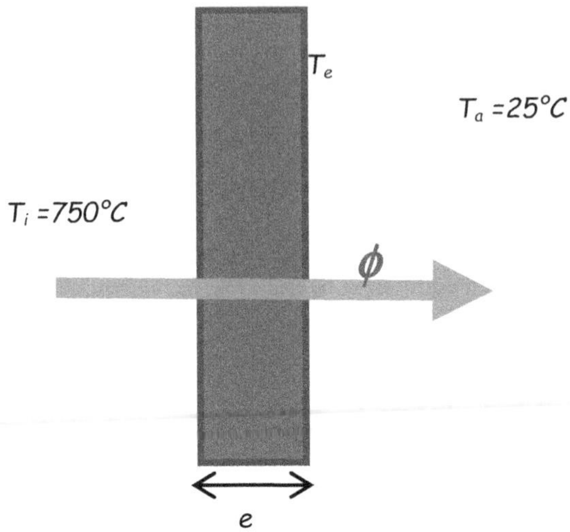

Figure 3.10. *Wall without heat lagging*

1) Calculating the temperature, T_e

The unknown temperature, T_e, can be calculated from the expressions of the flux.

Indeed, the flux can be expressed in two ways:

– As a function of the extreme temperatures, T_i and T_a: $\phi = \dfrac{T_i - T_a}{\Sigma R}$;

hence the following expression of the flux density, φ: $\varphi = \dfrac{T_i - T_a}{\dfrac{e}{\lambda} + \dfrac{1}{h}}$.

– As a function of the external temperatures, T_e and T_a: $\phi = h\ S\ (T_e - T_a)$.

T_e is obtained by equaling the two flux-density expressions,

i.e.: $h(T_e - T_a) = \dfrac{T_i - T_a}{\dfrac{e}{\lambda} + \dfrac{1}{h}}$.

Hence: $T_e = T_a + \dfrac{T_i - T_a}{\dfrac{eh}{\lambda} + 1}$.

NUMERICAL APPLICATION.–

$T_i = 750$ °C $\quad T_a = 25$ °C $\quad \lambda = 27$ W/m °C $\quad h = 40$ W/m^2 °C

$$T_e = 25 + \frac{750 - 25}{\dfrac{0.35 * 40}{27} + 1}$$ ***Te* = 502 °C**

2) Calculating the heat losses

The simplest method consists of calculating ϕ from the expression of the convective flux:

$$\phi = h\ S\ (T_e - T_a).$$

NUMERICAL APPLICATION.–

$T_a = 25$ °C $\quad T_e = 502.44$ °C $\quad h = 40$ W/m^2 °C $\quad S = 20$ m^2

$\phi = 40\ (20)\ (502.44 - 25)$ $\quad$ **ϕ = 382 kW**

3) Annual losses in monetary terms

Let *P* be the cost of losses, expressed in Dh/year.

We have: $P = \phi\ D_F\ P_E$.

NUMERICAL APPLICATION.–

$\phi = 381.95$ kW

Unit price of electricity: $P_E = 2.2$ Dh/kWh

Annual operating time: $D_F = 7{,}900$ hrs/year

P = 381.95 x 7,900 x 2.2 $\quad$ ***P* = 6.64 million Dh/year**

This realistic example shows that the annual cost of heat losses can be very high *if no heat insulation is used.*

B) With a heat-lagging thickness, e_c

1) Expression of the flux

With heat lagging, the density, $\varphi_c(e_c)$, of the heat lost can be expressed as a function of the thickness, e_c, of the heat lagging as follows:

$$\varphi_c(e_c)=\frac{T_i-T_a}{\frac{e}{\lambda}+\frac{e_c}{\lambda_c}+\frac{1}{h}}.$$

2) Overall annual cost as a function of e_c

The overall annual cost function is as follows: $C_T(e_c) = P_c(e_c) + C(e_c)$, where:

$P_c(e_c)$ is the annual cost of the heat losses with heat lagging, given by:

$$P_c(e_c) = \varphi_c(e_c)\, S\, D_F\, P_E$$

$C(e_c)$ is the annual cost of the heat insulation, given by:

$$C(e_c)=\frac{P_u(e_c)\, S}{D_v}$$

D_v is the lifetime of the heat lagging and $P_u(e_c)$ the cost per square meter of the heat lagging, given by:

$$P_u(e_c)=\alpha e_c$$

By replacing $P_c(e_c)$ and $C(e_c)$ with their expressions in the overall annual cost function, we obtain:

$$C_T(e) = \varphi_c(e_c)\, S\, D_F\, P_E + \frac{P_u(e)\, S}{D_v},$$

with $\varphi_c(e_c)=\frac{T_i-T_a}{\frac{e}{\lambda}+\frac{e_c}{\lambda_c}+\frac{1}{h}}$ and $P_u(e_c) = \alpha e_c$,

i.e.: $$C_T(e_c)=\frac{S\, D_F P_E (T_i-T_a)}{\frac{e}{\lambda}+\frac{e_c}{\lambda_c}+\frac{1}{h}}+\frac{\alpha S}{D_v}e_c.$$

3) Optimum heat-lagging thickness

This thickness, e_c^*, is obtained for: $\left.\dfrac{dC_T(e_c)}{de_c}\right|_{e_c=e_c^*} = 0.$

Yet: $$\frac{dC_T(e_c)}{de_c} = -\frac{1}{\lambda_c}\frac{S\,D_F P_E(T_i - T_a)}{\left(\dfrac{e}{\lambda} + \dfrac{e_c}{\lambda_c} + \dfrac{1}{h}\right)^2} + \frac{\alpha\,S}{D_v}.$$

e_c^* is then given by the solution of the equation:

$$\left(e_c^* + \frac{\lambda_c}{h} + \frac{\lambda_c}{\lambda}e\right)^2 = \frac{D_v D_F P_E \lambda_c}{\alpha\,\cdot}(T_i - T_a).$$

Hence: $$e_c^* = \sqrt{\frac{D_v D_F P_E \lambda_c}{\alpha}(T_i - T_a)} - \lambda_c\left(\frac{1}{h} + \frac{e}{\lambda}\right).$$

NUMERICAL APPLICATION.–

$T_i = 750$ °C $\quad T_a = 25$ °C $\quad \lambda = 27$ W/m°C $\quad \lambda c = 0.04$ W/m°C

$\alpha = 9{,}600$ Dh/m^3 $\quad h = 40$ W/m^2 °C

Overall lateral surface area of the four walls: $S = 20$ m^2

Thickness of the lateral walls of the furnace: $e = 35$ cm

Price of electricity: $P_E = 2.210^{-3}$ Dh/Wh

Annual operating time: $D_F = 7{,}900$ hrs/year

Lifetime of the heat lagging: $D_v = 5$ years

$$e_c^* = \sqrt{\frac{5(7\,900)(2{,}2\,10^{-3})0.04}{9\,600}(750-25)} - 0.04\left(\frac{1}{40} + \frac{0.35}{27}\right)$$

***ec** = 51 cm**

4) The savings made

With P_c being the cost of the losses with heat lagging, expressed in Dh/year, the savings are given by: $E = P - P_c$.

The losses without heat insulation have been calculated at P = *6.64 million Dh/year*. They correspond to the following expression: $P = \phi D_F P_E$,

with: $\phi = \dfrac{S(T_i - T_a)}{\dfrac{e}{\lambda} + \dfrac{1}{h}}$; i.e.: $P = \dfrac{S D_F P_E (T_i - T_a)}{\dfrac{e}{\lambda} + \dfrac{1}{h}}$.

With heat lagging, these losses will be reduced to: $P_c(e_c) = \phi_c(e_c) S D_F P_E$,

with: $\phi_c(e_c) = \dfrac{T_i - T_a}{\dfrac{e}{\lambda} + \dfrac{e_c}{\lambda_c} + \dfrac{1}{h}}$; i.e.: $P_c(e_c) = \dfrac{S D_F P_E (T_i - T_a)}{\dfrac{e}{\lambda} + \dfrac{e_c}{\lambda_c} + \dfrac{1}{h}}$.

Hence the expression of the savings made as a result of installing the heat insulation:

$$E = S D_F P_E (T_i - T_a)\left[\frac{1}{\dfrac{e}{\lambda} + \dfrac{1}{h}} - \frac{1}{\dfrac{e}{\lambda} + \dfrac{1}{h} + \dfrac{e_c}{\lambda_c}}\right]$$

NUMERICAL APPLICATION.–

T_i = 750 °C T_a = 25 °C P_E = 2.210^{-3} Dh/Wh e = 35 cm

e_c = 51 cm S = 20 m^2 D_F = 7,900 hrs/year

λ_c = 0.04 W/m°C λ = 27 W/m°C h = 40 W/m^2°C

$$E = 20(7\ 900)(0.0022)(750 - 25)\left[\frac{1}{\dfrac{0.35}{27} + \dfrac{1}{40}} - \frac{1}{\dfrac{0.35}{27} + \dfrac{1}{40} + \dfrac{0.51}{0.04}}\right]$$

***E* = 3.4 million Dh/year**

3.3.2. *Heat-lagging cylindrical jackets*

We saw in the previous sections that for insulated plane walls the thermal resistance increases with the thickness, e_c, of the heat lagging. This result is intuitively logical. Indeed, the more heat lagging is added, the greater the resistance will be and the more the heat losses will be reduced.

However, this result is not always true in the case of cylindrical jackets. Indeed, for this type of wall, the heat losses flux is not always a strictly decreasing function of the heat-lagging thickness. This is due to the fact that for cylindrical jackets, the thermal resistance in some cases could present a minimum. This minimum is encountered for a heat-lagging radius known as the *critical radius*.

3.3.2.1. *Determination of the critical thickness of a heat lagging*

Let us consider a cylindrical pipe of internal radius r_1 and external radius r_2, heat-lagged with an insulator of thermal conductivity, λ_c, and thickness, $r_3 - r_2$ (see Figure 3.11).

The pipe transports a fluid with an average temperature of T_i, while the surrounding environment is at temperature T_e.

The thermal conductivity of the pipe and the convection transfer coefficients between the fluid and the interior of the pipe, h_i, and between the outer surface of the heat lagging and the surrounding environment, h_e, are assumed to be known.

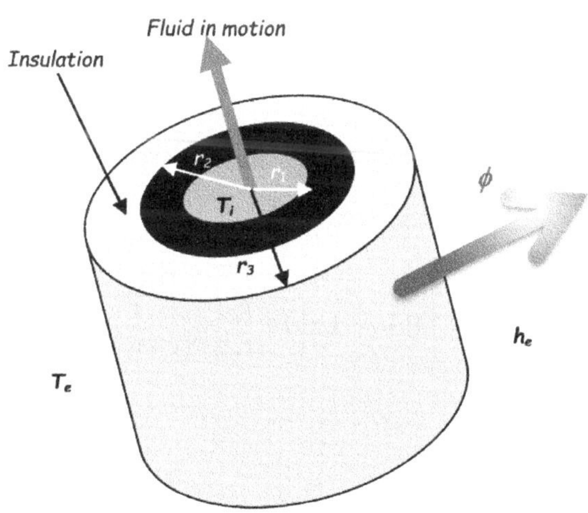

Figure 3.11. *Heat-lagged cylindrical pipe. For a color version of this figure, see www.iste.co.uk/benallou/energy1.zip*

As we have seen, the flux, ϕ, of heat losses to the outside (see section 2.13) is expressed as a function of the temperature difference ($T_i - T_e$), as follows:

$$\phi(r_3) = \frac{2\ \pi\ L\ (T_i - T_e)}{\frac{1}{r_1 h_i} + \frac{1}{\lambda} Ln\left(\frac{r_2}{r_1}\right) + \frac{1}{\lambda_C} Ln\left(\frac{r_3}{r_2}\right) + \frac{1}{r_3 h_e}}.$$

If e designates the thickness of the heat lagging, we have: $r_3 = r_2 + e$.

The flux expression can then be rewritten as a function of the thickness of the heat lagging material:

$$\phi(e) = \frac{A}{B + \frac{1}{\lambda_C} Ln\left(1 + \frac{e}{r_2}\right) + \frac{1}{(e + r_2) h_e}},$$

where A and B are constants given by:

$$A = 2\ \pi\ L\ (T_i - T_e) \text{ and } B = \frac{1}{r_1 h_i} + \frac{1}{\lambda} Ln\left(\frac{r_2}{r_1}\right).$$

Let us now analyze the variations of ϕ with e:

– for $e = 0$: $\phi(0) = \dfrac{A}{B + \frac{1}{r_2 h_e}}$;

– furthermore: $$\frac{d\phi}{de} = -A \frac{\frac{1}{\lambda_C r_2} \frac{1}{1 + \frac{e}{r_2}} - \frac{1}{h_e (e + r_2)^2}}{\left[B + \frac{1}{\lambda_C} Ln\left(1 + \frac{e}{r_2}\right) + \frac{1}{(e + r_2) h_e}\right]^2}$$

or: $$\frac{d\phi}{de} = A \frac{\frac{1}{(e + r_2)}\left(\frac{1}{h_e (e + r_2)} - \frac{1}{\lambda_C}\right)}{\left[B + \frac{1}{\lambda_C} Ln\left(1 + \frac{e}{r_2}\right) + \frac{1}{(e + r_2) h_e}\right]^2}.$$

We deduce therefrom that the sign of $\frac{d\phi}{de}$ is that of $-h_e e+\lambda_C-h_e r_2$ and that

$$\frac{d\phi}{de}=0, \text{ for } \frac{1}{\lambda_C}=\frac{1}{h_e\left(e+r_2\right)}.$$

$$\text{i.e.: } e=e_c=\frac{\lambda_C}{h_e}-r_2.$$

There is not much practical benefit in analyzing $e_c \leq 0$ as $\frac{d\phi}{de}$ will be negative $\forall\, e \in [0; \infty[$.

Consequently, ϕ will be decreasing over the whole interval $[0; \infty[$. This case is nevertheless discussed in Exercise 5.

Table 3.2 presents the variation of ϕ in the event that $e_c > 0$.

e	0	e_c	∞
ϕ'	+	0	–
ϕ	↗		↘

Table 3.2. *Variations of α as a function of* e

The curve of the flux of the heat losses, ϕ, as a function of the thickness, e, of the heat lagging is represented in Figure 3.12.

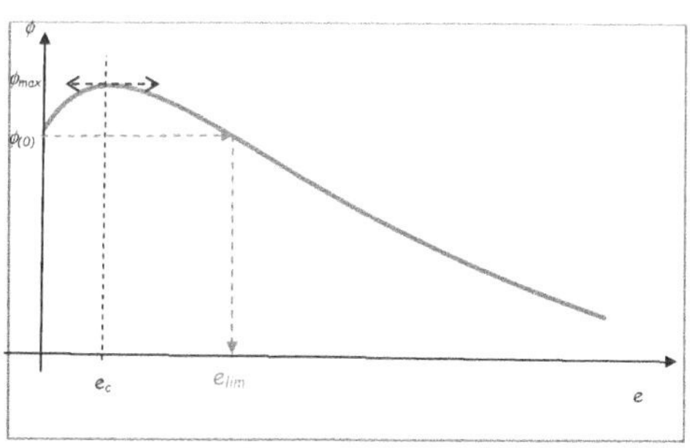

Figure 3.12. *Curve of the flux of the heat losses as a function of the thickness of the heat lagging. For a color version of this figure, see www.iste.co.uk/benallou/energy2.zip*

Thus:

– for $e < e_c$: the heat losses *with heat lagging* are greater than the losses without heat lagging. This is why e_c is called the *critical thickness of the heat lagging*;

– for $e > e_c$: using the heat lagging results in a reduction in energy losses, as hoped. Yet this reduction will only give heat losses below $\phi(0)$ when e exceeds a *boundary-layer thickness*, e_{lim}, defined by $\phi(e_{lim}) = \phi(0)$.

Consequently, when the wall to be heat-lagged is cylindrical, we will always need to make sure that the thickness of the heat lagging used is *greater than* e_{lim}: the *limit thickness*.

> NOTE.
>
> We can also define the *critical radius* of the pipe by: $r_c = r_2 + e_c$
>
> In the same way, we define the *limit radius* by: $r_{lim} = r_2 + e_{lim}$

The condition $e > e_{lim}$ implies that $e > e_c$, and therefore that: $e > \frac{\lambda_C}{h_e} - r_2$.

Thus, this condition yields: $(e + r_2) h_e > \lambda_C$, i.e.: $\frac{1}{\lambda_C} > \frac{1}{r_3 h_e}$.

Quite simply, this condition reveals that insulation thickness must be chosen such that the *conductive resistance* of the latter (which is proportional to $\frac{1}{\lambda_C}$) is greater than the convective resistance, which is proportional to $\frac{1}{(e+r_2)h_e}$.

The opposite condition, $e < e_{lim}$, would imply that $\phi(e) > \phi(0)$.

Physically speaking, this would correspond to a situation where heat flows more easily with heat lagging than without it. In this case, the thermal resistance also decreases with heat lagging. From a heat-loss point of view, applying insulation with a thickness smaller than the limit thickness, e_{lim}, is worse than leaving the pipe bare!

The flux, ϕ_{max}, at the critical point, is given by:

$$\phi_{max} = \frac{A}{B + \frac{1}{\lambda_C} Ln\left(\frac{\lambda_C}{h_e r_2}\right) + \frac{1}{\lambda_C}},$$

with: $A = 2\ \pi\ L\ (T_i - T_e)$ and $B = \frac{1}{r_1 h_i} + \frac{1}{\lambda} Ln\left(\frac{r_2}{r_1}\right)$.

As shown in Figure 3.12, only *thicknesses greater than the limit thickness* (i.e. insulation radii *greater than* the limit radius) are to be considered. Yet whilst the limit thickness guarantees energy losses below those obtained with a bare pipe, it does not provide any information with regard to minimizing energy losses.

The following subsection presents an approach to be used to determine the heat-lagging thickness through an optimization of the expected energy savings.

3.3.2.2. *Determining the optimum thickness of a heat lagging*

Similar to the case of plane walls, the optimum heat-lagging thickness is that which minimizes the *total annual operating cost* function. Let us recall that this function is defined, in terms of the heat-lagging thickness, *e*, by:

$$C_T(e) = P(e) + C(e),$$

where *P(e)* is the annual cost of the heat losses with heat lagging, given by:

$$P\ (e) = \phi(e)\ D_F\ P_E,$$

and *C(e)* is the annual cost of the heat lagging, given by:

$$C(e) = \frac{P_u(e)\ S}{D_v},$$

with:

D_v: the lifetime of the heat lagging

$P_u(e)$: the cost per m^2 of the heat lagging

S: the lateral surface covered by the heat lagging: $S = 2\pi\ (r_2 + e)\ L$

D_F: the annual operating time

P_E: the unit price of electricity

Thus, for a cylindrical pipe of internal radius, r_1, and external radius, r_2, heat-lagged with a thickness, e, of an insulation of thermal conductivity, λ_c, the *total annual operating cost* function is given by:

$$C_T(e) = D_F P_E \, \phi(e) + \frac{2\pi \, L \, (r_2 + e)}{D_v} P_u(e),$$

where: $\phi(e) = \dfrac{A \, D_F \, P_E}{B + \dfrac{1}{\lambda_C} \mathrm{Ln}\left(1 + \dfrac{e}{r_2}\right) + \dfrac{1}{(e + r_2) h_e}}$.

A and *B* are the constants above defined by.

$$A = 2 \, \pi \, L \, (T_i - T_e) \quad \text{and} \quad B = \frac{1}{r_1 h_i} + \frac{1}{\lambda} \mathrm{Ln}\left(\frac{r_2}{r_1}\right).$$

The optimum heat-lagging thickness, e^*, is that which cancels out $\dfrac{dC_T(e)}{de}$.

Yet: $\dfrac{dC_T(e)}{de} = D_F P_E \dfrac{d\phi(e)}{de} + \dfrac{2\pi L}{D_v}\left((r_2 + e)\dfrac{dP_u(e)}{de} + P_u(e)\right)$.

Moreover (see section 3.3.2.1 above):

$$\frac{d\phi}{de} = A \frac{\dfrac{1}{(e + r_2)}\left(\dfrac{1}{h_e (e + r_2)} - \dfrac{1}{\lambda_C}\right)}{\left[B + \dfrac{1}{\lambda_C} \mathrm{Ln}\left(1 + \dfrac{e}{r_2}\right) + \dfrac{1}{(e + r_2) h_e}\right]^2}$$

Replacing $\frac{d\phi(e)}{de}$ with its expression, we obtain:

$$\frac{dC_T(e)}{de} = D_F P_E A \frac{\frac{1}{(e+r_2)}\left(\frac{1}{h_e(e+r_2)} - \frac{1}{\lambda_C}\right)}{\left[B + \frac{1}{\lambda_C}\text{Ln}\left(1+\frac{e}{r_2}\right) + \frac{1}{(e+r_2)h_e}\right]^2} + \frac{2\pi L}{D_v}\left((r_2+e)\frac{dP_u(e)}{de} + P_u(e)\right).$$

Thus, e^* is given by the solution to the following equation with e:

$$\frac{1}{(e+r_2)}\left(\frac{1}{\lambda_C} - \frac{1}{h_e(e+r_2)}\right)$$

$$= \frac{2\pi L}{D_F P_E A D_v}\left((r_2+e)\frac{dP_u(e)}{de} + P_u(e)\right)\left[B + \frac{1}{\lambda_C}\text{Ln}\left(1+\frac{e}{r_2}\right) + \frac{1}{(e+r_2)h_e}\right]^2,$$

or alternatively, by replacing A with its expression and reorganizing:

$$D_F P_E D_v (T_i - T_e)\left(\frac{h_e(e+r_2) - \lambda_C}{\lambda_C h_e (e+r_2)^2}\right)$$

$$= \left[B + \frac{1}{\lambda_C}\text{Ln}\left(1+\frac{e}{r_2}\right) + \frac{1}{(e+r_2)h_e}\right]^2\left((r_2+e)\frac{dP_u(e)}{de} + P_u(e)\right).$$

Very often, P_u varies linearly with e:

$$P_u(e) = \alpha + \beta e$$

$$\frac{D_F P_E D_v (T_i - T_e)}{\alpha + \beta(2e+r_2)}\left(\frac{h_e(e+r_2) - \lambda_C}{\lambda_C h_e (e+r_2)^2}\right) = \left[B + \frac{1}{\lambda_C}\text{Ln}\left(1+\frac{e}{r_2}\right) + \frac{1}{(e+r_2)h_e}\right]^2.$$

Solving this equation will make it possible to determine the optimum thickness, e^*, of the heat lagging. An example is presented in the illustration below.

3.3.2.3. *Illustration: optimum heat-lagging thickness for a steam pipe*

Steam, overheated to T_i, is transported by a cylindrical pipe between the boiler and a production workshop. The temperature of the inner wall of the pipe wall can reasonably be assumed to be constant and equal to T_i across its whole length, L. Likewise, the temperature of the surrounding environment is constant and equal to T_e.

Questions

A) If no heat lagging was used to protect the pipe, you are asked to:

1) Calculate, in kW, the heat losses flux, ϕ_0.

2) Deduce therefrom, in Dh/year, the cost of the energy losses generated by transporting the steam between the boiler and the production workshop considered.

B) Given the significance of these losses, your management decides to heat-lag the pipe.

They ask you to:

1) Give, in this case (where the pipe inner wall temperature is known), the expression of the flux as a function of the thickness of the heat lagging.

2) Calculate the critical radius of the heat lagging.

3) Calculate the optimum thickness of the heat lagging.

4) Calculate, in monetary terms, the expected savings with a heat lagging with an optimum thickness.

Data:

– Pipe:
  Internal diameter: $d_i = 50$ mm
  External diameter: $d_e = 60$ mm
  $L = 78$ m
  $\lambda = 54$ W/m°C

– Heat lagging:
  $\lambda_c = 0.2265$ W/m °C
  Unit price per m²: $P_u(e) = 6{,}500 + 1.2\ 10^5\ e$
  Lifetime: $D_v = 5$ years

– Unit price of electricity: $P_E = 2.2$ Dh/kWh

– Convective transfer coefficient outer face surrounding: $h_e = 12$ W/m² °C

– Temperatures: $T_i = 360$ °C $T_e = 20$ °C

– Annual operating time: $D_F = 7{,}600$ hrs/year

Solutions

A) Without heat lagging

1) Calculating the flux, ϕ_0, of the heat losses

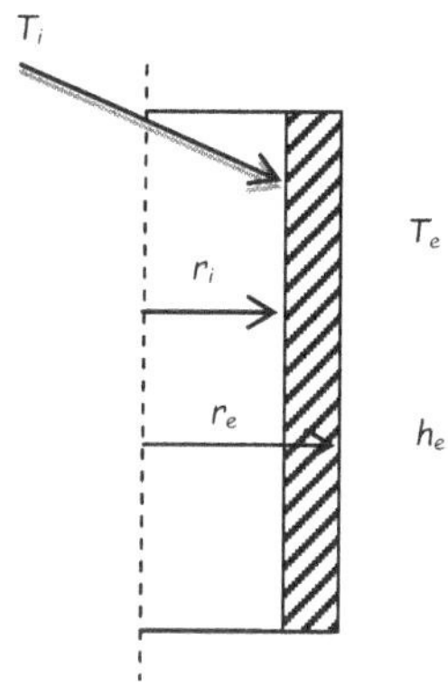

Figure 3.13. *Cross-section of the cylindrical pipe without heat lagging*

In this case, the temperature of the inner wall is known: T_i.

As a consequence, internal convection resistance is not considered in the expression of the heat flux; only resistances relative to conduction in the pipe wall and to external convection are to be taken into consideration. The flux ϕ_0 is then given by:

$$\varphi_0 = \frac{2\ \pi\ L\ (T_i - T_e)}{\frac{1}{\lambda}\text{Ln}\left(\frac{r_e}{r_i}\right) + \frac{1}{r_e h_e}}.$$

NUMERICAL APPLICATION.–

$T_i = 360$ °C $\quad T_e = 20$ °C $\quad r_i = 2.5\ 10^{-2}$ m $\quad r_e = 3\ 10^{-2}$ m

$L = 78$ m $\quad \lambda = 54$ W/m °C $\quad h_e = 12$ W/m^2 °C

$$\phi_0 = \frac{2\pi L(78)(360-20)}{\frac{1}{\lambda}\text{Ln}\left(\frac{3}{2.5}\right) + \frac{1}{3\ 10^{-2}\,(12)}}$$

$\boldsymbol{\phi_0}$ **= 59.9 kW**

2) Cost of the energy losses generated by transporting the steam

This cost is given by: $P_0 = \phi_0 D_F P_E$.

NUMERICAL APPLICATION.–

$\phi_0 = 59.9$ kW

Unit price of electricity: $P_E = 2.2$ Dh/kWh

Annual operating time: $D_F = 7{,}600$ hrs/year

Hence: $P_0 = 59.9\ (7{,}600)\ (2.2)$ **$P_0 \approx 1$ million Dh per year**

B) With heat lagging

1) Flux of the heat losses as a function of the thickness of the heat lagging

As the temperature of the inner wall is known, the flux is given by:

$$\phi(e) = \frac{2\pi\, L\, (T_i - T_e)}{\frac{1}{\lambda} Ln\left(\frac{r_e}{r_i}\right) + \frac{1}{\lambda_C} Ln\left(1 + \frac{e}{r_e}\right) + \frac{1}{(r_e + e) h_e}}.$$

To obtain a form equivalent to that mentioned in section 3.3.2.1 above, $\phi(e)$ may be presented in the form:

$$\phi(e) = \frac{A}{B + \frac{1}{\lambda_C} Ln\left(1 + \frac{e}{r_e}\right) + \frac{1}{(r_e + e) h_e}},$$

where A and B are given by:

$$A = 2\,\pi\, L\, (T_i - T_e) \text{ and } B = \frac{1}{\lambda} Ln\left(\frac{r_e}{r_i}\right)$$

2) Critical radius of the heat lagging

The critical radius, r_c, is given by: $rc = r_e + e_c$.

The critical thickness, e_c, is given by: $e_c = \frac{\lambda_C}{h_e} - r_e$.

Hence: $r_c = \dfrac{\lambda_C}{h_e}$.

NUMERICAL APPLICATION.–	
$\lambda_c = 0.2265$ W/m °C	$h_e = 12$ W/m^2 °C
$r_c = \dfrac{0,2265}{12}$	$r_c = \mathbf{18.88}$ **mm**

3) *Calculating the optimum thickness of the heat lagging*

Given that $P_u(e)$ is linear ($P_u(e) = \alpha + \beta e$), the optimum thickness of the heat lagging, e^*, is given by the solution to the following equation (see section 3.3.2.2 above):

$$\frac{D_F P_E D_v (T_i - T_e)}{\alpha + \beta(2e + r_e)} \left(\frac{h_e (e + r_e) - \lambda_C}{\lambda_C h_e (e + r_e)^2} \right) = \left[B + \frac{1}{\lambda_C} \text{Ln}\left(1 + \frac{e}{r_e}\right) + \frac{1}{(e + r_e) h_e} \right]^2,$$

with: $B = \dfrac{1}{\lambda} \text{Ln}\left(\dfrac{r_e}{r_i}\right)$;

i.e.:

$$\left[\frac{1}{\lambda} \text{Ln}\left(\frac{r_e}{r_i}\right) + \frac{1}{\lambda_C} \text{Ln}\left(1 + \frac{e}{r_e}\right) + \frac{1}{(e + r_e) h_e} \right]^2 = \frac{D_F P_E D_v (T_i - T_e)}{\alpha + \beta(2e + r_e)} \left(\frac{h_e (e + r_e) - \lambda_C}{\lambda_C h_e (e + r_e)^2} \right)$$

This equation can be rewritten in the following form: $f(e) - g(e) = 0$,

with: $f(e) = \left[\dfrac{1}{\lambda} \text{Ln}\left(\dfrac{r_e}{r_i}\right) + \dfrac{1}{\lambda_C} \text{Ln}\left(1 + \dfrac{e}{r_e}\right) + \dfrac{1}{(e + r_e) h_e} \right]^2$,

and: $g(e) = \dfrac{D_F P_E D_v}{\alpha + \beta(2e + r_e)} \left(\dfrac{h_e (e + r_e) - \lambda_C}{\lambda_C h_e (e + r_e)^2} \right) (T_i - T_e)$, or: $$g(e)=\gamma(e)\left(\frac{h_e(e+r_e)-\lambda_C}{\lambda_C h_e(e+r_e)^2}\right)(T_i-T_e),$$

with: $$\gamma(e)=\frac{D_F P_E D_v}{\alpha+\beta(2e+r_e)}.$$

NUMERICAL APPLICATION.–

λ = 54 W/m °C	λ_c = 0.2265 W/m °C	T_i = 360 °C
r_i = 2.5 10^{-2} m	r_e = 3 10^{-2} m	h_e = 12 W/m^2 °C
D_v = 5 years	α = 6,500 Dh/m^2	β = 1.2 10^5 Dh/m^3
D_F = 7,600 hrs/year	P_E = 0.0022 Dh/Wh	T_e = 20 °C

Figure 3.14 plots the *g(e)* and *f(e)* curves, as well as that representing the function, *f(e) - g(e)*. It shows that the solution is around 0.07 m.

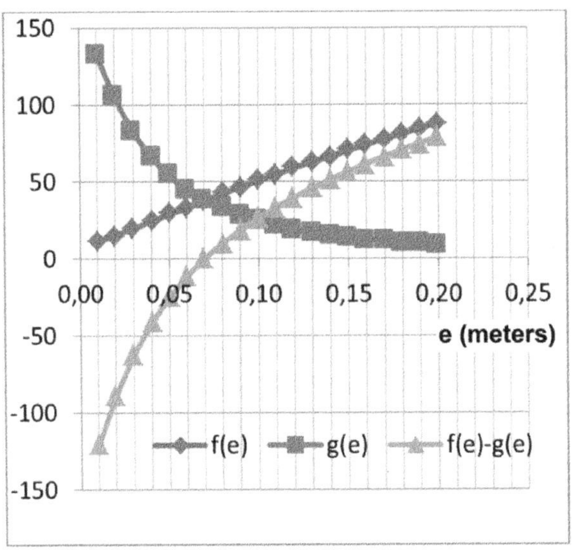

Figure 3.14. *Graphical determination of optimum heat-lagging thickness. For a color version of this figure, see www.iste.co.uk/benallou/energy2.zip*

Using the Newton–Raphson method, the solution is indeed found to be: e^* = 7 cm.

4) Savings

Using a heat lagging with an optimum thickness, the savings made are given by:

$$E = P_0 - D_F P_E \phi(e^*),$$

where:

– P_0 corresponds to the cost with no heat lagging:

$$\phi(e^*) = \frac{A}{B + \frac{1}{\lambda_C} \operatorname{Ln}\left(1 + \frac{e^*}{r_e}\right) + \frac{1}{(r_e + e^*) h_e}};$$

– A and B are given by: $A = 2\ \pi\ L\ (T_i - T_e)$ and $B = \frac{1}{\lambda}\ \operatorname{Ln}\left(\frac{r_e}{r_i}\right)$, i.e.:

$$E = P_0 - \frac{2\pi\ L D_F P_E (T_i - T_e)}{\frac{1}{\lambda} \operatorname{Ln}\left(\frac{r_e}{r_i}\right) + \frac{1}{\lambda_C} \operatorname{Ln}\left(1 + \frac{e^*}{r_e}\right) + \frac{1}{(r_e + e^*) h_e}}.$$

NUMERICAL APPLICATION.–

$P_0 = 1{,}001{,}733$ Dh/year	$D_F = 7{,}600$ hrs/year	$P_E = 0.0022$ Dh/W
$L = 78$ m	$T_i = 360$ °C	$T_e = 20$ °C
$\lambda = 54$ W/m °C	$r_i = 2.5\ 10^{-2}$ m	$r_e = 3\ 10^{-2}$ m
$\lambda_c = 0.2265$ W/m °C	$h_e = 12$ W/m^2 °C	
$e^* = 0.07$ m	$\phi(e^*) = 27.1$ kW	**E = 548,896 million Dh/year**

3.3.3. *Heat-lagging spherical containers*

Let us consider the multi-layer spherical container represented in Figure 3.15, which corresponds to a spherical storage reservoir which we wish to insulate.

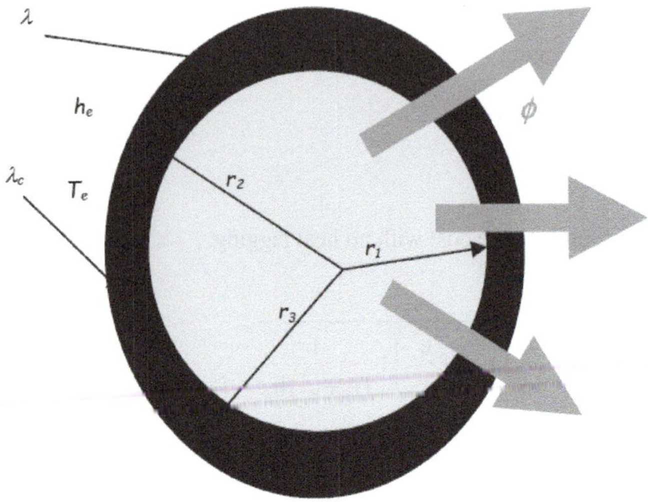

Figure 3.15. *Heat lagging a spherical reservoir*

This problem is similar to that for cylindrical jackets, explored above. Indeed, for spherical containers as well, the heat losses flux is not always a decreasing function of the heat-lagging thickness; as a result, the thermal resistance could, in some cases, present a minimum.

The flux, ϕ_c, of heat losses to the outside (see Chapter 2, section 2.9) is expressed as a function of the temperature difference $(T_i - T_e)$, as follows:

$$\phi_c(r_3) = \frac{4\pi(T_i - T_e)}{\frac{1}{r_1^2 h_i} + \frac{1}{\lambda}\frac{r_2 - r_1}{r_2 r_1} + \frac{1}{\lambda_c}\frac{r_3 - r_2}{r_3 r_2} + \frac{1}{r_3^2 h_e}},$$

or alternatively, with e designating the thickness of the heat lagging ($r_3 = r_2 + e$):

$$\phi_c(e) = \frac{4\pi(T_i - T_e)}{\frac{1}{r_1^2 h_i} + \frac{1}{\lambda}\frac{r_2 - r_1}{r_1 r_2} + \frac{1}{\lambda_c}\frac{e}{r_2(r_2 + e)} + \frac{1}{(r_2 + e)^2 h_e}}.$$

For the sake of convenience, ϕ_c is presented in the form $\phi_c(e)=\dfrac{K}{G(e)}$, with:

$$K=4\,\pi\,(T_i-T_e)$$

and: $G(e)=\dfrac{1}{r_1^2 h_i}+\dfrac{1}{\lambda}\dfrac{r_2-r_1}{r_2 r_1}+\dfrac{1}{\lambda_c}\dfrac{e}{r_2(r_2+e)}+\dfrac{1}{(r_2+e)^2 h_e}$.

We then have: $\phi'(e)=-\dfrac{K\,G'(e)}{G(e)^2}$.

Thus: $\phi'_c=0$ for $G'(e)=0$: $G'(e)=\dfrac{1}{\lambda_c r_2}\dfrac{r_2+e-e}{(r_2+e)^2}-\dfrac{1}{h_e}\dfrac{2}{(r_2+e)^3}$.

Thus: $\phi'_c=0$ for $\dfrac{1}{\lambda_c}=\dfrac{2}{h_e(r_2+e)}$.

The *critical thickness*, e_c, of the heat lagging is therefore given by: $e_c=\dfrac{2\lambda_c}{h_e}-r_2$.

The shape of the curve representing heat losses, ϕ_c, as a function of the heat lagging thickness, *e*, is then similar to that obtained in Figure 3.15 for the cylindrical jackets; *only the expression of* e_c *is different.*

Therefore, when $e_c>0$, care must be taken in choosing the insulation thickness to be used. In fact, it must be ensured that the heat-lagging thickness used is greater than the limit thickness, $e_{lim}>e_c$, defined by $\phi(e_{lim})=\phi(0)$.

Ultimately, as well as needing to satisfy the above condition ($e>e_{lim}>e_c$), the thickness of the heat lagging chosen for the spherical reservoirs should also minimize the *total annual operating cost*, as defined in section 3.3.1:

$$C_T(e)=D_F P_E \phi_c(e)+\frac{4\pi\,(r_2+e)^2}{D_V}P_u(e),$$

with:

D_F: the annual operating time

P_E: the unit price of energy

D_v: the lifetime of the heat lagging

$P_u(e)$: the cost per m^2 of the heat lagging,

$$\text{and: } \phi_c(e)=\frac{4\pi\ (T_i-T_e)}{\frac{1}{r_1^2 h_i}+\frac{1}{\lambda}\frac{r_2-r_1}{r_2 r_1}+\frac{1}{\lambda_c}\frac{e}{r_2(r_2+e)}+\frac{1}{(r_2+e)^2 h_e}}.$$

The optimum heat-lagging thickness, e^*, is that which cancels out $\frac{dC_T(e)}{de}$.

$$\text{Yet: } \frac{dC_T(e)}{de}=D_F P_E\frac{d\varphi_c(e)}{de}+\frac{4\pi}{D_v}(r_2+e)\left[2P_u(e)+(r_2+e)\frac{dP_u(e)}{de}\right].$$

$$\text{Moreover: } \frac{d\phi_c(e)}{de}=-K\frac{\frac{1}{\lambda_c(r_2+e)^2}-\frac{1}{h_e}\frac{2}{(r_2+e)^3}}{\left[\frac{1}{r_1^2 h_i}+\frac{1}{\lambda}\frac{r_2-r_1}{r_2 r_1}+\frac{1}{\lambda_c}\frac{e}{r_2(r_2+e)}+\frac{1}{(r_2+e)^2 h_e}\right]^2},$$

with: $K=4\ \pi\ \ (T_i-T_e)$.

Replacing $\frac{d\phi(e)}{de}$ with its expression, we obtain:

$$\frac{dC_T(e)}{de}=\frac{4\pi}{D_v}(r_2+e)\left[2P_u(e)+(r_2+e)\frac{dP_u(e)}{de}\right]-\frac{D_F P_E K\left[\frac{1}{\lambda_c(r_2+e)^2}-\frac{1}{h_e}\frac{2}{(r_2+e)^3}\right]}{\left[\frac{1}{r_1^2 h_i}+\frac{1}{\lambda}\frac{r_2-r_1}{r_2 r_1}+\frac{1}{\lambda_c}\frac{e}{r_2(r_2+e)}+\frac{1}{(r_2+e)^2 h_e}\right]^2}$$

Thus, by replacing K with its expression, e^* is given by the solution to the following equation with e:

$$\left[\frac{1}{r_1^2 h_i}+\frac{1}{\lambda}\frac{r_2-r_1}{r_2 r_1}+\frac{1}{\lambda_c}\frac{e}{r_2\left(r_2+e\right)}+\frac{1}{\left(r_2+e\right)^2 h_e}\right]^2=\frac{\left(T_i-T_e\right)D_F P_E D_v}{\left(r_2+e\right)^4}\frac{\left[\frac{r_2+e}{\lambda_c}-\frac{2}{h_e}\right]}{\left[2P_u\left(e\right)+\left(r_2+e\right)\frac{dP_u\left(e\right)}{de}\right]}$$

Very often, P_u varies linearly with e:

$$P_u(e)=\alpha+\beta e$$

$$\frac{\left(r_2+e\right)^4\left(2\alpha+\beta r_2+3\beta e\right)}{\left(T_i-T_e\right)D_F P_E D_v}\left[A+\frac{1}{\lambda_c}\frac{e}{r_2\left(r_2+e\right)}+\frac{1}{\left(r_2+e\right)^2 h_e}\right]^2=\left[\frac{r_2+e}{\lambda_c}-\frac{2}{h_e}\right],$$

with: $A=\frac{1}{r_1^2 h_i}+\frac{1}{\lambda}\frac{r_2-r_1}{r_2 r_1}$.

Solving this equation makes it possible to determine the optimum thickness, e^*, of the heat lagging.

3.3.3.1. *Illustration: thermal insulation of a spherical reservoir*

You have recently taken part in a specialist seminar focusing on rational use energy. You complete your report to your line manager with a proposal consisting of heat-lagging the large spherical fuel storage reservoir, in order to lose as little energy as possible.

The fuel inside the reservoir is heated using an electrical resistance to ensure that its temperature, T_i, always remains above T^*. The electricity is transformed into heat with an efficiency ε.

The storage reservoir is placed in open air. The surrounding temperature is assumed to be constant and equal to T_a.

Questions

In order to convince your manager of the merits of your proposal, you elaborate a design note comprising the following elements:

A) Presentation of the basic assumptions;

B) Determination of the heat losses when the reservoir is not heat-lagged:

1) Condition on the power, P, of the resistance to keep the fuel at a temperature $T_i \geq T^*$;

2) Determination of the annual electricity consumption in kWh/month and in Dh/year such that the fuel temperature constantly remains equal to T^*;

3) Cost of annual power consumption without heat lagging;

C) Analysis of the opportunity to heat-lag the reservoir:

1) Expression of the annual cost of electricity consumption as a function of the heat-lagging thickness;

2) Calculation of the critical radius of the heat lagging;

3) Calculation of the optimum thickness of the insulation to be used;

4) Estimation of the savings expected as a result of heat lagging.

Data:

Reservoir: $D_1 = 7.5$ m $\quad D_2 = 7.54$ m
$T_i^* = 86$ °C $\quad T_a = 21$ °C
$h_i = 32$ W/m^2 °C $\quad h_e = 22$ W/m^2 °C
$\lambda_c = 0.035$ W/m °C $\quad P_u(e) = 7{,}500 + 3.5\ 10^4\ e$ (Dh/m^2)
$D_v = 4.5$ years $\quad \lambda = 65$ W/m °C
Price of electricity: $P_E = 2.2$ Dh/kWh
Annual operating time: 365 days per year, 24 hours per day
$\varepsilon = 86\%$

Solutions

A) Assumptions:

One-dimensional conduction along direction r

B) Determination of the losses when the reservoir has no heat lagging:

1) Condition to retain on P so that $T_i \geq T^*$

If ϕ_0 indicates the energy flux without heat lagging:

$$\phi_0 = \frac{4\pi\left[T^* - T_a\right]}{\dfrac{4}{h_i D_1^2} + \dfrac{2(D_2 - D_1)}{\lambda\, D_1 D_2} + \dfrac{4}{h_e D_2^2}}.$$

Hence the condition on *P*: $P \geq \dfrac{4\pi\left[T^{*}-T_{a}\right]}{\varepsilon\left[\dfrac{4}{h_{i}D_{1}^{2}}+\dfrac{2\left(D_{2}-D_{1}\right)}{\lambda\, D_{1}D_{2}}+\dfrac{4}{h_{e}D_{2}^{2}}\right]}$.

2) Electricity consumption

When the temperature inside the reservoir equals T^*, the amount of energy lost, expressed in kWh per year, is given by: $Q_0 = \phi_0\left(\Delta t\right)$,

where:

ϕ_0 is expressed in kWatts

Δt is the number of hours per year.

Since electricity is transformed into heat with an efficiency ε, the annual electricity consumption corresponds to:

$$K_0 = \frac{24x365}{\varepsilon}\ \phi_0$$

3) Annual cost of electricity consumption

The cost of the annual losses thus stands at:

$C_0 = P_E K_0$=3.36 million Dh/year

C) Study of the opportunity to heat-lag the reservoir

1) Annual cost of electricity consumption as a function of the thickness

With a heat-lagging thickness, *e*, the term defining the thermal resistance becomes:

$$\frac{4}{h_i D_1^2}+\frac{2\left(D_2-D_1\right)}{\lambda\, D_1 D_2}+\frac{4\ e}{\lambda\, D_2\left(D_2+2\ e\right)}+\frac{4}{h_e\left(D_2+2\ e\right)^2}$$

The annual losses are thus given, in kWh/year, by:

$$Q(e)=\frac{4\pi\left[T^{*}-T\right]}{\varepsilon\frac{4}{h_iD_1^2}+\frac{2(D_2-D_1)}{\lambda\ D_1D_2}+\frac{4\ e}{\lambda\ D_2\left(D_2+2\ e\right)}+\frac{4}{h_e\left(D_2+2\ e\right)^2}}(\Delta t).$$

The annual cost of the losses is therefore:

$$C(e)=P_E\frac{Q(e)}{\varepsilon}.$$

The expression generating this cost as a function of *e* is thus:

$$C_T(e)=\frac{4\ \pi\ P_E}{\varepsilon}\frac{365\text{x}24}{\frac{4}{h_iD_1^2}+\frac{2(D_2-D_1)}{\lambda\ D_1D_2}+\frac{4\ e}{\lambda\ D_2\left(D_2+2\ e\right)}+\frac{4}{h_e\left(D_2+2\ e\right)^2}}.$$

2) Calculating the critical radius

The critical thickness is that which maximizes the total annual energy losses. It is therefore this thickness that optimizes the denominator, *D(e)*, with:

$$D\ (e)=\frac{4}{h_iD_1^2}+\frac{2\left(D_2-D_1\right)}{\lambda\ D_1D_2}+\frac{4\ e}{\lambda_cD_2\left(D_2+2\ e\right)}+\frac{4}{h_e\left(D_2+2\ e\right)^2}$$

$$\left.\frac{dD\ (e)}{de}\right|_{e=e_c}=0\ \text{yields:}\ \frac{4}{\lambda_cD_2}\frac{\left(D_2+2\ e_c\right)-2e_c}{\left(D_2+2\ e_c\right)^2}-\frac{4}{h_e}\frac{4\left(D_2+2\ e_c\right)}{\left(D_2+2\ e_c\right)^4}=0;$$

i.e.: $e_c=\frac{2\lambda_c}{h_e}-\frac{D_2}{2}.$

Consequently: $R_c=\frac{2\lambda_c}{h_e}.$

NUMERICAL APPLICATION.–

$\lambda c = 0.035$ W/m °C

$h_e = 22$ W/m^2 °C

Hence: $R_c = 0.32$ cm and $e_c = -3.67$ m

As the critical thickness is negative, there is no need to determine the limit thickness.

3) Optimum heat-lagging thickness

In this case, the total operating cost function is given by:

$$C(e) = \frac{4\pi P_E}{\varepsilon} \frac{\Delta t}{\dfrac{4}{h_i D_1^2} + \dfrac{2(D_2 - D_1)}{\lambda\, D_1 D_2} + \dfrac{4\,e}{\lambda_c D_2 (D_2 + 2\,e)} + \dfrac{4}{h_e (D_2 + 2\,e)^2}}.$$

The optimum thickness, e^*, is given by the solution to the following equation with e:

$$\left[\frac{1}{r_1^2 h_i} + \frac{1}{\lambda}\frac{r_2 - r_1}{r_2 r_1} + \frac{1}{\lambda_c}\frac{e}{r_2(r_2+e)} + \frac{1}{(r_2+e)^2 h_e}\right]^2$$

$$= \frac{(T_i - T_a) P_E D_v}{\varepsilon (r_2+e)^4} \frac{\left[\dfrac{r_2+e}{\lambda_c} - \dfrac{2}{h_e}\right]}{\left[2P_u(e) + (r_2+e)\dfrac{dP_u(e)}{de}\right]} (\Delta t).$$

In the present case, P_u varies linearly with e: $P_u(e) = \alpha + \beta\, e$

$$\frac{(r_2+e)^4 (2\alpha + \beta r_2 + 3\beta e)}{P_E D_v}\left[\frac{1}{r_1^2 h_i} + \frac{1}{\lambda}\frac{r_2 - r_1}{r_2 r_1} + \frac{1}{\lambda_c}\frac{e}{r_2(r_2+e)} + \frac{1}{(r_2+e)^2 h_e}\right]^2$$

$$= \left[\frac{r_2+e}{\lambda_c} - \frac{2}{h_e}\right]\frac{\Delta t}{\varepsilon}.$$

e^* is therefore the solution of the equation $f(e) = g(e)$, where f and g are given by:

$$f(e) = \left[\frac{r_2 + e}{\lambda_c} - \frac{2}{h_e}\right]\frac{\Delta t}{\varepsilon},$$

and:

$$g(e) = \frac{(r_2 + e)^4 (2\alpha + \beta r_2 + 3\beta e)}{P_E D_v}\left[\frac{1}{r_1^2 h_i} + \frac{1}{\lambda}\frac{r_2 - r_1}{r_2 r_1} + \frac{1}{\lambda_c}\frac{e}{r_2 (r_2 + e)} + \frac{1}{(r_2 + e)^2 h_e}\right]^2.$$

The resolution of the equation, $f(e) = g(e)$ was conducted on a spreadsheet:

e (in cm)	$f(e)$	$g(e)$	$f(e) - g(e)$
0.00	1,096,257	89	1,096,168
10.00	1,125,360	144,283	981,077
20.00	1,154,463	630,893	523,570
21.00	1,157,373	702,545	454,828
22.00	1,160,283	778,841	381,442
23.00	1,163,194	859,901	303,293
24.00	1,166,104	945,846	220,258
25.00	1,169,014	1,036,801	132,213
26.00	1,171,924	1,132,891	39,034
26.10	1,172,216	1,142,787	29,429
26.20	1,172,507	1,152,736	19,771
26.30	1,172,798	1,162,737	10,060
26.40	1,173,089	1,172,792	296
26.401	1,173,092	1,172,893	199
26.402	1,173,094	1,172,994	101
26.403	1,173,097	1,173,095	3
26.404	1,173,100	1,173,195	-95

Table 3.3. *Spreadsheet calculation of optimum heat-lagging thickness*

Figure 3.16 shows that the solution is: $e^* = 26.4$ cm.

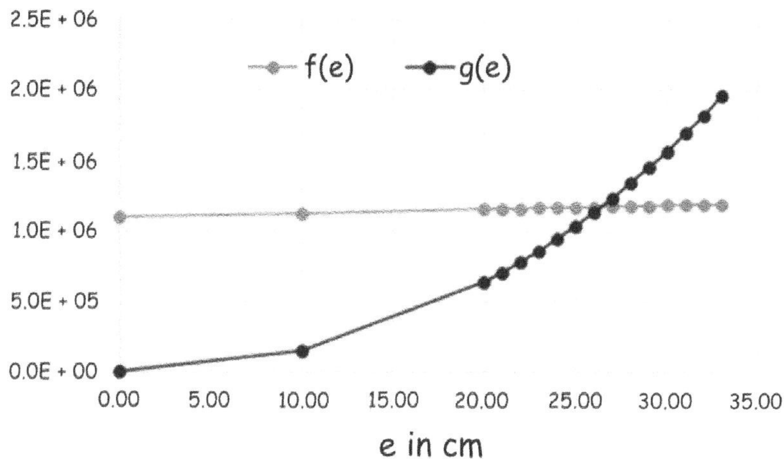

Figure 3.16. *Graphical determination of optimum thickness. For a color version of this figure, see www.iste.co.uk/benallou/energy2.zip*

4) Expected savings with heat lagging

The cost of the annual energy losses with heat lagging is $C(e^*)$, given by:

$$C\left(e^*\right) = \frac{4\,\pi\,P_E}{\varepsilon} \frac{\Delta t}{\dfrac{4}{h_i D_1^2} + \dfrac{2\left(D_2 - D_1\right)}{\lambda\, D_1 D_2} + \dfrac{4\,e^*}{\lambda_e D_2\left(D_2 + 2\,e^*\right)} + \dfrac{4}{h_e\left(D_2 + 2\,e^*\right)^2}}.$$

NUMERICAL APPLICATION.–

$D_1 = 7.5$ m $\quad D_2 = 7.54$ m $\quad \lambda = 65$ W/m °C $\quad e^* = 26.403$ cm

$T_i = 86$ °C $\quad T_a = 21$ °C $\quad h_i = 32$ W/m^2 °C $\quad h_e = 22$ W/m^2 °C

$\lambda_e = 0.035$ W/m °C $\quad P_E = 2.2$ Dh/kWh $\quad \Delta t = 365 \times 24$ $\quad \varepsilon = 86\%$

i.e.: ***C(e*)* = 561,727 Dh/year**

Hence the savings that can be made using the heat lagging: ***E* = C0 - C(e*)**

$C_0 = 3{,}361{,}742$ Dh/year

$C_e^* = 561{,}727$ Dh/year

***E* = 2,800,015 Dh/year**

4

Conduction Applications in the Reduction of Heat Losses in Construction

4.1. Introduction

For our comfort, buildings need to be heated or air-conditioned according to the climate conditions. This results in a difference in temperature between the inside and outside of buildings, and therefore a driving potential difference that will lead to the existence of a heat flux. This flux is *lost energy*.

The economic crises that followed the succession of oil shocks from 1973 onwards have led to the development of research in the field of sustainable use of energy, encompassing all sectors of consumption. The construction industry has benefited from this tremendous forward-looking dynamic.

Moreover, growing awareness of the cause-and-effect relationship between energy consumption patterns and climate change has been recorded internationally since the first Earth Summit held in Stockholm (Sweden) in 1972. This brought about a new consumer philosophy: sustainable development. Energy consumption henceforth needed to be accompanied by efforts to preserve, as much as possible, the environment and its natural resources in order for them to be passed on to future generations with the least alteration possible. In this concept, energy came to be central to environmental conservation.

Drawing on this basis of a sustainable-development vision, various government guidelines were thus established, with a view to ensuring that building designs observe the principles of sustainable development, with certain countries even introducing *thermal building regulations* as legislative frameworks for construction. These regulations include requirements as to the thermal resistance of building envelopes, for example. These imperatives involve using thermal insulation products for walls. Often designed in the form of double partitions (double walls), these are subject to heat-lagging requirements, designed to meet the required criteria in terms of thermal resistance. Transparent building walls (glazing) must meet similar constraints. This is reflected in the installation of double glazing, which minimizes heat exchange with the outside. As a result, a number of thermal insulation solutions have been developed for buildings and a whole range of innovative glazing has appeared on the market.

This chapter is devoted to applying conduction equations in the calculation of heat losses through building walls in order to establish the calculation methods that will make it possible to *verify compliance with thermal building regulations*.

The regulations of course vary from one country to the next, but the perspective and objectives are always the same: designing *smart buildings* in terms of energy consumption.

The following section looks at how thermal building regulations have evolved, and offers some advice as to the standards to be observed for thermal resistance, according to the geographical zone concerned.

Methods of calculating heat loss through flat walls are then re-explored in order to adapt them to the calculation of thermal resistances of building walls composed of single or double partitions, both with and without heat lagging. These methods are also used to calculate heat loss through single glazing and double glazing, where the panes are separated by an inert gas or simply air.

The last section of this chapter presents a method for the economic analysis of building thermal insulation projects.

4.2. Thermal building regulations

In accordance with the guidelines laid down by the United Nations Conferences on Environment and Development (Stockholm, Sweden, 1972; Nairobi, Kenya, 1982; Rio de Janeiro, Brazil, 1992; Johannesburg, South Africa, 2002 and Rio de Janeiro, Brazil, 2012), different countries have put in place guidelines or regulations that must be followed in the design and construction of buildings. The objective is to minimize heat losses in the various structures.

For example, the 2002 European Energy Performance of Buildings Directive (EPBD) requires Member States to be equipped with adequate tools and equipment in order to limit the energy consumption of buildings. This directive introduces minimum requirements for new constructions.

With this in mind, back in 2002, France for example, put in place thermal regulations (RT 2002) applicable to all new constructions. The 2002 version of these regulations was revised in 2005 and again in 2012. The next revision is scheduled for 2020.

In a similar approach, Morocco has introduced thermal regulations (*Official Bulletin* no. 6306 of 6/11/2014), applicable from 6 November 2015. These limit the flux densities of heat losses through the envelope of the building concerned. The latter must be lower than the reference densities defined in Table 4.1 for each climatic region listed in Figure 4.1.

Zone 1	40	Zone 4	4
Zone 2	46	Zone 5	1
Zone 3	48	Zone 6	5

Table 4.1. *Thermal performance of residential buildings by Moroccan climatic zone (in kWh/m²/year)*

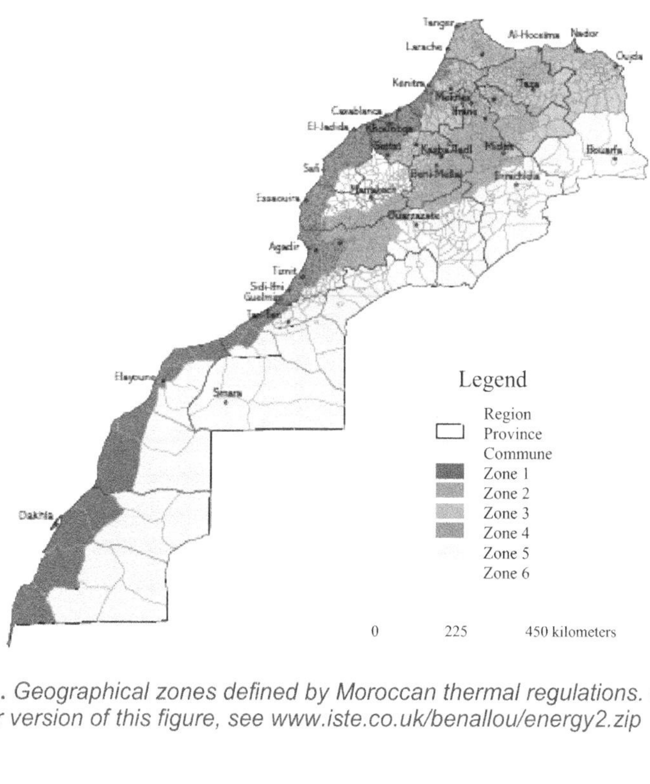

Figure 4.1. *Geographical zones defined by Moroccan thermal regulations. For a color version of this figure, see www.iste.co.uk/benallou/energy2.zip*

In addition, the Moroccan thermal regulations define the overall heat transfer coefficients and, if applicable, the reference thermal resistances for different walls, considering four intervals for the window-to-wall ratio (WWR)[1]: ≤ 15%, 16 - 25%, 26 – 35% and 36 – 45%. As an example, Table 4.2 shows the benchmarks used for a building with a window-to-wall ratio between 26 and 35%.

	Reference coefficient, U, in W/m^2 °K			Reference resistance in W^{-1}m^2 °K
Zones	Roofs	Outer walls	Glazings	Floors
1	≤ 0.75	≤ 1.20	≤ 3.30	NR(*)
2	≤ 0.65	≤ 0.70	≤ 3.30	NR(*)
3	≤ 0.60	≤ 0.70	≤ 2.60	> 0.75
4	≤ 0.55	≤ 0.60	≤ 2.60	≥ 1.25
5	≤ 0.55	≤ 0.60	≤ 2.60	≥ 1.00
6	≤ 0.55	≤ 0.60	≤ 2.60	≥ 1.00

(*) NR: no requirement.

Table 4.2. *Benchmarks of* U *and* R, *by zone, for WWR between 26 and 35%*

Similarly, the French thermal regulations define the energy performance of a given building compared to a reference building that is geometrically identical, but which has unit thermal characteristics predefined by the regulations. The most significant requirement of these regulations concerns energy losses: the transfer coefficient calculated through the envelope of the building considered (U_{bui}) must be lower than the reference coefficient $(U_{bui})_{ref}$ (see Table 4.3 and Figure 4.2).

	$(U_{bui})_{ref}$ in W/m^2 °K	
Wall type	Zones H_1 and H_2	Zone H_3
Walls	0.36	0.40
Zone 2	0.27	0.36
Zone 3	0.20	0.25

Table 4.3. *Overall heat transfer coefficients by region, defined by French regulations*

1 WWR: window-to-wall ratio:

$$\text{WWR} = \frac{\text{Total windows and bay windows area (including frames)}}{\text{Total external walls area}}$$

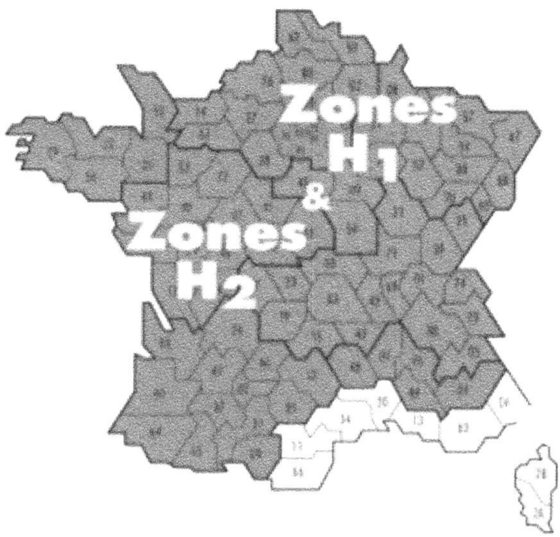

Figure 4.2. *Geographical zones defined by French thermal regulations. For a color version of this figure, see www.iste.co.uk/benallou/energy2.zip*

The tolerances accepted by French thermal regulations may in no event exceed 25%:

– for an individual house, $(U_{bui})_{rmax} = (U_{bui})_{ref} + 20\%$;

– for a residential building, $(U_{bui})_{rmax} = (U_{bui})_{ref} + 25\%$.

The heat-loss flux density is calculated for the average daily temperature. Of course, this average density also depends on the thermal conductivity of the laggings used and the glazings chosen. The minimum performances of the thermal insulations per building envelope type are presented by region in Table 4.4.

	Thermal resistance, in $W^{-1}m^2\,°K$	
Wall type	Zones H_1 and H_2	Zone H_3
Wall insulation	3.15	2.35
Floor insulation	2.25	2.25
Loft and roof insulation	6	5

Table 4.4. *Reference thermal resistances by zone in France*

The 2012 revision of the French thermal regulations (RT 2012) aims to divide the energy consumption of new buildings by three, that is, a maximum consumption of 50 kWh/m^2/year.

As for RT 2005, RT 2012 weights maximum consumption (50 kWh/m^2/year) according to the geographical zone, the altitude and the living space. All energy uses are covered; namely:

– Heating;

– Cooling;

– Domestic hot water production;

– Ventilation and auxiliaries.

Standard RT 2012 is the latest revision of French thermal standards, later to be replaced by RT 2020, which is currently in preparation. This provides for (renewable) energy self-production in a building, at a rate greater than its energy consumption. The house planned for in RT 2020 must thus produce at least as much energy as it consumes. Where excess energy is generated, it can be injected into the public grid. This is known as the *positive-energy house (PEH)* or *positive-energy building* (*PEB*).

Possible technical solutions for PEHs and PEBs include:

– eliminating *thermal bridges*[2];

– better insulated partitions;

– roofs, walls and windows functioning as energy accumulators, storing any excess solar radiation received during the day either to recover it at night within the building, or to convert it into electricity.

Thus, RT 2012 involves the use of at least one of the following renewable sources:

– *Solar thermal collectors* to produce domestic hot water;

– *Photovoltaic solar panels* to produce electricity;

– Connection to a *heating network* with more than 50% supplied by renewable energies;

– Use of *heat pumps*;

– Use of *wood boilers*.

---

2 Thermal bridges are structural building elements presenting little or no insulation. They tend to be supporting elements that generally link two partitions such as the roof and the wall, or a floor slab and a wall. Energy losses through thermal bridges account for 30 to 40% of a building's total heat losses.

4.3. Calculating losses through building partitions

Depending on the architectural project, the different partitions of a building can be diverse in nature, comprising walls, slabs (floors and roofs), but also windows and doors. The latter are composed of glazings that are transparent to some extent, often with a view to optimizing natural lighting or quite simply for architectural aesthetic reasons. In addition, modern architectural designs incorporate entire facades composed of glazings covered to a lesser or greater degree in metal jackets designed to offer protection and at times fashionable decorative designs.

Analysis of building heat losses therefore needs to focus on the different surfaces in contact with the outside. There are three types: walls, glazings and thermal bridges.

4.3.1. *Expressing the flux of energy losses*

To express the losses through building walls, we will use the results developed in the previous chapter to calculate the fluxes passing through single or composite plane walls. In this sub-section, the results are adapted to the specific notations usually used in building energy efficiency calculations.

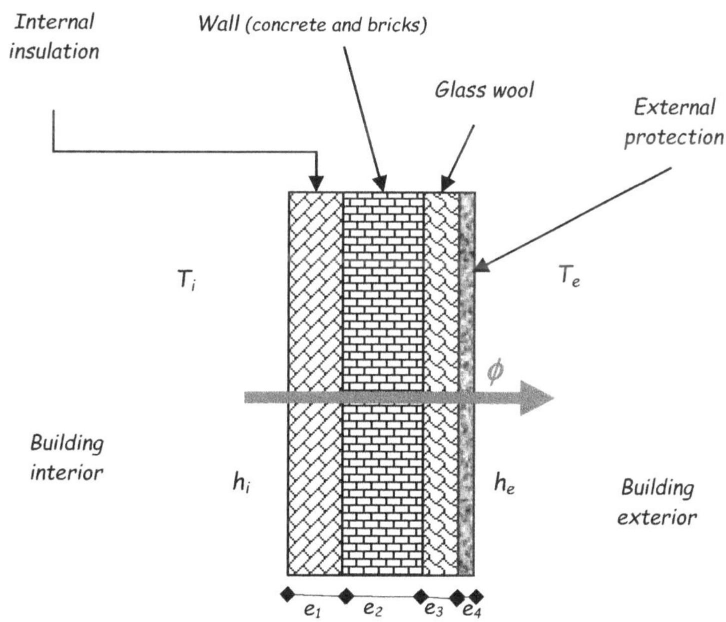

Figure 4.3. *Multi-component wall*

Let us recall that for a multi-component wall such as the one represented in Figure 4.3, the flux is expressed as a function of the conductive and convective resistances as follows:

$$\phi = \frac{T_i - T_e}{\frac{1}{h_i A} + \frac{e_1}{\lambda_D A} + \frac{e_2}{\lambda_2 A} + \frac{e_3}{\lambda_3 A} + \frac{e_4}{\lambda_4 A} + \frac{1}{h_e A}},$$

where:

A is the transfer area

h_i and h_e are the convection transfer coefficients with respect to the inside of the building and the outside surrounding environment, respectively

T_i and T_e are the temperatures inside and outside the building, respectively

e_i $1 \leq i \leq 4$ are the thicknesses of the different walls

λ_i $1 \leq i \leq 4$ are the thermal conductivities of these walls

More generally, for a wall comprising n conductive resistances and m convective resistances, the flux is given by:

$$\phi = \frac{A(T_i - T_e)}{\sum_{i=1}^{n} \frac{e_1}{\lambda_i} + \sum_{j=1}^{m} \frac{1}{h_j}}$$

4.3.2. *Notations specific to building energy efficiency calculations*

In building energy efficiency calculations, we prefer to use an overall heat transfer coefficient, noted U, such that the flux can be expressed as follows:

$$\phi = UA\,(T_i - T_e)$$

where:

ϕ is the heat-loss flux

A is the transfer area

T_i-T_e is the driving potential difference

T_i is the temperature inside the building

T_e is the temperature outside the building

By identifying this equation with the previous flux expression, we obtain the relation that defines the overall coefficient, U, as a function of the different resistances, that is:

$$U = \frac{1}{\sum_{i=1}^{n} \frac{e_i}{\lambda_i} + \sum_{j=1}^{m} \frac{1}{h_j}}$$

The overall coefficient, U, therefore corresponds to the inverse of the sum of the convective and conductive resistances. It is expressed in W/m^2K.

In building energy efficiency calculations, it is common to have different overall coefficients for each piece of masonry or woodwork. For woodwork, for example, we will note using:

- U_g the overall coefficient of the glazing alone;
- U_f the overall coefficient of the joinery frame alone;
- U_w the overall coefficient of the entire joinery of the window.

These overall coefficients (U_w, U_f and U_g) are often wrongly referred to as "heat coefficients". They should in fact be specifically described as "overall heat transfer coefficients".

Moreover, given that the structure of a building is composed of different types of partitions (single walls, double or composite walls, floor slabs, roof slabs, thermal bridges, etc.), we will look to determine the overall coefficient, U_{wall}, of the structure considered. This coefficient must integrate the contributions of the different layers making up the partition concerned.

4.3.3. *Calculating losses through composite partitions: walls, floors and roofs*

Walls, floors and roof slabs may, generally speaking, be considered as partitions composed of several layers: concrete, insulation, protection, paint, coatings, etc.

The heat-loss flux is given by: $\phi = UA\,(T_i - T_e)$.

Calculating the heat-loss flux therefore requires knowledge of the overall transfer coefficient, U. Yet, in most situations, this coefficient is not known. It is determined from more basic data that define the thermal resistance of the different layers making up the partition.

Let us consider, for example, a double building wall: the space between the walls is filled with a lagging of thermal conductivity, λ_c (see Figure 4.4).

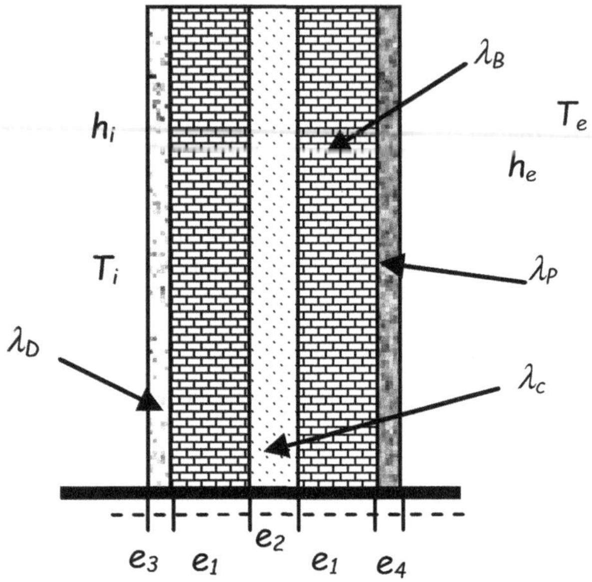

Figure 4.4. *Composite wall with internal and external convections*

The inside of this partition is covered with a decorative layer made of wood of thickness, e_3, and of thermal conductivity, λ_D, whilst the outside of the wall is coated with a protective, decorative layer of thickness, e_4, and of thermal conductivity, λ_P.

The overall coefficient of this partition is given by:

$$U = \frac{1}{\frac{1}{h_i} + \frac{e_3}{\lambda_D} + \frac{2e_1}{\lambda_B} + \frac{e_2}{\lambda_c} + \frac{e_4}{\lambda_P} + \frac{1}{h_e}}$$

4.4. Calculating losses through glass walls

Glass walls are not impenetrable to heat. Whatever their quality, thickness or number of layers, they do not represent a perfect insulator: a fair amount of energy will nevertheless be able to propagate outwards through these windows.

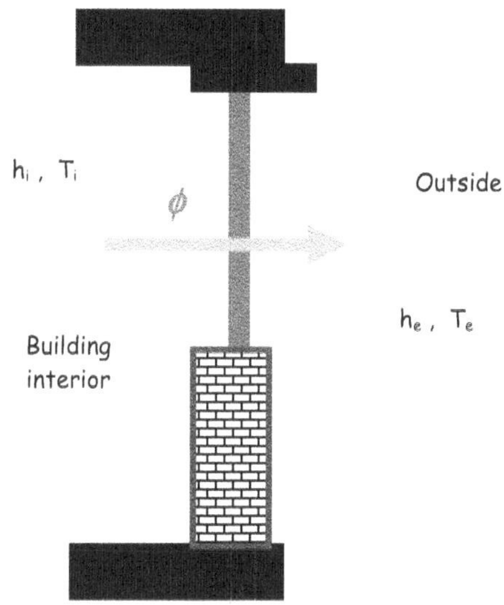

Figure 4.5. *Losses through the window of a building*

Indeed, as shown in Figure 4.5, the contact of glazings with outside air generates a heat transfer that essentially occurs through conduction or convection, although there is also radiation present.

In practical calculations concerning common situations where the temperatures considered are within the usual fields, radiation is negligible compared to the conduction and convection transfers. Convective transfers are characterized by the two coefficients, h_i and h_e. Conduction through the glazing can be differentiated by its thermal conductivity, λ.

The *standardized values* of h_e and h_i, to be used in calculating heat losses through glass walls, are:

– *regardless of the season*: $h_i = 8$ W/m^2 °C;

– *for summer*: $h_e = 23$ W/m^2 °C;

– *for winter*: $h_e = 34$ W/m^2 °C.

For a transfer area, *A*, and a glazing thickness, *e*, the heat-loss flux, ϕ, will therefore depend on the resistances to internal and external convections, and on the conductive resistance of glass; these resistances are given by:

Internal convective resistance: $R_i = \frac{1}{h_i A}$

External convective resistance: $R_e = \frac{1}{h_e A}$

Conductive resistance: $R_c = \frac{e}{\lambda A}$

The expression of the energy flux, ϕ, is written:

$$\phi = \frac{A(T_i - T_e)}{\frac{1}{h_i} + \frac{e}{\lambda} + \frac{1}{h_e}}$$

The *overall exchange coefficient*, *U*, for single glazing is then:

$$U = \frac{1}{\frac{1}{h_i} + \frac{e}{\lambda} + \frac{1}{h_e}}$$

As a result, the flux is obtained in the simple form: $\phi = UA(T_i - T_e)$

Commercially, the insulation performance of glazings is expressed as a function of their overall exchange coefficients, *U*.

For single glazings used in buildings, the overall exchange coefficient is between 5 and 6 W/m^2 °C.

We will then use for single glazing: $U = 5.5$ W/m^2 °C.

Double glazing is often used to improve the insulation performance of glass walls. Double glazing consists of two window panes separated by dry air, vacuum or an inert gas such as Argon. Thus, the space between the two panes creates an *additional thermal resistance* (see Figure 4.6). This leads to a significant improvement to the overall exchange coefficient. Generally, the double-glazing coefficient is *halved*, which implies that *the flux will also be twice as small.*

For *double glazings, we will use*: $U = 2.8$ W/m^2 °C.

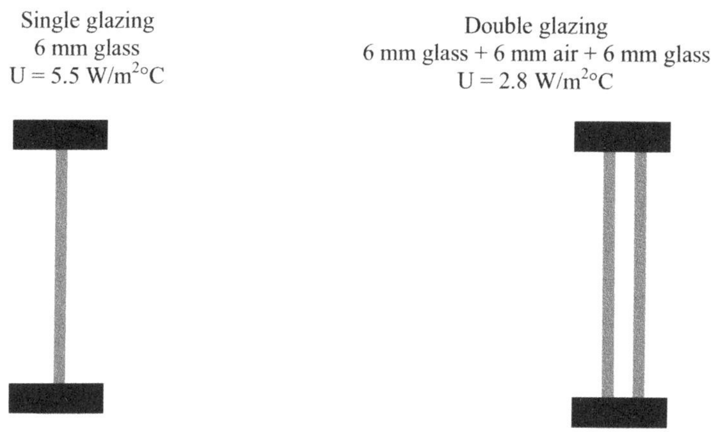

Figure 4.6. *Single- and double-glazed windows*

This improvement can be taken further if the specifications of the building considered so require, by doubling the distance between the window panes (12 mm instead of 6 mm) and by using a special (low-emitting) quality of glass. It is thus possible to virtually reduce by half the overall exchange coefficient, U. Once again, this implies that the flux will be twice as small.

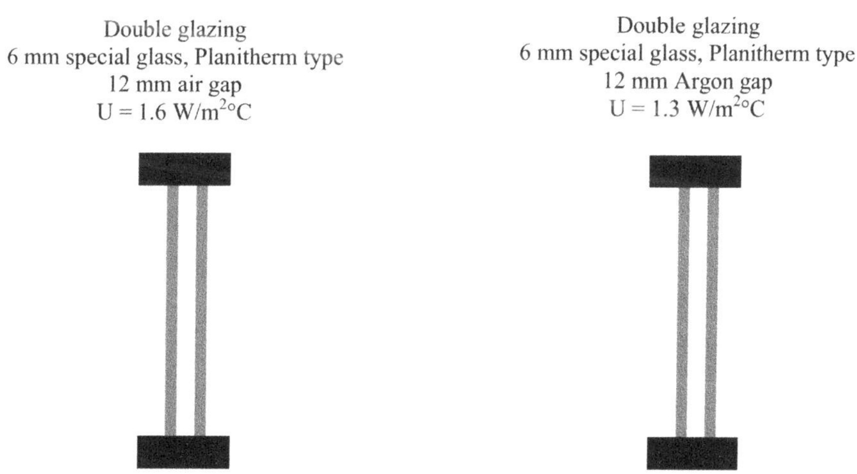

Figure 4.7. *Double glazings with air-filled and Argon-filled cavity*

Table 4.5 summarizes the values of the overall coefficient to be used for the different glazings available commercially.

Glazing type	Single, 6 mm	Double, 6 mm with 6-mm air gap	Double, 6 mm special with 12-mm air gap	Double, 6 mm special with 12-mm Argon-filled gap
U (W/m^2 °C)	5.5	2.8	1.6	1.3

Table 4.5. *Overall exchange coefficients for different glazings*

4.4.1. *Illustration: minimum thermal resistances for the walls of a hotel to be constructed*

A tourism development group is looking to construct a luxury hotel in Marrakech. The architect's plans show a building of length L, width w and height h, comprising:

- double-partition walls, the detailed plan of which is presented in Figure 4.8;
- different-sized windows;
- bay windows.

The dimensions and the number of windows and bay windows are given in Table 4.6.

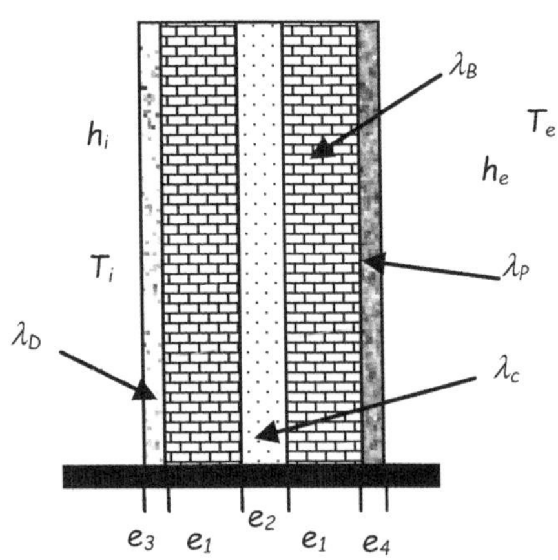

Figure 4.8. *Detailed plan of partitions*

Questions

1) Calculate the window-to-wall ratio for this building.

2) What are the maximum values that the overall coefficients of the roof, outer walls and glass walls can have in order to satisfy local thermal regulations?

3) What is the minimum thermal resistance that the floor needs to have?

4) What would the annual energy losses be, in MWh/year, by adopting the maximum values of the overall coefficients and the minimum resistance value for the floor?

5) What do these losses represent in MDh/year?

Data:

Sash type	Dimensions with frame (m)		Number
	Height	Width	
Windows	1.8	3.2	476
Windows	1.8	2.4	1,904
Windows	0.95	1.2	238
Bay windows	7.2	67	140

Table 4.6. *Dimensions of windows and bay windows*

$L = 455$ m

$w = 67$ m

$h = 278$ m

$\Delta\theta = 7.5$ °C (average over 24 hours)

electricity unit price: $P_E = 2.5$ Dh/kWh

annual operating time: $D_F = 8{,}760$ hours/year

Solutions

1) Window-to-wall ratio

By definition, this ratio is given by:

$$\text{WWR} = \frac{\text{Total windows and bay windows area (including frames)}}{\text{Total external walls area}}$$

NUMERICAL APPLICATION.–

Type of frame	Dimensions including frame: Height (m)	Width (m)	Number	S (m²)
Windows	1.8	3.2	476	2741.76
Windows	1.8	2.4	1904	8225,28
Windows	0.95	1.2	238	271.32
Bay windows	7.2	67	140	67536
			Total glazings	78774.36

Hence:

$$\mathrm{WWR} = \frac{78\,774.36}{2(455+67)278} \qquad \mathbf{WWR = 27.14\%}$$

2) Maximum possible values of the overall coefficients

The project is located in Marrakech, that is, in zone 5, which is defined by Moroccan thermal regulations.

In this zone, the thermal regulations plan for the following maximum transfer coefficients are:

– for walls: $U_{max} = 0.60$ W/m² °K;

– for glazings: $U_{max} = 2.60$ W/m² °K;

– for roofs: $U_{max} = 0.55$ W/m² °K.

3) Minimum thermal resistance that the floor needs to have

For zone 5, Moroccan thermal regulations plan for a thermal resistance of the floor that is as a minimum equal to: $R_{min} = 1$ W^{-1}m^2 °K.

4) Annual energy losses

By adopting the maximum overall-coefficient values and the minimum resistance value for the floor, the different fluxes are given by:

– flux through walls: $\phi^{\text{Walls}} = U^{\text{Walls}}_{\text{Max}} A^{\text{Walls}} (\Delta\theta)$

with: $U^{\text{Walls}}_{\text{Max}} = 0.60\ \text{W/m}^2\,°\text{K}$ and $A^{\text{Walls}} = 2(L+1)h$;

– flux through glass walls: $\phi^{\text{Panes}} = U_{\text{Max}}^{\text{Panes}} A^{\text{Panes}} (\Delta\theta)$

with: $U_{\text{Max}}^{\text{Panes}} = 2.60 \ \text{W}/\text{m}^2\,°\text{K}$ and $A^{\text{Panes}} = $ Total glazings;

– flux through roof: $\phi^{\text{Roof}} = U_{\text{Max}}^{\text{Roof}} A^{\text{Roof}} (\Delta\theta)$

with $U_{\text{Max}}^{\text{Roof}} = 0.55 \ \text{W}/\text{m}^2\,°\text{K}$ and $A^{\text{Roof}} = \text{L} \times 1$;

– flux through floor: $\phi^{\text{Floor}} = \dfrac{1}{R_{\text{Min}}^{\text{Floor}}} A^{\text{Floor}} (\Delta\theta)$

with: $R_{\text{Min}}^{\text{Floor}} = 1 \ \text{W}^{-1}\text{m}^2\,°\text{K}$ and $A^{\text{Floor}} = \text{L} \times 1$.

The heat-loss flux is then the sum of all of the components:

$$\phi = \phi^{\text{Walls}} + \phi^{\text{Panes}} + \phi^{\text{Roof}} + \phi^{\text{Floor}}$$

The losses are obtained from the flux value: $Q = \phi\, D_F$.

NUMERICAL APPLICATION.–

$L = 455$ m $w = 67$ m $h = 278$ m $\Delta\theta = 7.5$ °C

$D_F = 8{,}760$ hours/year

Hence: $A^{Walls} = 290{,}232 \ \text{m}^2$ $\quad$ $Total\ glazings = 78{,}774.36 \ \text{m}^2$

$A^{Floor} = A^{Roof} = 30{,}485 \ \text{m}^2$

$\phi^{Walls} = 1.31 \ \text{MW}$ $\quad$ $\phi^{Panes} = 1.54 \ \text{MW}$ $\quad$ $\phi^{Roof} = 0.13 \ \text{MW}$

$\phi^{Floor} = 0.23 \ \text{MW}$

$\phi = 3.21$ MW and **Q = 28,002 MWh/year**

5) Annual cost of energy losses

The cost of the energy losses is given by: $C_E = P_E Q$.

i.e.: $C_E = 70$ MDh/year.

4.5. Optimizing energy choices for building heat insulation

One problem is often posed during thermal analyses as part of architectural projects: how to choose which lagging and glazings to use. It is important to firstly emphasize that this choice must satisfy the guidelines of the thermal regulations of the country in question. Yet the multiple nature of the alternatives that are technically possible means that determining which solution to retain must essentially occur based on economic considerations, linked to the extent of the investment to be made to enable implementation.

As we will see below, this choice is conditioned by several parameters. Within this framework, it will be necessary to respond convincingly to the essential questions:

– Which glazing should be used: single or double?

– If double glazing is envisaged, what distance would need to be adopted between the window panes?

– Do we need to consider panes separated by air, vacuum or Argon?

– Which lagging should be used?

– What will its thickness need to be?

– What is the additional investment introduced for each alternative within the overall cost of constructing the building considered?

In principle, taking the lagging with the lowest thermal conductivity and the glazing with the lowest overall coefficient should lead to the lowest heat losses. However, lagging prices increase as thermal conductivity decreases and the cost of the best-insulating double glazings can become exorbitant compared to the budget component of the architectural project selected. Therefore, the answer to the questions asked depends on the following parameters:

– the different glazing solutions available commercially;

– the types of laggings available on the market;

– the cost of each glazing option: single or double, with or without separating gas;

– the lagging cost per m^2 of for different thicknesses;

– the lifetime of these laggings;

– the electricity or energy cost in general.

Therefore, the choice needs to be made after conducting an economic analysis by taking into account these different options. As with the problem posed by lagging in industrial installations (see previous chapter), this is an optimization problem. We will seek a compromise between the cost of heat losses and the cost of the additional investment required in order to put in place laggings and energy-efficient glazings.

A *total annual building operating cost function* is defined as being the sum of the cost of the residual heat losses over a year and the annual costs of the lagging, C_K, and the glazing, C_V.

C_V, which depends on the glazing option chosen, is given as a function of *its unit price*, $P_u(e_V)$, *its lifetime*, D_{vV}, *and the glazed area*, A_V, for each glazing option:

$$C_V(e_V) = \frac{P_u(e_V)\ A_V}{D_{vV}}$$

Likewise, C_K depends on the thickness, e_K, of the lagging, which is generally that of the double partition. It is given as a function of *its unit price*, $P_u(e_K)$, *its lifetime*, D_{Vk}, *and its lagged area*, A_k:

$$C_k(e_k) = \frac{P_u(e_k)\ A_k}{D_{vK}}$$

The *annual cost of heat losses*, P(e), is composed of two terms: the cost of losses through the lagged walls, $P_K(e_K)$, and those relating to the glazings, $P_V(e_V)$. In both cases, the cost of the losses is obtained by multiplying the annual flux of the losses by the energy price, that is:

$$P_K(e_K) = U_K(e_K)A_K(T_i - T_e)D_F P_E$$

$$P_V(e_V) = U_V(e_V)A_V(T_i - T_e)D_F P_E,$$

where:

A_K and A_V are the lagged and glazed transfer areas, respectively;

U_K and U_V are the overall coefficients of the lagged and glazed surfaces, respectively;

e_K and e_V are the thicknesses of the lagging and the glazing, respectively;

D_F is the annual operating time of the unit to be lagged, generally 8,760 hours/year;

P_E is the energy unit price (Dh/kWh or Dh/kcal).

$C_T(e)$ is then given by:

$$C_T(e_K, e_V) = \left[U_K(e_K)A_K + U_V(e_V)A_V\right]D_F P_E(T_i - T_e) + \frac{P_u(e_k)\ A_k}{D_{vK}} + \frac{P_u(e_V)\ A_V}{D_{vV}}$$

Therefore, the total cost function admits two variables. The optimization problem is relatively complex, but it is clear that the first term of $C_T\ (e_K,\ e_V)$ decreases with e_K and e_V (since the losses decrease with the thickness); inversely, the second term increases with e_K and e_V, since the lagging cost is higher for large thicknesses. The sum thus admits an optimum, which will determine the economic thickness of the lagging to be used and the glazing option to be retained.

Nevertheless, let us note that for the most common buildings, the lagging employed is generally determined by the availability and humidity-resistance conditions. Expanded polystyrene is most often preferred in construction.

Furthermore, to facilitate execution, the lagging thickness is generally set by the gap of the double partition. Consequently, the only options remaining to be studied are those relating to the glazing types. This helps simplify the analysis. An example of such an analysis, conducted to determine the choice of glazing type to be used for the windows of a building, is presented in section 4.5.1.

4.5.1. *Illustration: energy losses through the windows of a building*

The architectural plans of a large building present different windows, as well as bay windows looking out onto balconies. The dimensions and the number of windows and bay windows are given in Table 4.7.

The architect in charge of the project asks you to provide him with arguments to convince the contractor to use double glazing, despite the additional investment that this implies.

Questions

With this in mind, you undertake to:

1) Determine (in Dh/year) the cost of energy losses through the glass walls, if these were composed of 6-mm single glazing.

2) Determine this cost in the event that the glass walls were to be composed of:

i) two panes, 6-mm thick, separated by a 6-mm air cavity;

ii) two low-emitting special panes, each 6-mm thick, separated by a 12-mm air cavity;

iii) two low-emitting special panes, each 6-mm thick, separated by a 12-mm Argon-filled cavity.

Given that the different solutions examined above correspond to different investment costs, you advise the architect to use a parameter which informs about investment profitability in order to convince his customer. You suggest he uses the "average payback time", owing to its simplicity. Indeed, this profitability analysis parameter corresponds to the time needed for the saving made thanks to the glazing to equal the additional investment cost. Put another way:

$$\tau = \frac{I_s}{E_\alpha},$$

where:

τ is the average payback time, in years;

I_s is the additional investment brought about by the solution being considered;

E_a is the annual saving made using the solution being considered.

3) Calculate, for each of cases (i), (ii) and (iii) mentioned in the second question, the values of I_s, E_a and τ.

4) What type of glazing would you recommend to the architect?

Sash type	Dimensions	Number
Windows	1.20 m x 0.80 m	105
Windows	3.20 m x 0.80 m	45
Bay windows	4.80 m x 2 m	45

Table 4.7. *Dimensions and number of windows and bay windows*

Glazing type		U (W/m^2 °C)	Price (Dh/m^2)
Single	6 mm	5.5	1,100
Double	6 mm glass + 6 mm air + 6 mm glass	2.8	2,500
Double	6 mm special glass + 12 mm air + 6 mm special glass	1.6	3,900
Double	6 mm special glass + 12 mm Argon + 6 mm special glass	1.3	4,200

Table 4.8. *Costs of glazings available on the market*

Data:

$\Delta\theta = 15$ °C
electricity unit price: $P_E = 2.5$ Dh/kWh
annual operating time: $D_F = 8{,}760$ hours/year

Solutions

1) Cost of energy losses with 6-mm single glazing

If ϕ_l designates the heat-loss flux through glass walls, the cost of these losses is given by:

$$C = \phi_l D_F P_{E,,}$$

where: $\phi_l = U_l\, A\, \Delta\theta$.

i.e.: $C = U_l\, A\, D_F\, P_E\, \Delta\theta$.

NUMERICAL APPLICATION.–

$A = 105\ (1.2 \text{ x } 0.8) + 45\ (3.2 \text{ x } 0.8) + 45\ (4.8 \text{ x } 2)$

$A = 648$ m^2

$D_F = 8{,}760$ hours/year

$P_E = 2.5$ Dh/kWh

$\Delta\theta = 15$ °C

$U_l = 5.5\ 10^{-3}$ kW/m² °C

Hence: $C = A\ D_F\ P_E\ \Delta\theta U_l$ = 648 x 8,760 x 2.5 x 15 x U_l = 212,868 $10^3\ U_l$

C = 1.171 million Dh/year

2) Cost of energy losses in the case of double glazing

If ϕ_i designates the heat-loss flux through double-glazed walls, the corresponding costs are given by:

$$C_i = A\ D_F\ P_E\ \Delta\theta U_i$$

NUMERICAL APPLICATION.–

C_i = 212,868,103 U$_i$

Case	Glazing type		U_i (kW/m² °C)	C_i (million Dh/year)
i)	Double	6 mm glass + 6 mm air + 6 mm glass	2.8 10^{-3}	0.596
ii)	Double	6 mm special glass + 12 mm air + 6 mm special glass	1.6 10^{-3}	0.341
iii)	Double	6 mm special glass + 12 mm Argon + 6 mm special glass	1.3 10^{-3}	0.277

Table 4.9. *Costs of heat losses according to the double-glazing option*

3) Calculating the investment, saving and payback time

If P_i designates the unit price of the glazing of solution (i), the corresponding investment is given by:

$$I_i = A\ P_i$$

Hence the additional investment due to solution (i): $I_{si} = I_i - I_0$.

Where I_i is the investment corresponding to the basic solution, i.e. the solution with single glazing.

Furthermore, the annual saving that would be made by introducing solution (i) is given by:

$$E_{ai} = C - C_i$$

The payback time corresponding to solution (i) is then:

$$\tau_i = \frac{I_{si}}{E_{ai}} = \frac{A\ P_i - I_0}{C - C_i}$$

NUMERICAL APPLICATION.–

$A = 648\ m^2$

$I_0 = P_0 \times A$ $\quad I_0 = 1{,}100 \times A$ $\quad I_0 = 1{,}100 \times 648$ $\quad I_0 = 712{,}800$ Dh

C = 1,170,774 Dh/year

Case	Glazing type		$A\ P_i$(Dh)	I_{si} (Dh)	C_i (Dh/year)	E_{ai} (Dh/year)	τ (years)
i)	Double	6 mm glass + 6 mm air + 6 mm glass	1,620,000	907,200	596,030	574,744	1.58
ii)	Double	6 mm special glass + 12 mm air + 6 mm special glass	2,527,200	1,814,400	340,589	830,185	2.19
iii)	Double	6 mm special glass + 12 mm Argon + 6 mm special glass	2,721,600	2,008,800	276,728	894,046	2.25

Table 4.10. *Calculating payback times for the different options*

4) Glazing to be recommended to the architect

This is the glazing that gives the lowest payback time, τ. We will therefore recommend a *double glazing of 6 mm glass + 6 mm air + 6 mm glass*.

Indeed, the investment corresponding to this solution will be recovered in under 19 months.

4.6. Reading: financing energy renovations, innovative schemes

Whilst they differ slightly from one country to another, national thermal building regulations all point in the same direction. They establish minimum energy efficiency requirements for new buildings, but do not impose obligations for existing buildings. Instead, often attractive incentives and innovative financing schemes are offered, but owners of older homes are not required to adopt them. Indeed, analysis of the different thermal regulations shows that whilst the regulatory constraints related to energy efficiency are strict for new constructions, incentives are favored when it comes to existing buildings.

Among these measures, mechanisms have been put in place to encourage owners of old buildings to carry out Energy Performance Diagnoses (EPD) with a view to carrying out renovations. These diagnoses, which are 100% subsidized in the majority of cases, allow homeowners to find out in what ways their housing deviates from their country's thermal regulations. It is up to them to decide to initiate improvement works, but attractive funding schemes have been developed to encourage such efforts.

In Germany, for example, subsidies and loans at preferential rates have been put in place by the state-owned investment bank, KfW, for the conducting of energy renovation work on existing buildings. This scheme resulted in the renovation of more than 42,000 homes in 2011 alone.

In Sweden, a major public construction program, launched in the 1960s, has made it possible to build more than one million homes in 10 years, especially in the suburbs of large cities. However, this operation did not focus any attention on energy efficiency since, at the time of its implementation, the new awakening around energy matters had not yet generated sufficient awareness. Currently, these homes are largely out of step with the country's thermal regulations and their energy rehabilitation represents one of the biggest challenges in the field. Yet an incentive mode of financing these renovations remains to be developed.

The United Kingdom and Germany are among the few countries to have adopted measures requiring energy renovation works to be undertaken, even for older buildings.

As such, Germany imposes a minimum of 10 to 15% of renewable energies in the energy mix retained during housing renovation works.

As of 2016 in the United Kingdom, homes or offices not presenting an acceptable level of energy performance can no longer be rented out. But, in parallel, the United Kingdom has launched, in application of its Energy Act 2011, a very large incentive program called the Green Deal, with 700 million Euros of public funds to ensure the energy upgrade of 3.5 million homes using an innovative mode of financing. In fact, thermal insulation work and, more generally, work to improve the energy efficiency of old buildings, are financed through energy bills. A loan taken out by an individual can be repaid gradually (over 25 years), thanks to the savings made on energy expenditure. For example, if the energy bill of a house was €300 a month prior to renovations, and if the renovation project allowed this bill to be reduced to €220 a month, then the €80 a month thus saved would give the possibility of repaying the loan taken out in order to conduct the renovation in question. A Green Deal provider provides the individual with guidance from the very beginning of the project. They are an energy provider that make the initial estimate, supervise the work and finance it.

The financing scheme developed by the Green Deal is extremely attractive, but it has certain drawbacks; in particular, the high interest rate on loans and the difficulty relating to the transfer of loan repayments when selling a property (apartment, house, etc.). Anyone purchasing a property that has benefited from a renovation under the Green Deal scheme will be required to continue to repay the loan taken out by the seller, via the energy bill.

5

Conduction with Energy Generation

5.1. Introduction

In several practical situations, we are led to analyze conduction problems in materials inside which energy is generated. As well as transferring energy by conduction, these materials also produce energy themselves. We then have an example of conduction with energy generation. This is the case of industrial reactors where exothermic reactions (positive heat generation) or endothermic reactions (negative heat generation) take place. Another example is that of the fuel bars of nuclear reactors, where fission reactions are accompanied by significant energy generation. In such contexts, analysis of how energy is propagated in the materials considered is a key element in equipment design.

This chapter is dedicated to the study of conduction with heat generation. We will discuss the most common cases in practice, including generation in plane conductors, generation in cylindrical bars and heat propagation in rectangular fins, a configuration that is very common in electronic radiators used to dissipate the heat generated in the operation of electronic chips (see Volume 3, Chapter 5).

5.2. Plane conductor with generation

Figure 5.1 presents a plane conductor where energy with a power density, P, is generated.

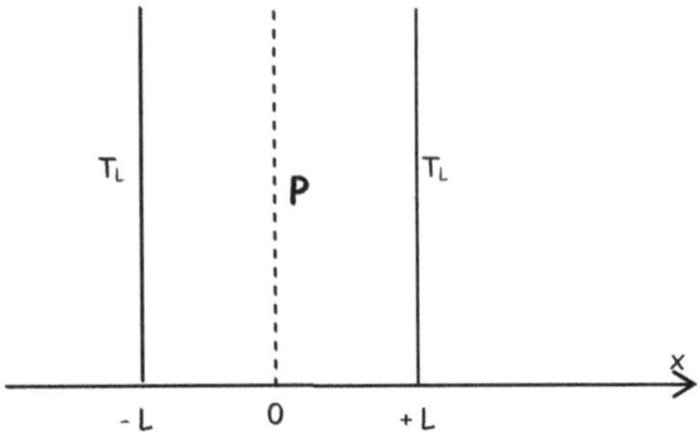

Figure 5.1. *Plane conductor with an energy generation of power P*

In this case, we have a one-dimensional conduction in the *x* direction. The conduction equation is given by:

$$\frac{d^2T}{dx^2} + \frac{P}{\lambda} = 0$$

Integrating this equation leads to the temperature profile in the material considered. A first integration gives:

$$\frac{dT}{dx} = -\frac{P}{\lambda}x + K_1$$

A second integration makes it possible to obtain *T(x)*:

$$T(x) = -\frac{P}{2\lambda}x^2 + K_1x + K_2$$

Constants K_1 and K_2 are determined from the boundary conditions. The most common are:

CL_1: $x = -L$; $T = T_L$

CL_2: $x = L$; $T = T_L$

In this case, we obtain: $K_1 = 0$ and $K_2 = T_L + \frac{PL^2}{2\lambda}$.

Hence the parabolic temperature profile: $T(x) = -\frac{P}{2\lambda}x^2 + T_L + \frac{PL^2}{2\lambda}$.

or: $T(x)-T_m = -\frac{P}{2\lambda}x^2$, where: $T_m = \frac{PL^2}{2\lambda}+T_L$.

The flux density is obtained by differentiating the temperature profile:

$$\varphi(x) = -\lambda \frac{dT}{dx}$$

$$\varphi(x) = \lambda \frac{P}{2\lambda}(2x) = P\ x$$

The flux density at the wall is therefore: $\varphi(L) = P\ L$.

5.2.1. *Illustration: generation in a plane conductor*

A plane conductor of thickness e and thermal conductivity λ is found to be the seat of a heat generation with a power density G that is constant but unknown.

The faces of the conductor are in contact with a fluid, the temperature of which can be assumed to be constant and equal to T_a.

The heat transfer coefficient between the conductor and the fluid is noted h_a.

Questions

1) Considering that conduction occurs in the x direction only, determine the conduction equation and the associated boundary conditions.

2) Deduce therefrom the temperature profile, $T(x)$, in the interval $\left[-\frac{e}{2}, \frac{e}{2}\right]$, as a function of the wall temperature, T^*, G, e and λ.

3) Calculate the flux density at the wall.

4) Give the expression of T^* as a function of G, e and λ.

5) Show that $T(x)$ passes by a maximum and give the expression of T_{Max} as a function of T^*, G, e and λ.

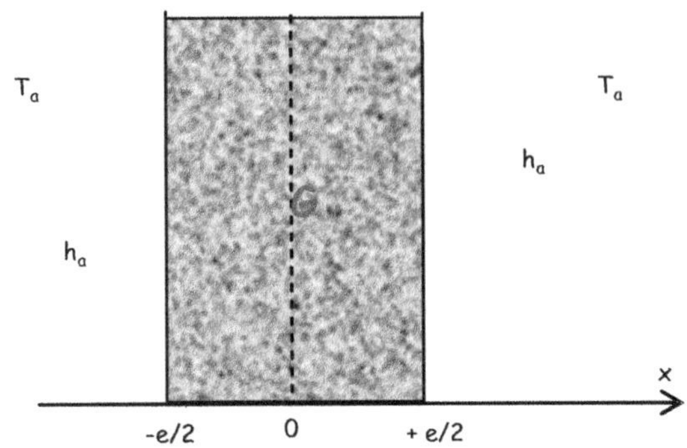

Figure 5.2. *Plane conductor with heat generation*

6) Using the expressions established above, verify that the following condition is satisfied:

$$T_{Max} > T^* > T_a$$

7) What must the value of h_a be so that $T_{Max} < 650$ °C?

8) Calculate T^* for $h_a = 15$ W/m^2 °C.

<u>*Data:*</u>

$e = 20$ cm $\quad$ $G = 75$ kW/m^3

$\lambda = 25$ W/m °C $\quad$ $T_3 = 40$ °C

Solutions

<u>*1) Equation of conduction and associated boundary conditions*</u>

With a single propagation direction and a generation power density, G, the equation of conduction is given by:

$$\frac{d^2T}{dx^2} + \frac{G}{\lambda} = 0,$$

with the boundary conditions:

$$CL_1\text{: } x = \frac{e}{2} \Rightarrow T = T^*$$

CL_2: $x = \frac{e}{2} \Rightarrow T = T^*$

2) *Temperature profile*

Integrating the equation of conduction gives:

$$T(x) = -\left(\frac{G}{2\lambda}\right) x^2 + \alpha x + \beta$$

Using the boundary conditions we obtain:

$$x = \frac{e}{2} \Rightarrow T^* = -\left(\frac{G}{8\lambda}\right) e^2 + \frac{\alpha}{2} e + \beta$$

$$x = \frac{e}{2} \Rightarrow T^* = -\left(\frac{G}{8\lambda}\right) e^2 - \frac{\alpha}{2} e + \beta$$

By subtracting one member after another we arrive at:

$$0 = \alpha e \Rightarrow \alpha = 0$$

The equation giving the temperature profile then becomes:

$$T(x) = -\left(\frac{G}{2\lambda}\right) x^2 + \beta$$

Using boundary condition CL_2, we arrive at:

$$T^* = -\left(\frac{G}{8\lambda}\right) e^2 + \beta$$

Hence: $\beta = T^* + \left(\frac{G}{8\lambda}\right) e^2$.

We therefore obtain the temperature profile as a function of T^*, G, e and λ:

$$T(x) = T^* + \left(\frac{G}{2\lambda}\right) \left[\left(\frac{e}{2}\right)^2 - x^2\right]$$

3) Calculating the flux density at the wall

At the wall, the flux density is given by:

$$\varphi_{x=\frac{e}{2}} = -\lambda\left(\frac{dT}{dx}\right)_{x=\frac{e}{2}}$$

Yet: $\dfrac{dT}{dx} = -\dfrac{G}{\lambda}x$.

Hence: $\varphi_{x=\frac{e}{2}} = -\lambda\left(-\dfrac{G}{\lambda}\right)\left(\dfrac{e}{2}\right)$.

i.e.: $\varphi_{x=\frac{e}{2}} = \dfrac{G}{2}e$.

NUMERICAL APPLICATION.–

$e = 20$ cm $\qquad G = 75$ kW/m^3

$$\varphi_{x=\frac{e}{2}} = \frac{75}{2}(0.2)$$

$$\varphi_{x=\frac{e}{2}} = 7.5 \text{ kW / m}^2$$

4) Expressing T as a function of G, e, T_a and h_a*

To determine T*, it is necessary to take into account the fact that at the wall there is an equality of the conductive and convective fluxes.

Hence: $-\lambda\left(\dfrac{dT}{dx}\right)_{x=\frac{e}{2}} = h_a\left(T^* - T_a\right)$

or: $-\lambda\left(-\dfrac{G}{2\lambda}\,e\right) = h_a\left(T^* - T_a\right) \Rightarrow \dfrac{G}{2}\,e = h_a\left(T^* - T_a\right)$

i.e.: $T^* = T_a + \dfrac{G\,e}{2h_a}$.

5) Expressing T as a function of T*, G, e and λ.*

The profile of *T(x)* is parabolic and:

$$\frac{d^2T}{dx^2} = -\frac{G}{\lambda},$$

where *G* and λ are positive: $\frac{d^2T}{dx^2} < 0$.

T(x) therefore admits a maximum, T_{Max}.

Moreover, $\frac{dT}{dx} = -\frac{G}{\lambda}x \Rightarrow \frac{dT}{dx} = 0$ for $x = 0 \Rightarrow T_{Max}$ is obtained for $x = 0$.

Consequently: $T_{Max} = T(0) = T^* + \frac{G\,e^2}{8\lambda}$.

6) Verifying the condition $T_{Max} > T^* > T_a$

$$\forall x \in \left[-\frac{e}{2}, \frac{e}{2}\right] \quad \text{we have:} \quad T(x) - T^* = \left(\frac{G}{2\lambda}\right)\left[\left(\frac{e}{2}\right)^2 - x^2\right]$$

$$\forall x \in \left[-\frac{e}{2}, \frac{e}{2}\right]: \frac{e}{2} > x \Rightarrow \left(\frac{e}{2}\right)^2 - x^2 > 0$$

Next, $\forall x \in \left[-\frac{e}{2}, \frac{e}{2}\right]$ $T(x) - T^* > 0 \Rightarrow T(x) > T^*$.

Yet: $T^* = T_a + \frac{G\,e}{2h_a}$: $T^* > T_a \Rightarrow T(x) > T^* > T_a$.

Moreover: $\forall x \in \left[-\frac{e}{2}, \frac{e}{2}\right]$: $T_{Max} > T(x)$

i.e.: $\forall x \in \left[-\frac{e}{2}, \frac{e}{2}\right]$: $T_{Max} > T^* > T_a$.

7) Value of h_a so that $T_{Max} < 650$ °C

We have: $T_{Max} = T^* + \dfrac{G\,e^2}{8\lambda}$ and $T^* = T_a + \dfrac{G\,e}{2h_a}$.

Hence:

$$T_{Max} = T_a + \frac{G\,e}{2h_a} + \frac{G\,e^2}{8\lambda}$$

$$T_{Max} > 650 \;\Rightarrow\; T_a + \frac{G\,e}{2h_a} + \frac{G\,e^2}{8\lambda} > 650$$

Hence: $h_a > \dfrac{1}{2}\dfrac{G\,e}{650 - T_a - \dfrac{G\,e^2}{8\lambda}}$.

NUMERICAL APPLICATION.–

$e = 20$ cm $G = 75$ kW/m^3 $\lambda = 25$ W/m °C $T_3 = 40$ °C

$$h_a > \frac{1}{2}\frac{75\,000\,(0.2)}{650 - 40 - \dfrac{75\,000\,(0.2)^2}{8(25)}}$$

$\boldsymbol{h_a > 12.61}$ **W/m² °C**

8) Calculating T^* for $h_a = 15$ W/m² °C

$$T^* = T_a + \frac{G\,e}{2\,h_a}$$

NUMERICAL APPLICATION.–

$e = 20$ cm $G = 75$ kW/m^3 $Ta = 40$ °C $h_a = 15$ W/m^2 °C

$$T^* = 40 + \frac{75\,000\,(0.2)}{2(15)}$$

$\boldsymbol{T^* = 540}$ **°C**

5.3. Cylindrical conductor with generation

Very often, exothermic catalytic chemical reactions are conducted in tubular reactors. Similarly, the core of a nuclear reactor consists of bars (rods) in which the fuel is placed. In such a context, conducting reactions in an acceptable manner, that is to say without runaway, requires the energy produced during operation to be extracted. We are thus led to study conduction in cylindrical spaces where extensive energy generation takes place.

Figure 5.3 presents a cylindrical conductor in which energy is generated with a power density, *P*.

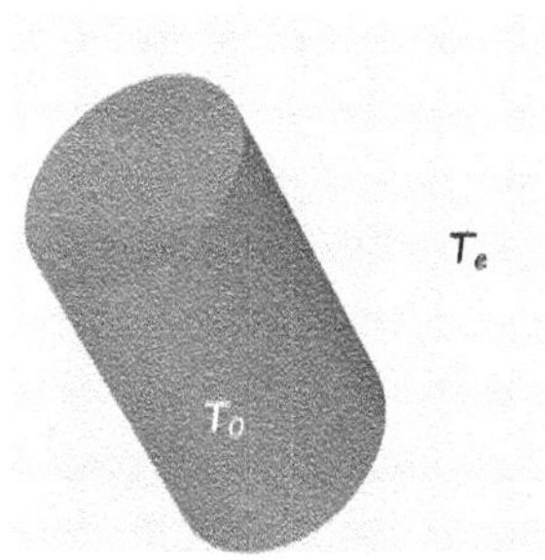

Figure 5.3. *Cylindrical conductor with energy generation*

In this type of situation, the bar is placed in a medium of temperature, T_e. This medium exchanges by convection with the wall of the bar, which is at T_0.

Thus, the energy transfer of interest to us occurs in the radial direction, *r*.

We are therefore looking at one-dimensional conduction here.

The conduction equation in simplified cylindrical coordinates, in the case of a one-dimensional conduction, is:

$$\frac{d^2T}{dr^2}+\frac{1}{r}\frac{dT}{dr}+\frac{P}{\lambda}=0,$$

with the boundary conditions:

CL_1: $r = R$; $T = T_R$

CL_2: $r = 0$; finite T

Taking into account the boundary conditions, again in this case integrating the equation of conduction also gives a parabolic temperature profile admitting a maximum, T_{Max}.

This profile corresponds to:

$$T(r) = T_R + \frac{P\,R^2}{\lambda}\left[1 - \left(\frac{r}{R}\right)^2\right]$$

It will be noted that this profile admits a maximum, i.e.:

$$T_{Max} = T(0) = T_R + \frac{PR^2}{\lambda}$$

The flux density is obtained by differentiating the expression of the temperature profile:

$$\varphi(r) = -\lambda \frac{dT(r)}{dr}$$

$$T(r) = T_R + [T_{Max} - T_R]\left[1 - \left(\frac{r}{R}\right)^2\right]$$

$$\frac{dT(r)}{dr} = -\frac{2[T_{Max} - T_R]}{R^2} r$$

Hence: $\varphi(r) = 2\lambda \dfrac{T_{Max} - T_R}{R^2}\, r$.

The flux density profile is therefore linear. The flux at the wall is:

$$\varphi(R) = \frac{2\lambda}{R}[T_{Max} - T_R]$$

Thus, the flux density at the abscissa, *r*, is expressed as a function of the flux at the wall by:

$$\frac{\varphi(r)}{\varphi(R)} = \left(\frac{r}{R}\right)$$

NOTE.– Conduction in an annular space with energy generation.

For this case, another type of boundary condition is used:

CL_1: $r = r_1$; $T = T_1$

CL_2: $r = r_2$; $T = T_2$

In this configuration, the temperature profile is given by:

$$\frac{T-T_2}{T_1-T_2} = \frac{P \cdot r_2^{\,2}}{4\lambda(T_1-T_2)}\left\{\left[1-\left(\frac{r_1}{r_2}\right)^2\right]\frac{\mathrm{Ln}\left(\frac{r}{r_2}\right)}{\mathrm{Ln}\left(\frac{r_2}{r_1}\right)} - \left[1+\left(\frac{r}{r_2}\right)^2\right]\right\}\frac{\mathrm{Ln}\left(\frac{r}{r_2}\right)}{\mathrm{Ln}\left(\frac{r_2}{r_1}\right)}$$

5.3.1. *Illustration: thermal technology in the core of a nuclear reactor*

The core of the nuclear research reactor, Triga Mark I, is composed of n tubular fuel rods, each of radius R and length L. The power released by this reactor is ψ.

To ensure its cooling, the reactor core is placed in a "well" of water, in which significant forced circulation of the cooling water is maintained. This makes it possible to extract the heat generated in each of the fuel rods.

It is assumed that the energy produced is conveyed from the inside of the rods to the bath water, in the radial direction only.

Each rod consists of a fuel cylinder protected by a metal sheath of thickness e.

Let us also imagine that the power, ψ, released by this reactor is the total of the powers, ψ_i, generated by each of the rods.

Questions

1) Calculate the power density, P, of each fuel rod.

2) Give the differential equation governing conduction in a fuel rod.

3) Indicate the differential equation governing conduction in the metal cladding.

4) Assuming that we know the temperature at the fuel/cladding interface, $T(R)=T_R$, establish the temperature profile in the fuel.

5) Give, as a function of P, the expression of the maximum temperature that could be reached in the fuel rod.

6) To assure the mechanical stability of the material constituting the fuel-rod cladding, the temperature at the fuel/cladding interface, T_R, must not exceed a given limit, defined by T_{Lim}. We will therefore choose $T_R \leq T_{Lim} - T^*$.

In these conditions, what would be the maximum temperature that could be reached by T_{Max}?

This temperature will be noted ${}^{Max}T_{Max}$.

7) We will then take $T_R = T_{Lim} - T^*$:

a) Calculate the flux densities at half-radius (φ at $r = R/2$) and at the fuel/sheath interface: $\varphi(R)$;

b) Deduce therefrom the temperature of the outer face of the sheath, T_{R+e};

c) Deduce therefrom the average water temperature in the well, T_a.

Let us assume that there is perfect contact between the different interfaces.

We will use the data and diagrams communicated by the design office, provided in the appendices below.

Data:

$n = 10$	$R = 2.5$ cm	$L = 70$ cm
$\psi = 100$ kW	$h_a = 2{,}300$ W/m^2 °C	$\lambda = 46$ W/m °C
$\lambda_g = 65$ W/m °C	$T_{Lim} = 650$ °C	
$e = 1$ cm	$T^* = 215$ °C	

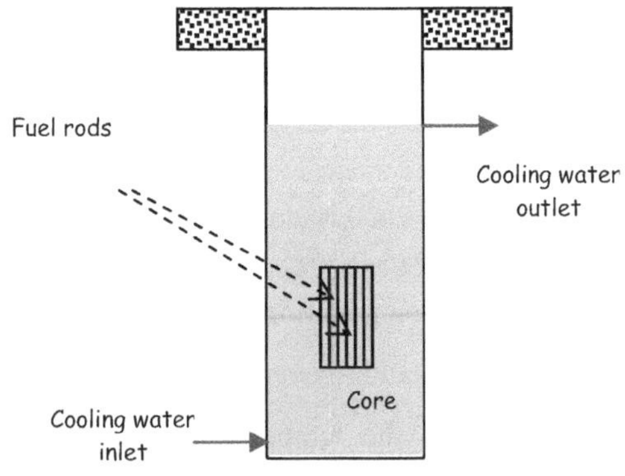

Figure 5.4. *Reactor core in the cooling well*

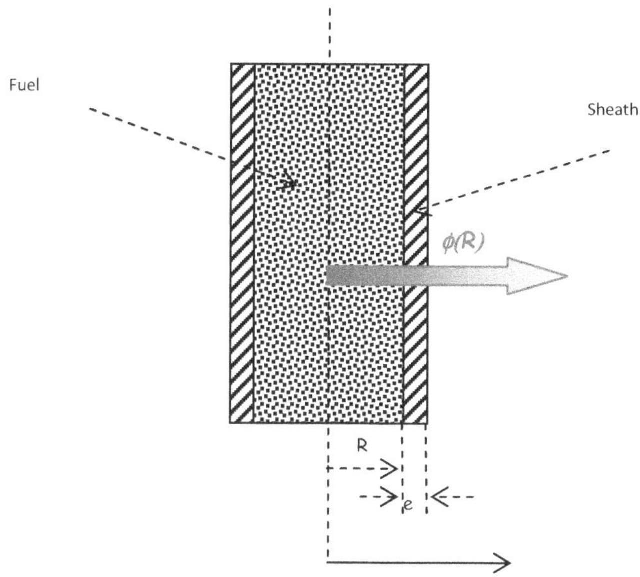

Figure 5.5. *Longitudinal cross-section of a fuel rod*

Answers

1) Calculating the power density

With the reactor having a power, π, and n fuel rods, we have: $\pi_i = \dfrac{\pi}{n}$.

Moreover: $P = \dfrac{\psi_i}{V} = \dfrac{\psi}{nV}$.

i.e.: $P = \dfrac{\psi}{n\,\pi\,R^2 L}$.

<table><tr><td>

NUMERICAL APPLICATION.–

$\psi = 100$ kW $n = 10$ $R = 2.5$ cm $L = 70$ cm

$$P = \frac{100\,000}{10\;\pi\;(0.025)^2\,(0.70)}$$

$\boldsymbol{P = 7{,}275}$ **kW/m^3**

</td></tr></table>

2) Differential equation for conduction in a fuel rod

We are in the presence of one-dimensional conduction with energy generation.

P being the generation power density, the equation of conduction is given by:

$$\frac{d^2T}{dr^2}+\frac{1}{r}\frac{dT}{dr}+\frac{P}{\lambda}=0$$

3) Differential equation governing conduction in the cladding

In this case, there is no energy generation.

The conduction equation in simplified cylindrical coordinates with a single propagation direction is appropriate in this context:

$$\frac{d^2T}{dr^2}+\frac{1}{r}\frac{dT}{dr}=0$$

4) Temperature profile in fuel

The differential equation governing conduction in the fuel is incorporated with the following boundary conditions:

$r = R$; $T = T_R$

$r = 0$; T is finite;

which lead to the following profile: $T(r)=T_R+\frac{P\,R^2}{\lambda}\left[1-\left(\frac{r}{R}\right)^2\right]$.

5) Maximum temperature in the fuel rod

The maximum of $T(r)$ is reached for $r = 0$.

Hence: $T_{Max}=T(0)=T_R+\frac{P\,R^2}{\lambda}$.

6) Maximum of T_{Max}

T_{Max} depends on T_R: $T_{Max}=T_R+\frac{P\,R^2}{\lambda}$.

The maximum of T_{Max} will therefore be recorded for the maximum of T_R; namely: $^{Max}T_{Max} = {}^{Max}T_R + \frac{P\,R^2}{\lambda}$.

Yet: $T_R \leq T_{Lim} - T^*$.

Thus, the maximum value of T_R is: $^{Max}T_R = T_{Lim} - T^*$.

Hence: $^{Max}T_{Max} = T_{Lim} - T^* + \frac{P\,R^2}{\lambda}$.

NUMERICAL APPLICATION.–

$P = 72{,}759$ W/m^3 $R = 2.5$ cm $\lambda = 46$ W/m °C $T_{Lim} = 650$ °C

$T^* = 215$ °C

$\boldsymbol{^{Max}T_{Max} = 534}$ **°C**

7) $T_R = T_{Lim} - T^*$

$T_R = T_{Lim} - T^* \Rightarrow T_{Max}$ is then at its maximum, $\Rightarrow T_{Max} = {}^{Max}T_{Max}$.

a) Calculating the flux densities

i) Calculating at half-radius

For r and T_{Max}, we have: $\varphi(r) = 2\lambda \frac{T_{Max} - T_R}{R^2}\, r$.

Hence, for $\frac{R}{2}$ and $^{Max}T_{max}$: $\varphi\left(\frac{R}{2}\right) = \lambda \frac{^{Max}T_{Max} + T^* - T_{Lim}}{R}$

ii) Calculation at the fuel/cladding interface

Likewise, for R and $^{Max}T_{max}$: $^{Max}\varphi(R) = \frac{2\lambda}{R}\left[{}^{Max}T_{Max} + T^* - T_{Lim}\right]$.

NUMERICAL APPLICATION.–

$R = 2.5$ cm $\lambda = 46$ W/m °C $^{Max}T_{Max} = 534$ °C $T_{Lim} = 650$ °C

$T^* = 215$ °C

$$\phi\left(\frac{R}{2}\right) = 46\frac{534+215-650}{0.025}$$

i.e.: $\varphi\left(\frac{R}{2}\right) = 182\ \text{kW}$

and: $\varphi(R) = 2\varphi\left(\frac{R}{2}\right) = 364\ \text{kW}$

b) Calculating temperature T_{R+e}

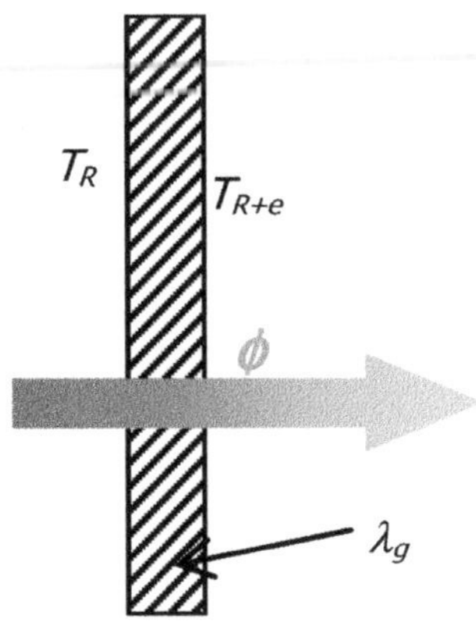

Figure 5.6. *Temperature,* T_{R+e}, *of the outer face of the sheath*

Let us now consider conduction in the metal sheath, the inner face of which is at T_R and the outer face at T_{R+e}.

In this case, there is conduction without generation. The flux expression is then given by:

$$\phi(R+e) = 2\pi L\lambda_g \frac{T_R - T_{R+e}}{\text{Ln}\left(\frac{R+e}{R}\right)}$$

Yet, this flux, $\phi(R+e)$, which has passed through the metal cladding, is the same as that generated in the fuel, expressed at $r = R$; namely:

$$\phi(R+e) = 2\pi RL\phi(R) = 2\pi RL\left[2\lambda\frac{T_{Max} - T_R}{R^2}R\right]$$

Hence: $\phi(R+e) = 4\pi L\lambda(T_{Max} - T_R)$.

Or, by replacing $\phi(R+e)$ with its expression obtained from the conduction in the fuel cladding:

$$2\pi L\lambda_g \frac{T_R - T_{R+e}}{\mathrm{Ln}\left(\frac{R+e}{R}\right)} = 4\pi L\lambda(T_{Max} - T_R)$$

Hence, if we substitute for $T_R = T_{Lim} - T^*$, we obtain:

$$T_{R+e} = T_{Lim} - T^* - 2\frac{\lambda}{\lambda_g}(T_{Max} + T^* - T_{Lim})\mathrm{Ln}\left(\frac{R+e}{R}\right)$$

NUMERICAL APPLICATION.–

$T_{Lim} = 650$ °C $T^* = 215$ °C $T_{Max} = 534$ °C

$\lambda = 46$ W/m °C $\lambda_g = 65$ W/m °C $R = 2.5$ cm

$e = 1$ cm

$$T_{R+e} = T_{Lim} - T^* - 2\frac{\lambda}{\lambda_g}(T_{Max} + T^* - T_{Lim})\mathrm{Ln}\left(\frac{R+e}{R}\right)$$

$\boldsymbol{T_{R+e} = 388}$ **°C**

c) Calculating temperature T_a

To calculate T_a, we use the convective flux expression:

$$\phi(R+e) = 2\pi(R+e)Lh_a\left[T(R+e) - T_a\right]$$

Yet, this flux, $\phi(R+e)$, is the same as that generated in the fuel, expressed at $r = R$:

$$\phi(R+e) = 4\pi L\lambda(T_{Max} - T_R)$$

By equaling the two flux expressions, we obtain:

$$2\pi(R+e)Lh_a[T_{R+e} - T_a] = 4\pi L\lambda(T_{Max} - T_R)$$

Hence: $T_a = T_{R+e} - \dfrac{2\lambda}{(R+e)h_a}(T_{Max} - T_R)$.

NUMERICAL APPLICATION.–

$T_R = 435$ °C	$T_{R+e} = 435$ °C	$T_{Max} = 534$ °C	
$\lambda = 46$ W/m °C	$h_a = 730$ W/m^2 °C	$R = 2.5$ cm	$e = 1$ cm

$$T_a = T_{R+e} - \frac{2\lambda}{(R+e)h_a}(T_{Max} - T_R)$$

$\boldsymbol{T_a = 32}$ **°C**

5.4. Conduction in rectangular fins

In practice, the flux transferred between a solid surface and the surrounding environment is often limited with regard to the requirements of the energy generating process. One of the most common examples is the case of power electronic circuits, whose operation generates a significant amount of heat. In order for these circuits to function correctly, the heat generated must be conveyed outwards.

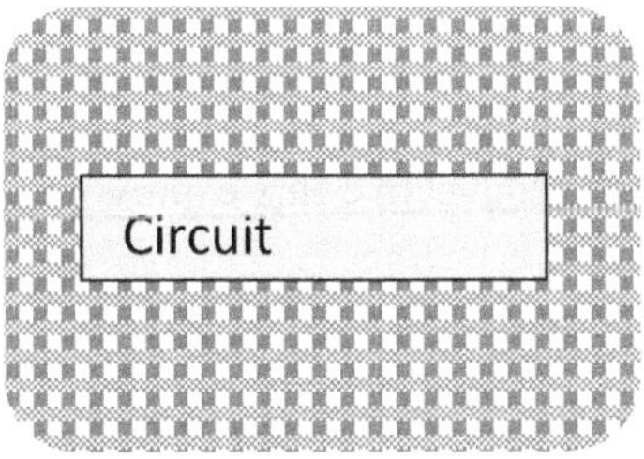

Figure 5.7. *Electronic circuit*

When designing such circuits, it is therefore necessary to seek configurations that maximize the heat flux between the electronic circuit and the outside. Yet, we know that this flux is proportional to the transfer area and to the temperature gradient between the circuit and the outside. As the latter element is not a degree of freedom, only the transfer area can be a design parameter.

But as the demands of the market are focused on the miniaturization of electronic equipment, these circuits are required to be designed in ever smaller sizes. This induces reduced transfer areas, which are very often not sufficient to ensure that the energy generated by the operation of the electronic circuit is adequately transferred to the outside.

As a result we will seek to artificially increase the transfer area without significantly expanding the circuit size. To do so, we add radiators known as heat sinks. These are generally composed of fins, enabling the transfer area to be extended (see Figure 5.8).

Other examples of this type of design can be found in the automotive industry (car radiators), in cold production (air-conditioning evaporators) and in certain heat exchangers used in processes where compactness is a parameter taken into account in the design.

In all of these situations, the solid concerned (electronic circuit wall, car radiator surface, heat exchanger tubes, etc.) is extended by means of generally metallic fins. The latter are firmly fastened or welded to the wall in question so as to virtually create an extension. They penetrate into the fluid (air, in the case of electronic circuits), thus making it possible to increase the transfer areas and thus to improve the fluxes transferred.

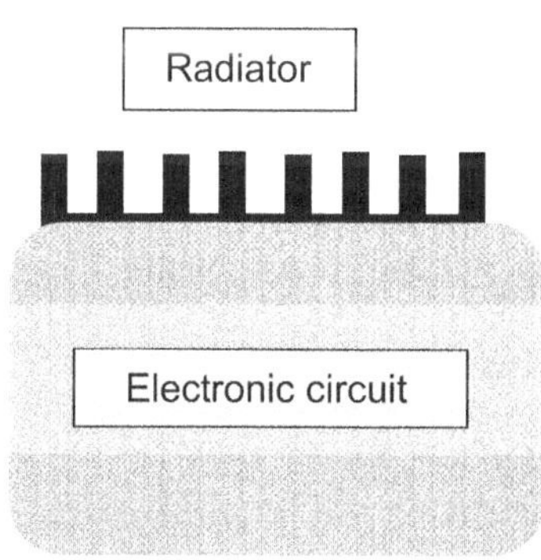

Figure 5.8. *Circuit with heat sink*

Consider the rectangular fin represented in Figure 5.9. This fin is embedded in a wall of temperature T_0 and is in contact with a fluid at temperature T_∞.

The length of the fin is L and its thickness is noted e.

Clearly, this fin will improve heat exchange in two ways:

– By prolonging the conduction produced in the solid;

– By increasing the area available for heat transfer between the solid and the fluid.

We suggest we determine, in steady state, the flux transferred from the fin to the fluid at T_∞. In this case, the heat propagates essentially in the z direction. As a result we will use the heat equation in rectangular coordinates with a single propagation direction, z, namely:

$$\frac{d^2T}{dz^2} + \frac{P}{\lambda} = 0$$

This case presents the following specific feature: convective exchanges across the contact surface between the fin and the fluid can be interpreted as an energy generation presenting a negative power density, P. The power density is given, as a function of z, by:

$$P(z) = -\frac{h\ (p\ dz)\ \left[T(z) - T_\infty\right]}{(S\ dz)}$$

Indeed, the convective transfer area between the fluid and the fin element of perimeter p and length dz is given by: $S_E = p\ dz$, where the perimeter, p, corresponds to: $p = 2\ (\ell + e)$.

The thickness, e, of the fin is assumed to be low enough for there to be no temperature variation within the same cross-section of the fin.

The fin cross-section, S, is given by: $S = e \times \ell$.

The flux exchanged by convection is thus obtained as a function of the convective heat transfer coefficient, h, between the fin and the bath:

$$\varphi_{conv} = h\ p\ dz\ \left[T(z) - T_\infty\right]$$

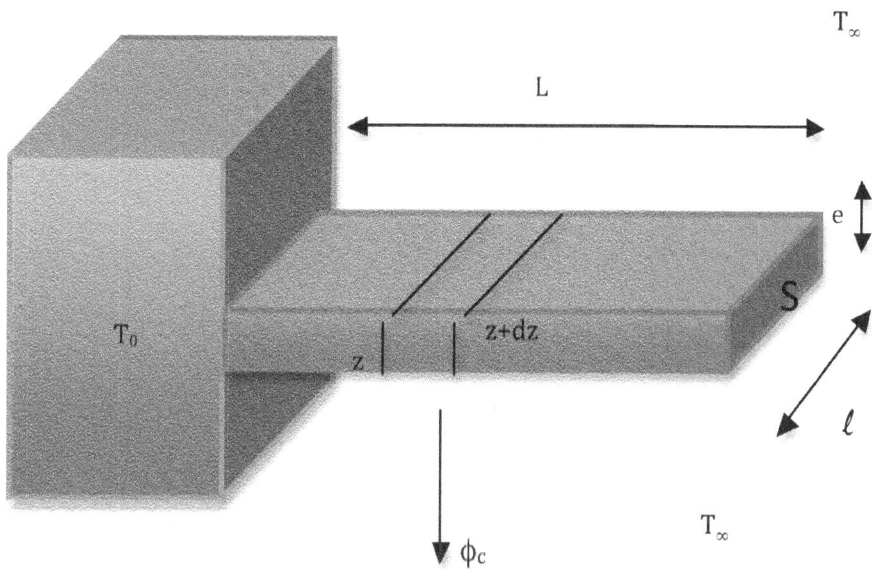

Figure 5.9. *Rectangular fin*

Consequently, the flux lost through convection, per unit volume of the fin, is given by:

$$P(z) = -\frac{\varphi_{conv}}{S\ dz} = -\frac{h\ (p\ dz)\ \left[T(z) - T_\infty\right]}{(S\ dz)}$$

By replacing *P* with its expression in the heat equation r, we obtain:

$$\frac{d^2T}{dz^2} - \frac{h\ p}{\lambda\ S}\left(T - T_\infty\right) = 0$$

NOTE.– This equation can be established without the need to interpret the energy exchanged by convection as a negative heat generation. Indeed, if we go back to the balance approach, the same equation can be obtained by establishing the differential energy balance on the volume element defined by *S dz*. This type of development is covered in Exercise 18 (Chapter 7).

The boundary conditions considered for this case are:

$z = 0$; $T = T_0$

$z = L$; $T = T_\infty$

By applying the change in variable: $\theta = T - T_\infty$ and by defining the following parameters: $\omega^2 = \dfrac{h\,p}{\lambda\,S}$, the differential equation becomes:

$$\frac{d^2\theta}{dz^2} - \omega^2\theta = 0,$$

the general solution of which is given by:

$$\theta = K_1\, e^{\omega z} + K_2\, e^{-\omega z},$$

with the following boundary conditions:

$z = 0$; $\theta = T_0 - T_\infty = \theta_0$

$z = L$; $\theta = 0$

Hence the constant of integration values:

$$K_1 = \frac{\theta_0}{1 - e^{2\omega L}}$$

$$K_2 = -\frac{\theta_0\, e^{2\omega L}}{1 - e^{2\omega L}}$$

Ultimately: $\theta = \dfrac{\theta_0}{1 - e^{2\omega L}}\left[e^{\omega z} - e^{\omega(2L-z)}\right]$.

or, returning to the initial variable: $\dfrac{T - T_\infty}{T_0 - T_\infty} = \dfrac{e^{\omega z} - e^{\omega(2L-z)}}{1 - e^{2\omega L}}$.

The flux transferred by convection, from the fin to the fluid, can therefore be calculated and it is to be noted that this value is equal to the conductive flux at $z = 0$, i.e.:

$$\phi_{conv} = \phi_{z|_{z=0}} = -\lambda\ S \left(\frac{dT}{dz}\right)_{|z=0}.$$

Yet: $\left(\frac{dT}{dz}\right)_{|z=0} = -\omega\ e^{2\omega L}\ (T_0 - T_\infty)$.

Hence: $\phi_{conv} = \sqrt{\lambda\ h\ p\ S}\ e^{2\ \omega\ L}(T_\circ - T_\infty)$.

5.4.1. *Illustration: gain in efficiency through use of a fin*

The nest of fins making up a car radiator can be represented by a single plane conductor onto which a fin is crimped, the dimensions and characteristics of which are given below. The fin and the plane wall are composed of the same material, which means that they have the same heat conductivity. The fin/wall assembly is at uniform temperature, T_0, and it is immersed in a fluid, the temperature of which is maintained at T_∞.

Questions

1) Calculate the flux transferred by convection between the fin and the fluid.

2) Calculate the gain in transfer efficiency obtained by adding the fin.

Data:

$L = 2.5$ cm	$e = 1$ mm	$\ell = 1$ cm
$\lambda = 60$ kcal/hm °C	$h = 15$ kcal/hm^2 °C	
$T_0 = 80$ °C	$T_\infty = 40$ °C	

Solutions

1) Calculating the flux transferred by the fin

$$\phi_c = \sqrt{2\lambda\ h\ (l + e)\ e\ l}\ \ (T_\circ - T_\infty)e^{2\omega L},$$

with: $\omega^2 = \frac{hp}{\lambda S} = \frac{2h\ (e + l)}{\lambda\ e\ l}$

i.e.: $\omega^2 = \dfrac{2*15(0.01+0.10)}{60*0.01*0.10} = 550$

or: $\omega = 23.45$.

Hence: $\phi_c = \sqrt{2*60*15*0.11*0.001}*40*e^{2*7,42*0,025}$

i.e.: $\phi_c = 1.82$ kcal/hr.

2) Gain in transfer efficiency

This value, ϕ_c, needs to be compared to the flux that would have been transmitted by convection, with no fin, by the same contact surface, $S = e*\ell$. This flux is given by:

$\phi = h\,S\,(T_0 - T_\infty)$

$\phi = h\,(e*\ell)\,(T_0 - T_\infty)$

$\phi = 15 \times 0.001 \times 40 = 0.01$ kcal/hr

ϕ_c is therefore more than three hundred times' greater than ϕ.

6

Conduction in Transient State

6.1. Introduction

The study of conduction in transient state is important for examining the thermal behavior of materials in the event of a sudden change in the external conditions. This occurs, for example, in the following situations:

– in the case of the heat treatment of metals (also known as "metal tempering") when metal parts are abruptly immersed in a cooling bath;

– during the start-up of installations when the materials making up the equipment are suddenly subject to temperature variations;

– as a space shuttle enters the atmosphere, with the materials making up the shuttle becoming immersed in a different temperature medium, etc.

In all of these situations, we are faced with having to resolve conduction problems in transient state. The first step in this resolution is to write the conduction equations that correspond to the situation being considered (plane, cylinder or sphere) and to put in place the boundary conditions in order to prepare the resolution.

Since the objective is to follow the spatial and temporal evolutions of the temperature in the solid being considered, we will then have a mathematical description in terms of partial differential equations of order at least equal to three. Indeed, in the simplest case, where we would consider a single direction for energy propagation (x or r, as applicable), the conduction equation will take one of the following forms:

– *In simplified rectangular coordinates* (a single propagation direction: *x*):

$$\frac{1}{\alpha}\frac{\partial \theta}{\partial t}=\frac{\partial^2 \theta}{\partial x^2}$$

– *In simplified cylindrical coordinates* (a single propagation direction: *r*):

$$\frac{1}{\alpha}\frac{\partial \theta}{\partial t}=\frac{\partial^2 \theta}{\partial r^2}+\frac{1}{r}\frac{\partial \theta}{\partial r}$$

– *In simplified spherical coordinates* (a single propagation direction: *r*):

$$\frac{1}{\alpha}\frac{\partial \theta}{\partial t}=\frac{\partial^2 \theta}{\partial r^2}+\frac{2}{r}\frac{\partial \theta}{\partial r}$$

We will therefore need to resolve the partial differential equation corresponding to the situation considered (plane, cylindrical or spherical solid), using the appropriate set of boundary conditions.

6.2. Methods for resolving the conduction equation

Let us recall that the heat equation, in the case of a three-dimensional conduction, is given, in rectangular coordinates[1], by:

$$\frac{\partial^2 \theta}{\partial x^2}+\frac{\partial^2 \theta}{\partial y^2}+\frac{\partial^2 \theta}{\partial z^2}+\frac{P}{\lambda}=\frac{1}{\alpha}\frac{\partial \theta}{\partial t}$$

The analytical resolution of this equation is generally too complex or even impossible, except in some very simple cases, where solutions can be found by applying Laplace transforms or integration methods by separation of variables (Lagrée, 2010).

In sections 6.5 and 6.6, we present the developments that lead to exact or approximate analytical solutions in the case of one-dimensional problems. However, for plane, cylindrical and spherical conductors with a single heat propagation direction, solutions are expressed in the form of Fourier series, which are

1 In this section we look at developments in rectangular coordinates. Cylindrical and spherical coordinates are obtained through similar treatments.

relatively complex to implement in general. Nevertheless it is demonstrated (see section 6.6) that in certain contexts corresponding to commonly encountered practical situations, the Fourier series can easily be represented by their first term, with acceptable accuracy. The approximations thus obtained were graphically documented by Heisler (1956), leading to the graphical resolution method presented in section 6.7. This method has been and continues to be used to develop, without too much effort, orders of magnitude that give an idea of the solutions sought.

Yet in the most general cases where the use of analytical solutions is not conceivable, we tend to instead resort to numerical resolution methods. One of these is based on a discretization of the conduction equation using the finite-difference method (Carslaw, 1959). The following section presents a numerical method employed in the complex case of problems where several heat propagation directions are considered.

Note that numerical methods of resolving the heat equation are practical for dealing with multi-directional conduction problems or analyzing heat transfer situations where a certain level of accuracy is required. Indeed, with the currently available calculation means, these methods can lead to representations whose accuracy depends to a certain degree on the user. Above all, they provide the opportunity to address the complexity of considering multiple propagation directions.

6.3. Discretizing the heat equation

The different partial derivatives appearing in the conduction equation are developed according to the following finite-difference approximations (Dusinberre, 1961); namely:

$$\frac{\partial \theta}{\partial t} = \frac{\theta_{t+dt} - \theta_t}{\Delta t}$$

$$\frac{\partial \theta}{\partial x} = \frac{\theta_{x+dx} - \theta_x}{\Delta x} \qquad \frac{\partial^2 \theta}{\partial x^2} = \frac{\theta_{x+dx} - 2\theta_x + \theta_{x-dx}}{(\Delta x)^2}$$

$$\frac{\partial^2 \theta}{\partial y^2} = \frac{\theta_{y+dy} - 2\theta_y + \theta_{y-dy}}{(\Delta y)^2} \qquad \frac{\partial^2 \theta}{\partial z^2} = \frac{\theta_{z+dz} - 2\theta_z + \theta_{z-dz}}{(\Delta z)^2}$$

To simplify the writing of equations intended for computer programming, it is assumed that the space can be described by a grid in the *x*, *y* and *z* directions. This grid is defined by the *increments* Δx, Δy and Δz.

Thus, a point *M* of space, *M(x, y, z)*, can be characterized by the indices *i*, *j* and *k*, such that:

$$i = \frac{x}{\Delta x}, \; j = \frac{y}{\Delta y} \text{ and } k = \frac{z}{\Delta z}$$

As a result, any point *M(x, y, z)* in the space can therefore be represented by its indices *i*, *j* and *k*. We will then write *M (i, j, k)* or $M_{i,j,k}$.

The temperature at any *M(x, y, z)* point of the solid in question can also be written in the form $\theta_{i,j,k}$.

Moreover, we will adopt the same discretization with respect to time, with an increment, Δt, such that the temperature, $\theta_{i,j,k}$, at instant *t* ($t = n\,\Delta t$) can be represented using the index *n* as follows: $\theta^n_{i,j,k}$.

Likewise, the temperature, $\theta_{i,j,k}$, at instant $t + n\,\Delta t$ will be written as follows: $\theta^{n+1}_{i,j,k}$.

$\theta^{n+1}_{i,j,k}$ will therefore correspond to the temperature at the point of coordinates *(i, j, k)*, at the instant defined by $n + 1$.

Thus, with these notations, the finite-difference approximations of the various partial derivatives are written, at each point *(i, j, k)* in the space and at each instant, *n*, as follows:

$$\left[\frac{\partial^2\theta}{\partial x^2}\right]^n = \frac{\theta^n_{i-1,j,k} - 2\theta^n_{i,j,k} + \theta^n_{i+1,j,k}}{(\Delta x)^2}$$

$$\left[\frac{\partial^2\theta}{\partial y^2}\right]^n = \frac{\theta^n_{i,j-1,k} - 2\theta^n_{i,j,k} + \theta^n_{i,j+1,k}}{(\Delta y)^2}$$

$$\left[\frac{\partial^2\theta}{\partial z^2}\right]^n = \frac{\theta^n_{i,j,k-1} - 2\theta^n_{i,j,k} + \theta^n_{i,j,k+1}}{(\Delta z)^2}$$

$$\left[\frac{\partial\theta}{\partial t}\right]^n = \frac{\theta^{n+1}_{i,j,k} - \theta^n_{i,j,k}}{\Delta t}$$

We can note that all of the approximations of these differentials highlight the instant *n*, except for the derivative $\left[\frac{\partial\theta}{\partial t}\right]^n$, where the term $\theta^{n+1}_{i,j,k}$ shows that it is the temperature $\theta_{i,j,k}$, expressed at instant $n + 1$. Also note that instant $n + 1$ represents time $t + \Delta t$, that is to say, the immediate future.

By inserting these approximations into the general equation of conduction, we obtain ($\Delta x = \Delta y = \Delta z$):

$$\theta^{n+1}_{i,j,k} = (Fo_\Delta)\left[\theta^n_{i-1,j,k} + \theta^n_{i+1,j,k} + \theta^n_{i,j-1,k} + \theta^n_{i,j+1,k} + \theta^n_{i,j,k-1} + \theta^n_{i,j,k+1}\right] + \left[1 - 6(Fo_\Delta)\right]\theta^n_{i,j,k} + \frac{(Fo_\Delta)\cdot P}{\lambda}(\Delta x)^2,$$

where Fo_Δ designates the *discrete Fourier number* given by:

$$Fo_\Delta = \alpha\frac{(\Delta t)}{(\Delta x)^2}$$

This equation is the explicit discrete representation of the general heat conduction equation. It can be noted that, instead of the partial differential form, discretization allows us to obtain linear equations between the temperatures, $\theta^{n+1}_{i,j,k}$, and the temperatures, $\theta^n_{i,j,k}$.

When this equation is written for all the points of the space, we obtain a linear system of equations where some of the $\theta_{i,j,k}$ values are the unknowns. Solving this system of equations makes it possible to determine the unknown $\theta_{i,j,k}$ values.

6.4. Implementing the discrete heat equation

The discrete representation presented is an extremely practical approximation for numerically resolving the conduction equation in any situation. Note that this representation makes it possible to determine the unknown temperature, $\theta_{i,j,k}^{n+1}$; that is to say at instant $(n + 1)\Delta t$, as a function of the temperatures, $\theta_{i,j,k}^{n}$, $\theta_{i-1,j,k}^{n}$, $\theta_{i+1j,k}^{n}$, $\theta_{i,j-1,k}^{n}$, $\theta_{i,j+1,k}^{n}$, $\theta_{i,j,k-1}^{n}$ and $\theta_{i,j,k+1}^{n}$, which are all temperatures at instant $t = n\Delta t$ and therefore assumed to be known. In other words, the discrete representation of the conduction equation presented makes it possible to determine future temperatures (at instant $t+\Delta t$) as a function of the temperatures at time t.

Thus, starting from a known initial profile, in other words the initial temperatures, $\theta_{i,j,k}^{0}$, $\theta_{i-1,j,k}^{0}$, $\theta_{i+1j,k}^{0}$, $\theta_{i,j-1,k}^{0}$, $\theta_{i,j+1,k}^{0}$, $\theta_{i,j,k-1}^{0}$ and $\theta_{i,j,k+1}^{0}$, which are known for all of the points of coordinates *i, j, k* of the solid, it should be possible to determine, step by step, the evolution of the temperature profile over time.

6.4.1. *Resolution algorithm*

The iterative algorithm presented in the steps described below enables the numerical resolution of conduction problems in transient state.

Step 0: define a mesh for the solid with increments Δx, Δy and Δz so that each point of the solid is defined with its three increments, *i*, *j* and *k*. In general, we take an identical step in the three directions of the space: $\Delta x = \Delta y = \Delta z$.

Step 1: assign to each point of the mesh an initial value of the temperature $\theta_{i,j,k}^{0}$ consistent with the initial conditions of the solid and the imposed boundary conditions.

Step 2: write the equation of conduction in its "finite difference" form for all of the points of the solid considered, and in the second members of the equations obtained, replace the temperatures of each point of the mesh with the $\theta_{i,j,k}^{0}$ values assigned in Step 1.

Step 3: arrive at a linear system of equations.

Step 4: as all the $\theta_{i,j,k}^{0}$ temperatures in the second members of the system of equations are known, the temperatures at instant 1 can be easily calculated: $\theta_{i,j,k}^{1}$.

Step 5: steps 1 to 4 are repeated, replacing the $\theta^0_{i,j,k}$ temperatures with $\theta^1_{i,j,k}$ in order to calculate the $\theta^2_{i,j,k}$ values.

Next Steps: steps 1-4 are repeated, replacing the $\theta^{n-1}_{i,j,k}$ temperatures with $\theta^n_{i,j,k}$ in order to calculate the $\theta^{n+1}_{i,j,k}$ values.

An example of the implementation of this algorithm is shown in section 6.4.4.1.

6.4.2. *Choosing the Δx and Δt increments*

Obviously, the accuracy of the results that will be obtained by this numerical method will depend closely on the values chosen for the Δx and Δt increments. In principle, the smaller Δx and Δt are, the more accurate the results will be (Süli and Mayers, 2003). However, small Δx and Δt values cause a large number of points in the mesh, meaning a large number of calculations at each step. Moreover, given that there is a truncation error associated with each calculation, it is inevitable that at a certain point in time, the errors will outweigh the gain in accuracy generated by smaller Δx and Δt steps. Consequently, the increments selection is limited: they can not be chosen as small as desired since the gain in accuracy sought will be counterbalanced by accumulations of calculation errors, by a greater number of equations to be resolved, by a larger order of the system of equations, and lastly by a greater complexity in the resolution of this system.

Different studies into the convergence of calculations carried out via this method show that discretization steps Δx and Δt must be chosen with respect to the accuracy sought, while observing the following criteria, in order to guarantee the stability of the resolution:

– *one-dimensional* scenarios require: $Fo_\Delta \leq 1/2$;

– *two-dimensional* scenarios require: $Fo_\Delta \leq ¼$;

– *three-dimensional* scenarios require: $Fo_\Delta \leq 1/6$.

6.4.3. *Simplifications in the case of stationary state*

In stationary state, the conduction equation for the most general case is given by:

$$\frac{\partial^2\theta}{\partial x^2}+\frac{\partial^2\theta}{\partial y^2}+\frac{\partial^2\theta}{\partial z^2}+\frac{P}{\lambda}=0$$

By injecting the discrete expressions of the partial derivatives into the conduction equation, we obtain:

$$\frac{\theta_{i+1,j,k}-2\theta_{i,j,k}+\theta_{i-1,j,k}}{(\Delta x)^2}+\frac{\theta_{i,j+1,k}-2\theta_{i,j,k}+\theta_{i,j-1,k}}{(\Delta y)^2}+\frac{\theta_{i,j,k+1}-2\theta_{i,j,k}+\theta_{i,j,k-1}}{(\Delta z)^2}+\frac{P}{\lambda}=0$$

Or, by taking $\Delta x = \Delta y = \Delta z$:

$$\theta_{i+1,j,k}+\theta_{i-1,j,k}+\theta_{i,j+1,k}+\theta_{i,j-1,k}+\theta_{i,j,k+1}+\theta_{i,j,k-1}-6\theta_{i,j,k}+\frac{P}{\lambda}(\Delta x)^2=0$$

NOTE.– The discrete form of the equation of conduction in stationary state can be obtained from the transient-state equation by taking:

$Fo_\Delta = 1$ and $\theta^{n+1}_{i,j,k} = \theta^{n}_{i,j,k}$

The discrete form of the conduction equation in stationary state can prove extremely useful for the study of *multi-dimensional* problems in steady state.

In stationary state the equation is resolved by adapting the algorithm presented in section 6.4.1: time not being an issue, we will start by choosing a mesh: definition of Δx.

Subsequently, the known $\theta_{i,j}$ temperatures are determined and assigned to the corresponding points of the mesh. The unknown temperatures are numbered and placed on the corresponding points of the mesh.

We then write the discrete conduction equation in its stationary form for all of the points of the mesh, including for any points whose temperatures are unknown. We obtain a linear system of equations, which comprises the same number of equations as there are unknowns.

Solving this linear system of equations makes it possible to determine the unknown temperatures.

6.4.4. *Simplifications in the two-dimensional case*

In the case of a two-dimensional conduction, the discrete form of the conduction equation is given by:

– *In transient state*:

$$\theta^{n+1}_{i,j}=(Fo_\Delta)\left[\theta^{n}_{i-1,j}+\theta^{n}_{i+1,j}+\theta^{n}_{i,j-1}+\theta^{n}_{i,j+1}\right]+\left[1-4(Fo_\Delta)\right]\theta^{n}_{i,j}+\frac{(Fo_\Delta)\cdot P}{\lambda}(\Delta x)^2,$$

where Fo_Δ designates the discrete Fourier number given by: $Fo_\Delta = \alpha \frac{(\Delta t)}{(\Delta x)^2}$.

– *In steady state*:

$$\theta_{i+1,j} + \theta_{i-1,j} + \theta_{i,j+1} + \theta_{i,j-1} - 4\theta_{i,j} + \frac{P(\Delta x)^2}{\lambda} = 0$$

6.4.4.1. *Illustration: two-dimensional temperature profile in a copper plate*

Consider a copper plate measuring 3 cm by 4 cm, of low thickness. Conduction in this plate can therefore be considered to be two-dimensional.

Each of the two opposite sides of the plate is at a different constant temperature, but the peaks are maintained at 350 °C. In other words, sides AB and CD are at 400 °C but points A, B, C and D are at 350 °C. Likewise, sides AC and BD are kept at 300 °C, with the exclusion of points A, B, C and D, which reach 350 °C.

Questions

1) Taking $\Delta x = \Delta y = 1$ cm, proceed with the discretization of the plate and place the known and unknown temperatures.

2) Determine the system of equations that needs to be satisfied by the unknown temperatures in steady state.

3) Calculate the temperature profile in the plate in steady state.

Data:

We will assume that there is no generation:

$C_{p\,Cu} = 383$ J/kg °C $\qquad$ $\lambda = 386$ W/m °C $\qquad$ $\rho_{Cu} = 8{,}954$ kg/m^3

Solutions

1) Discretizing the plate

$\Delta x = \Delta y = 1$ cm

The plate is then discretized and the temperatures of the different points on the meshing are placed on Figure 6.1, including the unknown temperatures, θ_1, θ_2, θ_3, θ_4, θ_5 and θ_6.

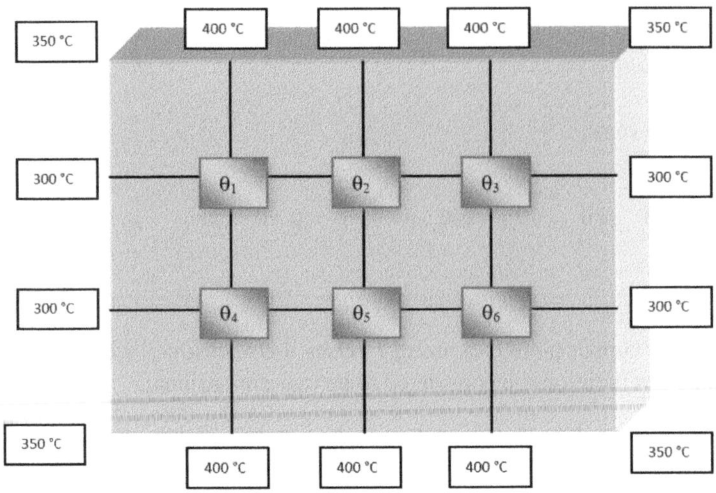

Figure 6.1. *Meshing of the plate and placing of known and unknown temperatures*

<u>*2) System of equations to be satisfied by the unknown temperatures*</u>

These temperatures need to satisfy the discretized conduction equation, with no heat generation:

$$\theta_{i+1,j} + \theta_{i-1,j} + \theta_{i,j+1} + \theta_{i,j-1} - 4\theta_{i,j} = 0$$

Or: $$\theta_{i,j} = \frac{\theta_{i+1,j} + \theta_{i-1,j} + \theta_{i,j+1} + \theta_{i,j-1}}{4}$$

Thus, the temperature at each point must be equal to the average of the four neighboring temperatures. The unknown temperatures therefore need to satisfy the following system of equations:

$$\begin{cases} -4\theta_1 + \theta_2 + \theta_4 = -700 \\ \theta_1 - 4\theta_2 + \theta_3 + \theta_5 = -400 \\ \theta_2 - 4\theta_3 + \theta_6 = -700 \\ \theta_1 - 4\theta_4 + \theta_5 = -700 \\ \theta_2 + \theta_4 - 4\theta_5 + \theta_6 = -400 \\ \theta_3 + \theta_5 - 4\theta_6 = -700 \end{cases}$$

3) Calculating the temperature profile in the plate in steady state

The system of equations is resolved by substitution:

$$\theta_1 = 4\theta_2 - \theta_3 - \theta_5 - 400 \qquad (1)$$

$$\theta_2 = 4\theta_3 - \theta_6 - 700 \qquad (2)$$

$$\theta_3 = -\theta_5 + 4\theta_6 - 700 \qquad (3)$$

$$\theta_4 = 4\theta_1 - \theta_2 - 700 \qquad (4)$$

$$\theta_5 = 4\theta_4 - \theta_1 - 700 \qquad (5)$$

$$\theta_6 = 4\theta_5 - \theta_4 - \theta_2 - 400 \qquad (6)$$

Hence:

$$(3) \Rightarrow 51{,}38298\ \theta_1 = 18351{,}0638$$

$$(2) \Rightarrow \theta_2 = \frac{120}{47}\theta_1 - \frac{25400}{47}$$

$$(4) \Rightarrow \theta_4 = 4\theta_1 - \theta_2 - 700$$

$$(5) \Rightarrow \theta_5 = 4\theta_4 - \theta_1 - 700$$

$$(6) \Rightarrow \theta_6 = 4\theta_5 - \theta_4 - \theta_2 - 400$$

$$(1) \Rightarrow \theta_3 = 4\theta_2 - 4\theta_4 + 300$$

Hence the solution:

$\theta_1 =$	357.14
$\theta_2 =$	371.43
$\theta_3 =$	357.14
$\theta_4 =$	357.14
$\theta_5 =$	371.43
$\theta_6 =$	357.14

Table 6.1. *Calculating unknown temperatures in steady state*

6.4.5. *Simplifications in the one-dimensional case*

In the case of a one-dimensional conduction, such as that considered in the Heisler method, the discrete form of the conduction equation is given by:

– *In transient state*:

$$\theta_i^{n+1} = \theta_i^n + \mathrm{Fo}_\Delta\left[\theta_{i+1}^n - 2\theta_i^n + \theta_{i-1}^n\right],$$

where Fo_Δ designates the discrete Fourier number given by: $\mathrm{Fo}_\Delta = \alpha\dfrac{(\Delta t)}{(\Delta x)^2}$.

– *In steady state*:

$$\theta_{i+1}^n = 2\theta_i^n - \theta_{i-1}^n$$

6.5. Developing precise analytical solutions in the one-dimensional case

The analytical solutions developed in this section relate to cases of one-dimensional conductions in plane, cylindrical or spherical conductors subject to convective boundary conditions. These problems are represented by the following equations:

– *For plane conductors*:

$$\frac{1}{\alpha}\frac{\partial\theta}{\partial t} = \frac{\partial^2\theta}{\partial x^2},$$

with: $$\begin{cases} t = 0\ ;\ \forall x\ :\ \theta(x,\ 0) = \theta_i \\ x = \pm s\ ;\ \forall t : -\lambda\left(\dfrac{\partial\theta}{\partial x}\right) = \pm h(\theta - \theta_e) \end{cases}.$$

– *For cylindrical conductors*:

$$\frac{1}{\alpha}\frac{\partial\theta}{\partial t} = \frac{\partial^2\theta}{\partial r^2} + \frac{1}{r}\frac{\partial\theta}{\partial r},$$

with: $$\begin{cases} t = 0\ ;\ \forall r\ :\ \theta(r,\ 0) = \theta_i \\ r = s\ ;\forall t : -\lambda\left(\dfrac{\partial\theta}{\partial r}\right) + h(\theta - \theta_e) = 0 \\ r = 0\ ;\ \forall t :\ \dfrac{\partial\theta}{\partial r} = 0 \end{cases}.$$

– *For spherical conductors*:

$$\frac{1}{\alpha}\frac{\partial\theta}{\partial t}=\frac{\partial^2\theta}{\partial r^2}+\frac{2}{r}\frac{\partial\theta}{\partial r},$$

$$\text{with:}\begin{cases}\text{Fo}=0\ ;\forall\bar{r};\ 0\le\bar{r}\le 1\ :\bar{T}(\bar{r},\ 0)=1\\ \forall\ \text{Fo}>0:\bar{r}=0\ :\ \dfrac{\partial\bar{T}}{\partial\bar{r}}=0\\ \forall\ \text{Fo}>0:\bar{r}=1\ :\ \dfrac{\partial\bar{T}}{\partial\bar{r}}+\text{Bi}\ \bar{T}(1,\text{Fo})=0\end{cases}.$$

We start by conducting a non-dimensionalization on the equations by introducing the following variable and function changes:

– *For plane conductors*:

$$\bar{x}=\frac{x}{s};\ \text{Fo}=\frac{\alpha}{s^2}t\ \text{ and }\ \bar{T}(\bar{x},\text{Fo})=\frac{\theta(x,t)-\theta_e}{\theta_i-\theta_e}.$$

– *For cylindrical conductors*:

$$\bar{r}=\frac{r}{s};\ \text{Fo}=\frac{\alpha}{s^2}t\ \text{ and }\ \bar{T}(\bar{r},\text{Fo})=\frac{\theta(r,t)-\theta_e}{\theta_i-\theta_e}.$$

– *For spherical conductors*:

$$\bar{r}=\frac{r}{s}\ ;\ \text{Fo}=\frac{\alpha}{s^2}t\ \text{ and }\ \bar{T}(\bar{r},\text{Fo})=\frac{\theta(r,t)-\theta_e}{\theta_i-\theta_e}.$$

The differential equations become ($\text{Bi}=\frac{hs}{\lambda}$):

– *For plane conductors*:

$$\frac{\partial\bar{T}}{\partial(\text{Fo})}=\frac{\partial^2\bar{T}}{\partial\bar{x}^2}$$

$$\text{with:}\begin{cases}\text{Fo}=0\ ;\forall\bar{x};\ -1\le\bar{x}\le 1\ :\bar{T}(\bar{x},\ 0)=1\\ \forall\ \text{Fo}>0:\bar{x}=\pm 1\ :\ -\left(\dfrac{\partial\bar{T}}{\partial\bar{x}}\right)_{\bar{x}\pm 1}=\pm\text{Bi}\ \bar{T}\end{cases}.$$

– *For cylindrical conductors*:

$$\frac{\partial \overline{T}}{\partial(Fo)} = \frac{1}{\overline{r}}\frac{\partial}{\partial \overline{r}}\left(\overline{r}\frac{\partial \overline{T}}{\partial \overline{r}}\right)$$

$$\text{with: } \begin{cases} Fo = 0 \; ;\forall \overline{r}; \; 0 \le \overline{r} \le 1 \; : \overline{T}(\overline{r}, \; 0) = 1 \\ \forall \; Fo > 0 : \overline{r} = 0 \; : \; \dfrac{\partial \overline{T}}{\partial \overline{r}} = 0 \\ \forall \; Fo > 0 : \overline{r} = 1 \; : \; \dfrac{\partial \overline{T}}{\partial \overline{r}} + Bi \; \overline{T}(1, Fo) = 0 \end{cases}.$$

– *For spherical conductors*:

$$\frac{\partial \overline{T}}{\partial(Fo)} = \frac{1}{\overline{r}}\frac{\partial}{\partial \overline{r}}\left(\overline{r}\frac{\partial \overline{T}}{\partial \overline{r}}\right)$$

$$\text{with: } \begin{cases} Fo = 0 \; ;\forall \overline{r}; \; 0 \le \overline{r} \le 1 \; : \overline{T}(\overline{r}, \; 0) = 1 \\ \forall \; Fo > 0 : \overline{r} = 0 \; : \; \dfrac{\partial \overline{T}}{\partial \overline{r}} = 0 \\ \forall \; Fo > 0 : \overline{r} = 1 \; : \; \dfrac{\partial \overline{T}}{\partial \overline{r}} + Bi\overline{T}(1, Fo) = 0 \end{cases}.$$

We will continue the development focusing on plane conductors, but the developments are similar for the other geometry types (cylinder or sphere).

To resolve the differential equation, we proceed by separating the variables. To do this, we write:

$$\overline{T}(\overline{x}, Fo) = f(\overline{x})g(Fo).$$

The differential equation becomes: $f(\overline{x}) \; g'(Fo) = g(Fo) \; f''(\overline{x})$.

Or: $\dfrac{g'(Fo)}{g(Fo)} = \dfrac{f''(\overline{x})}{f(\overline{x})}$.

This is not possible if: $\dfrac{g'(Fo)}{g(Fo)} = \dfrac{f''(\overline{x})}{f(\overline{x})} = Cste$, i.e.: $\dfrac{g'(Fo)}{g(Fo)} = A$ and $\dfrac{f''(\overline{x})}{f(\overline{x})} = A$.

Hence: *g(Fo)* is of the form, $e^{A\,Fo}$ ⇨ $A < 0$.

By writing: $A = -\omega^2$ we therefore obtain: $f''(\overline{x}) = -\omega^2 f(\overline{x})$

This equation admits a general solution that is written in the following form:

$$f(\overline{x}) = \cos(\omega\overline{x} + \phi_\omega)$$

The symmetry of the problem imposes: $\phi_\omega = 0$.

Thus, the solution sought is the sum of all of the solutions taking the following form:

$$\overline{T}_j(\overline{x}, Fo) = A_{\omega j}\, e^{-\omega_j^2 Fo} \cos(\omega_j \overline{x}),$$

namely: $\overline{T}(\overline{x}, Fo) = \sum_{j>0} A_{\omega j}\, e^{-\omega_j^2 Fo} \cos(\omega_j \overline{x})$.

Constants ω_j and $A_{\omega j}$ are determined from the boundary conditions. Indeed, the secondary boundary condition makes it possible to define the ω_j values:

$$\forall\ Fo > 0:\ \omega_j\ tg(\omega_j) = Bi$$

For a given Biot number ($Bi = \dfrac{hs}{\lambda}$), the ω_j values correspond to all of the solutions of the equation: $\omega_j\ tg(\omega_j) = Bi$

These are infinite in number.

Moreover, the first boundary condition makes it possible to determine the $A_{\omega j}$ values:

$$Fo = 0:\ \overline{T}(\overline{x}, 0) = 1 \Rightarrow \sum_{j>0} A_{\omega j} \cos(\omega_j \overline{x}) = 1$$

By multiplying the two members by $\cos\left(\omega_k \overline{x}\right)$ and by integrating the equation thus obtained between 0 and 1, we arrive at:

$$\sum_{j>0} A_{\omega j} \int_0^1 \cos\left(\omega_k \overline{x}\right)\cos\left(\omega_j \overline{x}\right)\, d\overline{x} = \int_0^1 \cos\left(\omega_k \overline{x}\right)\, dx$$

Note that the first member of this equation comprises integrals of the form $\int_0^1 \cos\left(\omega_k \overline{x}\right)\cos\left(\omega_j \overline{x}\right) dx$, which are all nil except for $k = j$.

Hence:

$$A_{\omega k} \int_0^1 \cos^2\left(\omega_k \overline{x}\right)\, d\overline{x} = \int_0^1 \cos\left(\omega_k \overline{x}\right)\, dx \Rightarrow A_{\omega k} = \frac{2\sin\left(\omega_k\right)}{\omega_k + \sin\left(\omega_k\right)\cos\left(\omega_k\right)}$$

Thus, the solution becomes:

$$\overline{T}\left(\overline{x}, Fo\right) = \sum_{j=1}^{\infty}\left[\frac{2\sin\left(\omega_j\right)}{\omega_j + \sin\left(\omega_j\right)\cos\left(\omega_j\right)} e^{-\omega_j^2 Fo} \cos\left(\omega_j \overline{x}\right)\right],$$

where the ω_j constants are the solutions of $\omega tg\left(\omega\right) = Bi$.

For cylindrical and spherical conductors, we obtain similar forms, which in the case of cylinders involve Bessel functions of the first kind of order 0 and 1: J_0 and J_1 (see Appendix). Table 6.2 summarizes the solutions obtained for each case.

Conductor	$\overline{T}\left(\overline{x}, Fo\right)$	Where the ω_j constants are the solutions of the equation
Plane: - of thickness 2s - of conductivity λ - the two faces of which are subject to a convection, *h*	$\overline{T}\left(\overline{x}, Fo\right) = \sum_{j=1}^{\infty}\left[\frac{2\sin\left(\omega_j\right)}{\omega_j + \sin\left(\omega_j\right)\cos\left(\omega_j\right)} e^{-\omega_j^2 Fo} \cos\left(\omega_j \overline{x}\right)\right]$	$\omega_j\, tg\left(\omega_j\right) = Bi$

Cylindrical: - of radius r - of conductivity λ - subject to a convection, h	$\overline{T}(\overline{r},Fo)=\sum_{j=1}^{\infty}\left[\frac{2}{\omega_j}\frac{J_1(\omega_j)}{J_0^2(\omega_j)+J_1^2(\omega_j)}e^{-\omega_j^2 Fo}J_0(\omega_j\overline{r})\right]$	$\omega_j\frac{J_1(\omega_j)}{J_0(\omega_j)}=Bi$
Spherical: - of radius r - of conductivity λ - subject to a convection, h	$\overline{T}(\overline{r},Fo)=\sum_{j=1}^{\infty}\left[\frac{2\left[\sin(\omega_j)-\omega_j\cos(\omega_j)\right]}{\omega_j-\sin(\omega_j)\cos(\omega_j)}e^{-\omega_j^2 Fo}\frac{\sin(\omega_j\overline{r})}{\omega_j\overline{r}}\right]$	$1-\omega_j\cot(\omega_j)=Bi$

Table 6.2. *Precise analytical solutions to the conduction equation*

$$\overline{T}=\frac{\theta-\theta_e}{\theta_i-\theta_e},\ \overline{x}=\frac{x}{s},\ \overline{r}=\frac{r}{s},\ Fo=\frac{\alpha}{s^2}t \text{ and } Bi=\frac{hs}{\lambda}$$

NOTE.– Note that the equations enabling the ω_j values to be determined [$\omega tg(\omega)=Bi$ for a plane conductor, $\omega_j\frac{J_1(\omega_j)}{J_0(\omega_j)}=Bi$ for a cylindrical conductor and $1-\omega_j\cot(\omega_j)=Bi$ for a spherical conductor], admit an infinity of solutions. Nevertheless, it should be pointed out that in most practical situations, the first solution (ω_1), or the first two solutions (ω_1 and ω_2) can provide a fair representation of the full analytical solution (Soutif, 1991). This is due to the presence of the term, $e^{-\omega_j^2 Fo}$, which creates an exponential decrease in the other terms of the series, particularly for very low *Fo* values.

Thus, as shown by the analysis performed in sections 6.6.3 and 6.6.4, for $Fo > 0.2$, it is enough for us to retain the first solution (ω_1), thus obtaining approximations presenting an accuracy of more than 98%. The following section presents a methodology that, based on the Biot number, allows us to establish the most appropriate analysis, along with the approximations to use, if any.

6.6. Approximate analytical solutions

In this section, we analyze the heat transfer with respect to the situation considered. This analysis allows us to show in which case approximations of the temperature field, $\theta(x, t)$ or $\overline{T}(\overline{x},Fo)$, can be used and what precisions would be obtained for each approximation. The different analyses are presented as a function of the Biot number via the definition of Bi intervals, which correspond to practical situations. In this instance, the reasoning is applied to a plane solid, but it can easily be adapted to the other geometry types.

6.6.1. *For* Bi = *0*

This is the case for athermic and adiabatic systems.

Physically: the situation $Bi = 0$ occurs when the convective transfer coefficient is nil: $h = 0$.

In such a situation, there is no convection and the temperature of the solid considered remains constant.

Analytically: with $Bi = 0$, the temperature field is given, for a plane conductor, by:

$$\overline{T}(\overline{x}, Fo) = \sum_{j=1}^{\infty}\left[\frac{2\sin(\omega_j)}{\omega_j + \sin(\omega_j)\cos(\omega_j)} e^{-\omega_j^2 Fo} \cos(\omega_j \overline{x})\right],$$

where the ω_j constants are the solutions to: $\omega_j \, tg(\omega_j) = 0$,

namely: $\omega_j\pi$.

Hence: $A_{\omega 1} = 1$ and the other values of $A_{\omega j} = 0$.

The temperature field then becomes: $\forall \ Fo > 0 \ \ \overline{T}(\overline{x}, Fo) = 1$;

or: $\forall t \ \ \theta = \theta_i$.

Thus, analytically we come to the result that proves obvious physically, that is: for a diathermic or adiabatic system (Bi = 0), the temperature of the solid considered remains constant.

6.6.2. *For 0* < Bi < *0.1*

This is the case for “thermally thin” systems.

Physically:

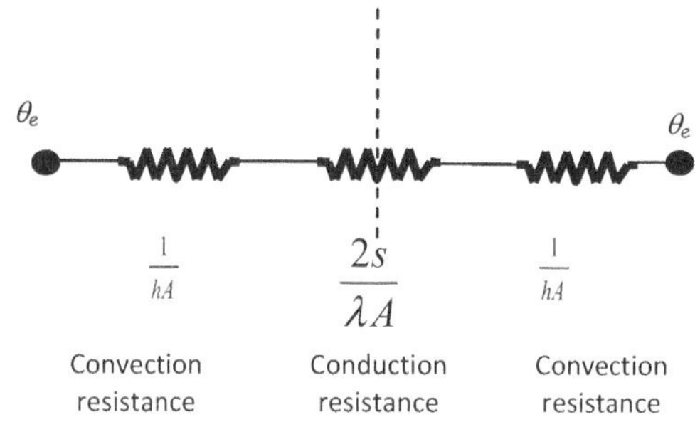

Figure 6.2. *Electrical analogy of a plane of thickness 2s, immersed in a medium at θe*

$Bi < 0.1 \Rightarrow \frac{1}{h} \rangle 10\left(\frac{s}{\lambda}\right) \Rightarrow$ on each side of the solid, the convection resistance $\left(\frac{1}{h\ A}\right)$ is more than ten times greater than the conduction resistance $\left(\frac{s}{\lambda\ A}\right)$.

In such situations, *convection* constitutes the limiting physical process conditioning the transfer of heat. Convection is said to *govern* heat transfer.

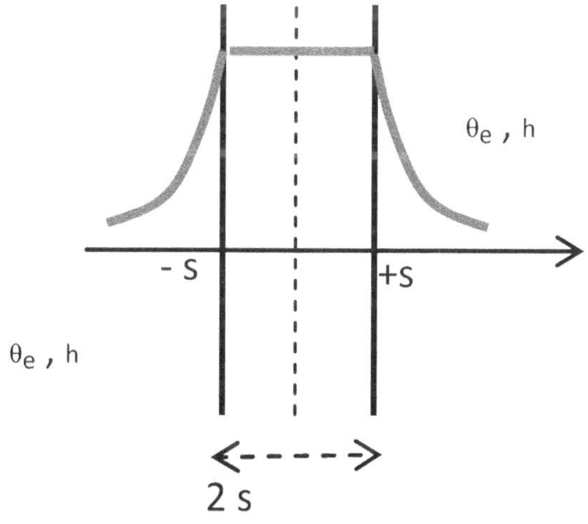

Figure 6.3. *Temperature profile inside a thermally-thin system*

Conduction, for it part, is virtually instantaneous. As soon as the temperature on the surface of the solid changes, the same will be true for the entire solid. This is reflected by a constant temperature profile inside the solid.

This is why we talk of "thermally thin systems".

Analytically: for such systems, the temperature is uniform in the solid, in other words, it does not depend on x, but solely on time.

The resolution in this case uses lumped system analysis, which is based on the balance approach.

Indeed, if the contact area between the plane and the bath is noted A and if θ designates the temperature of the solid at instant t, an energy balance on the plane solid, in transient state, gives:

Accumulation = Input – Output; (no generation)

or: $\rho\ V\ C_p \dfrac{d\theta}{dt} = 0 - h(2A)(\theta - \theta_e)$,

i.e.: (V = 2 s A): $2\ s\ A\ \rho\ C_p \dfrac{d\theta}{dt} = -2h\ A\ (\theta - \theta_e)$.

Hence, remembering that $Bi = \dfrac{h\ s}{\lambda}$ and that $\alpha = \dfrac{\lambda}{\rho C_p}$: $\dfrac{d\theta}{(\theta - \theta_e)} = -\dfrac{h}{\rho\ s\ C_p} dt$,

or: $\dfrac{d\theta}{(\theta - \theta_e)} = -\dfrac{\lambda\ Bi}{\rho\ s^2 C_p} dt$.

i.e.: $\dfrac{d\theta}{(\theta - \theta_e)} = -\dfrac{\alpha\ Bi}{s^2} dt$.

After integration, we then obtain: $Ln\left(\dfrac{\theta - \theta_e}{\theta_i - \theta_e}\right) = -\dfrac{\alpha\ Bi}{s^2} t$.

or, remembering that $Fo = \dfrac{\alpha}{s^2} t$: $Ln\left(\dfrac{\theta - \theta_e}{\theta_i - \theta_e}\right) = -\ Bi\ Fo$.

The temperature field is therefore exponential: $\frac{\theta-\theta_e}{\theta_i-\theta_e}=e^{-\frac{\alpha\,Bi}{s^2}t}$.

Note that the term $\frac{\alpha\,Bi}{s^2}$ has the dimension of the inverse of a time. Generally, we will note: $\tau=\frac{s^2}{\alpha\,Bi}$.

We then have: $\frac{\theta-\theta_e}{\theta_i-\theta_e}=e^{-\frac{t}{\tau}}$.

τ has the dimension of time. It is referred to as the system time constant.

τ indicates the pace at which a thin system cools down (or heats up). Figure 6.4 shows that more than 95% of the cooling task is assured after a time greater than or equal to 3 τ.

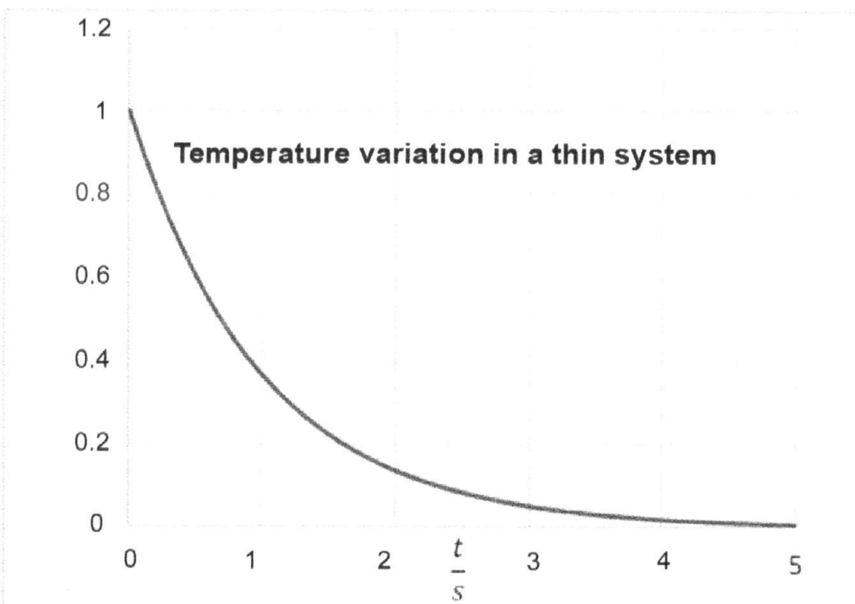

Figure 6.4. *Profile in a thin system of time constant, τ*

6.6.3. *For* Bi = *0.1*

We can no longer consider the system as thermally thin and, consequently, the lumped analysis can no longer be performed. The temperature is now a function of x and t: $\theta(x, t)$.

We therefore need to return to the general solution, given by:

$$\overline{T}(\overline{x},Fo)=\sum_{j=1}^{\infty}\left[\frac{2\sin(\omega_j)}{\omega_j+\sin(\omega_j)\cos(\omega_j)}e^{-\omega_j^2 Fo}\cos(\omega_j\overline{x})\right]$$

where the constants ω_j are the solutions of the equation: $\omega_j\ \mathrm{tg}(\omega_j)=\mathrm{Bi}$.

For $Bi = 0.1$, the first four solutions of the equation $\omega_j\ \mathrm{tg}(\omega_j)=\mathrm{Bi}$ are:

$\omega_1 = 0.311$

$\omega_2 = 3.173$

$\omega_3 = 6.300$

$\omega_4 = 9.435$

Thus, a good approximation of the temperature field is obtained by:

$$\overline{T}(\overline{x},Fo)\approx\sum_{j=1}^{4}A_{\omega j}\ e^{-\omega_j^2 Fo}\cos(\omega_j\overline{x})$$

where: $A_{\omega j}=\dfrac{2\sin(\omega_j)}{\omega_j+\sin(\omega_j)\cos(\omega_j)}$.

The $A_{\omega j}$ coefficients are calculated in Table 6.3:

j	ω_j	$A_{\omega j}$
1	0.311	1.0161
2	3.173	-0.0196
3	6.300	0.0053
4	9.435	-0.0022

Table 6.3. *The first four solutions of* $\omega_j tg\omega_j = 0.1$

We can thus note that the coefficient of the second term, A_{ω_2}, is *more than fifty times* smaller than A_{ω_1}.

Moreover, we can easily establish that for *Fo* > 0.2, the $A_{\omega j}e^{-\omega_j^2 Fo}$ terms become negligible as of *j* = 2. Indeed, Table 6.4 shows that, for *Fo* = 0.2, the term $A_{\omega 2}e^{-\omega_2^2 Fo}$ is of the order of 10^{-3}, whereas the first term is of the order of 1.

J	ω_j	$e^{-\omega_j^2 Fo}$	$A_{\omega j}e^{-\omega_j^2 Fo}$
1	0.3111	9.81E-01	9.97E-01
2	3.173	1.34E-01	-2.62E-03
3	6.300	3.57E-04	1.90E-06
4	9.435	1.85E-08	-4.01E-11

Table 6.4. *Contributions of the different* $e^{-\omega_j^2 Fo}$ *values for* Fo = *0.2*

The end result is clear: we obtain a good approximation of $\overline{T}(\overline{x}, Fo)$ by conserving solely the first term of the development:

$$\overline{T}(\overline{x}, Fo) \approx A_{\omega 1}\, e^{-\omega_1^2 Fo} \cos\left(\omega_1 \overline{x}\right)$$

i.e.: $\overline{T}(\overline{x}, Fo) \approx 1.0161\, e^{-0.09678\ Fo} \cos\left(0.3111\ \overline{x}\right)$.

6.6.4. *For* Bi *> 0.1*

In this case too, it is no longer possible to perform lumped analysis. But as in the previous example, we obtain a good approximation of the temperature field by stopping at the first four development terms.

Table 6.5 gives the first four solutions to the equation $\omega_j\ tg(\omega_j) = Bi$ for different values of *Bi* greater than 0.1.

$\overline{T}(\overline{x},Fo) \approx \sum_{j=1}^{4} A_{\omega j}\, e^{-\omega_j^2 Fo} \cos(\omega_j \overline{x})$, with: $A_{\omega j} = \frac{2\sin(\omega_j)}{\omega_j + \sin(\omega_j)\cos(\omega_j)}$								
Bi	ω_1	ω_2	ω_3	ω_4	$A_{\omega 1}$	$A_{\omega 2}$	$A_{\omega 3}$	$A_{\omega 4}$
1	0.863	3.4256	6.4373	9.5293	1.1191	-0.1517	0.0466	-0.0217
5	1.3138	4.0336	6.9096	9.8928	1.2402	-0.3442	0.1588	-0.0876
10	1.430	4.306	7.223	10.200	1.2621	-0.3935	0.2097	-0.1308
100	1.555	4.666	7.777	10.880	1.2731	-0.4240	0.2539	-0.1807
∞	$\frac{\pi}{2}$	$\frac{3\pi}{2}$	$\frac{5\pi}{2}$	$\frac{7\pi}{2}$	$\frac{4}{\pi}$	$-\frac{4}{3\pi}$	$\frac{4}{5\pi}$	$-\frac{4}{7\pi}$

Table 6.5. *Solutions limited to the first four terms for different Biot values*

Note that here the $A_{\omega j}$ values are of almost the same order of magnitude. Note on Table 6.6, however, that for $Fo > 0.2$, the $e^{-\omega_j^2 Fo}$ terms become negligible as of $j = 3$.

Bi	ω_1	ω_2	ω_3	ω_4	$e^{-\omega_1^2 Fo}$	$e^{-\omega_2^2 Fo}$	$e^{-\omega_3^2 Fo}$	$e^{-\omega_4^2 Fo}$
1	0.8603	3.4256	6.4373	9.5293	0.8624	0.0957	2.52E-04	1.30E-08
5	1.3138	4.0336	6.9096	9.8928	0.7081	0.0386	7.13E-05	3.16E-09
10	1.430	4.306	7.223	10.200	0.6643	0.0245	2.94E-05	9.19E-10
100	1.555	4.666	7.777	10.880	0.6166	0.0129	5.58E-06	5.23E-11
∞	1.5708	4.7124	7.8540	10.9956	0.6105	0.0118	4.39E-06	3.15E-11

Table 6.6. *Contributions of the different $e^{-\omega_j^2 Fo}$ values for* Fo *= 0.2 and different Biot values*

We thus obtain a good approximation of $\overline{T}(\overline{x},Fo)$ by solely retaining the first two terms of the development:

$$\overline{T}(\overline{x},Fo) \approx A_{\omega 1}\, e^{-\omega_1^2 Fo} \cos(\omega_1 \overline{x}) + A_{\omega 2}\, e^{-\omega_2^2 Fo} \cos(\omega_2 \overline{x})$$

This results in the expressions of the temperature fields for the different Biot number values considered: Bi = 1, Bi = 5, Bi = 10, Bi = 100 and $Bi \rightarrow \infty$. These expressions are set out in Table 6.7:

Bi	$\overline{T}(\overline{x},Fo)$
1	$\overline{T}(\overline{x},Fo) \approx 1.1191\ e^{-0.7401\ Fo}\cos(0.8603\overline{x})\ - 0.1517\ e^{-11.7347\ Fo}\cos(3.4256\overline{x})$
5	$\overline{T}(\overline{x},Fo) \approx 1.2402\ e^{-1.7261\ Fo}\cos(1.3138\overline{x})\ - 0.3442\ e^{-16.2697\ Fo}\cos(4.0336\overline{x})$
10	$\overline{T}(\overline{x},Fo) \approx 1.2621\ e^{-2.0449\ Fo}\cos(1.43\overline{x})\ - 0.3935\ e^{-18.5416\ Fo}\cos(4.306\overline{x})$
100	$\overline{T}(\overline{x},Fo) \approx 1.2731\ e^{-2.4180\ Fo}\cos(1.555\overline{x})\ - 0.4240\ e^{-21.7716\ Fo}\cos(4.666\overline{x})$
∞	$\overline{T}(\overline{x},Fo) \approx \frac{4}{\pi}e^{-\frac{\pi^2}{4}Fo}\cos\left(\frac{\pi}{2}\overline{x}\right) - \frac{4}{3\pi}e^{-\frac{9\pi^2}{4}Fo}\cos\left(\frac{3\pi}{2}\overline{x}\right)$

Table 6.7. *Approximate solutions to the heat equation*

From a practical point of view, for values not presented in Table 6.7, for each value of Bi > 0.1, we will proceed by determining the first two solutions, ω_1 and ω_2, to equation $\omega tg\omega = Bi$, then the two coefficients, $A_{\omega 1}$ and $A_{\omega 2}$, will be calculated. The expression of the temperature field will then come about in a straightforward manner.

The different temperature-field expressions developed in this section for plane solids are summarized in Table 6.8. This offers a more immediate view of the solutions than the graphical and numerical methods, whilst remaining accurate. Exercise 28 in Chapter 7 allows us to appreciate the simplicity of the calculation offered by these expressions.

Bi	$\overline{T}(\overline{x},Fo)$
0	The temperature remains constant
0 < Bi < 0.1	$\overline{T}(Fo) = e^{-Bi\ Fo}$
0.1	$\overline{T}(\overline{x},Fo) \approx 1.0161\ e^{-0.09678\ Fo}\cos(0.3111\ \overline{x})$
1	$\overline{T}(\overline{x},Fo) \approx 1.1191\ e^{-0.7401\ Fo}\cos(0.8603\overline{x})\ - 0.1517\ e^{-11.7347\ Fo}\cos(3.4256\overline{x})$

5	$\bar{T}(\bar{x},Fo) \approx 1.2402\ e^{-1.7261\ Fo}\cos(1.3138\bar{x})\ -0.3442\ e^{-16.2697\ Fo}\cos(4.0336\bar{x})$
10	$\bar{T}(\bar{x},Fo) \approx 1.2621\ e^{-2.0449\ Fo}\cos(1.43\bar{x})\ -0.3935\ e^{-18.5416\ Fo}\cos(4.306\bar{x})$
100	$\bar{T}(\bar{x},Fo) \approx 1.2731\ e^{-2.4180\ Fo}\cos(1.555\bar{x})\ -0.4240\ e^{-21.7716\ Fo}\cos(4.666\bar{x})$
∞	$\bar{T}(\bar{x},Fo) \approx \frac{4}{\pi}e^{-\frac{\pi^2}{4}Fo}\cos\left(\frac{\pi}{2}\bar{x}\right) - \frac{4}{3\pi}e^{-\frac{9\pi^2}{4}Fo}\cos\left(\frac{3\pi}{2}\bar{x}\right)$

Table 6.8. *Solutions for different fields of* Bi *for a plane solid*

6.7. Graphical method for solving the heat equation

The graphical method presented in this section appears to be very useful when instead of seeking exact temperature profiles we wish to obtain rapid estimates of these profiles in transient state. It is entirely suitable for the metal tempering calculations and may be applied to any of the geometries: plane, cylindrical or spherical conductor.

The problem considered is the same as that solved using the analytical method: the conductor, which is initially at uniform temperature, θ_i, is immersed in a medium of temperature, θ_e. As the convection heat transfer coefficient between the conductor and the medium are known, the method makes it possible to determine the temperature profile as a function of time inside the solid: $\theta(x,t)$.

Let us recall the definitions of the following parameters, the same that are used for analytical resolution:

– *s* designates:

- the half-thickness for a plane conductor;
- the radius for a cylindrical conductor;
- the radius for a spherical conductor;

– *s* is used to define the *reduced length,* $\bar{x} = \frac{x}{s}$ or $\bar{r} = \frac{r}{s}$, as applicable, with: $0 \le \bar{x} \le 1$ and $0 \le \bar{r} \le 1$;

– $\bar{T}$ designates the *reduced temperature*: $\bar{T} = \frac{\theta - \theta_e}{\theta_i - \theta_e}$;

– $\bar{T}(0,Fo)$ will indicate the *reduced temperature at the center of the solid*: $\bar{T}(0,Fo)=\dfrac{\theta(0,t)-\theta_e}{\theta_i-\theta_e}$;

– the Biot number: $Bi=\dfrac{hs}{\lambda}$;

– the Fourier number: $Fo=\dfrac{\alpha t}{s^2}$ with $\alpha=\dfrac{\lambda}{\rho C_p}$.

The graphical method presented here (Heisler, 1956; Gröber, Erk and Grigull, 1961) is based on the use of an approximation of the exact solution, which only retains the first term of the sum, $\bar{T}(\bar{x},Fo)$. As seen in sections 6.6.3 and 6.6.4, this approximation could be justified for Fourier numbers above 0.2.

The expressions of the approximate solutions thus obtained are presented in Table 6.8, for which the values of ω_l and $A_{\omega l}$ to be used have been calculated and tabulated in the Appendix (database), as a function of the Biot number and for each geometry type considered. Table 6.9 gives several values for significant Biot numbers.

Conductor	$\bar{T}(\bar{x},Fo)$ or $\bar{T}(\bar{r},Fo)$	$A_{\omega l}$	Where ω_l is the first solution of the equation
Plane: - of thickness 2s - of conductivity λ - the two faces of which are subject to a convection, *h*	$\bar{T}(\bar{x},Fo)=$ $A_{\omega_1}e^{-\omega_1^2 Fo}\cos(\omega_1\bar{x})$	$A_{\omega_1}=\dfrac{2\sin(\omega_1)}{\omega_1+\sin(\omega_1)\cos(\omega_1)}$	$\omega\ tg(\omega)=Bi$
Cylindrical: - of radius *r* - of conductivity λ - subject to a convection, *h*	$\bar{T}(\bar{r},Fo)=$ $A_{\omega_1}e^{-\omega_1^2 Fo}J_0(\omega_1\bar{r})$	$A_{\omega_1}=\dfrac{2}{\omega_1}\dfrac{J_1(\omega_1)}{J_0^2(\omega_1)+J_1^2(\omega_1)}$	$\omega\dfrac{J_1(\omega)}{J_0(\omega)}=Bi$
Spherical: - of radius *r* - of conductivity λ - subject to a convection, *h*	$\bar{T}(\bar{r},Fo)=$ $A_{\omega_1}e^{-\omega_1^2 Fo}\dfrac{\sin(\omega_1\bar{r})}{\omega_1\bar{r}}$	$A_{\omega_1}=\dfrac{2[\sin(\omega_1)-\omega_1\cos(\omega_1)]}{\omega_1-\sin(\omega_1)\cos(\omega_1)}$	$1-\omega\cot(\omega)=Bi$

Table 6.9. *The approximate solutions used to establish Heisler charts (case where Fo > 0.2)*

$$\overline{T} = \frac{\theta - \theta_e}{\theta_0 - \theta_e}; \quad \overline{x} = \frac{x}{s}; \; \overline{r} = \frac{r}{s}; \quad Fo = \frac{\alpha}{s^2} t \text{ and } Bi = \frac{hs}{\lambda}$$

	Plane		Cylinder		Sphere	
Bi	ω_1	$A\omega_1$	ω_1	$A\omega_1$	ω_1	$A\omega_1$
0.01	0.0998	1.0017	0.1412	1.0025	0.1731	1.0030
0.1	0.3111	1.0161	0.4417	1.0246	0.5423	1.0298
1	0.8603	1.1191	1.2558	1.2071	1.5708	1.2732
10	1.4289	1.2620	2.1795	1.5677	2.8363	1.9249
50	1.5400	1.2727	2.3272	1.6002	3.0788	1.9962
100	1.5552	1.2731	2.3809	1.6015	3.1102	1.9990
∞	1.5708	1.2732	2.4048	1.6021	3.1416	2.0000

Table 6.10. *Values of ω_1 and $A_{\omega 1}$ to be used for different Biot values and different geometries*

The values of Bessel functions J_0 and J_1 are tabulated in the Appendix, which also gives the expressions for calculating $J_0(x)$ and $J_1(x)$.

6.7.1. *Temperature profile at center of solid*

One of the most characteristic parts of the solid is its center, be it the central plane for a plane solid, the central axis of a cylinder or the center of a sphere. The graphical method presented in this section enables the temperature profile to be easily determined in this part of the solid: $\overline{T}(0, Fo)$.

$\overline{T}(0, Fo)$ is obtained from the expressions in Table 6.8 by taking $\overline{x} = 0$ or $\overline{r} = 0$, as applicable, and by noting that cos(0) = $J_0(0)$ =1, and that the limit of $\frac{\sin x}{x}$ when $x \rightarrow 0$ is also 1.

The expressions thus produced for $\overline{T}(0, Fo)$ are presented in Table 6.11:

Conductor	$\overline{T}(0, Fo)$	Where ω_1 is the first solution to the equation
Plane: - of thickness 2s - of conductivity λ - the two faces of which are subject to a convection, *h*	$\overline{T}(0,Fo) \approx \frac{2\sin(\omega_1)}{\omega_1 + \sin(\omega_1)\cos(\omega_1)} e^{-\omega_1^2 Fo}$	$\omega \; tg(\omega) = Bi$
Cylindrical: - of radius *r* - of conductivity λ - subject to a convection, *h*	$\overline{T}(0,Fo) \approx \frac{2}{\omega_1} \frac{J_1(\omega_1)}{J_0^2(\omega_1) + J_1^2(\omega_1)} e^{-\omega_1^2 Fo}$	$\omega \frac{J_1(\omega)}{J_0(\omega)} = Bi$
Spherical: - of radius *r* - of conductivity λ - subject to a convection, *h*	$\overline{T}(0,Fo) \approx \frac{2\left[\sin(\omega_1) - \omega_1 \cos(\omega_1)\right]}{\omega_1 - \sin(\omega_1)\cos(\omega_1)} e^{-\omega_1^2 Fo}$	$1 - \omega \cot(\omega) = Bi$

Table 6.11. *Approximate profile at center of solid:* $\overline{T}(0, Fo)$ *(case where* Fo > 0.2*)*

The charts in Figures 6.5 to 6.7 are graphical representations of the approximate profiles presented in Table 6.11, calculated for different Biot numbers, using the ω values from Table 6.10 and from section A.11 of the Appendix.

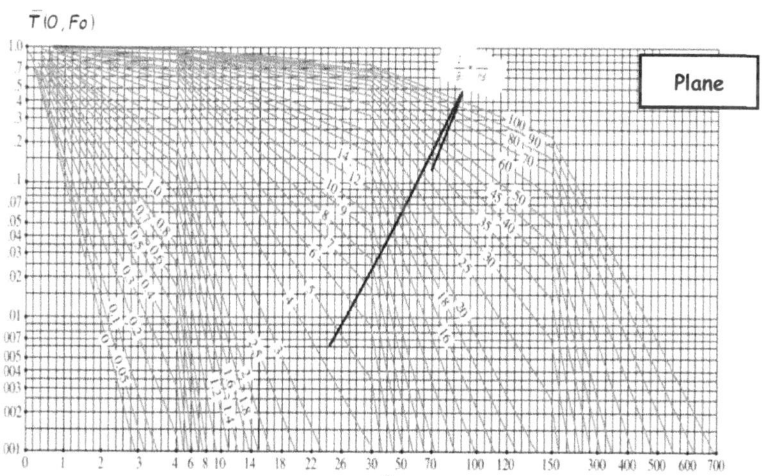

Figure 6.5. *The central temperatures as a function of* Fo *and of* $\frac{1}{Bi}$ *(plane conductor:* Fo > 0.2*). For a color version of this figure, see www.iste.co.uk/benallou/energy2.zip*

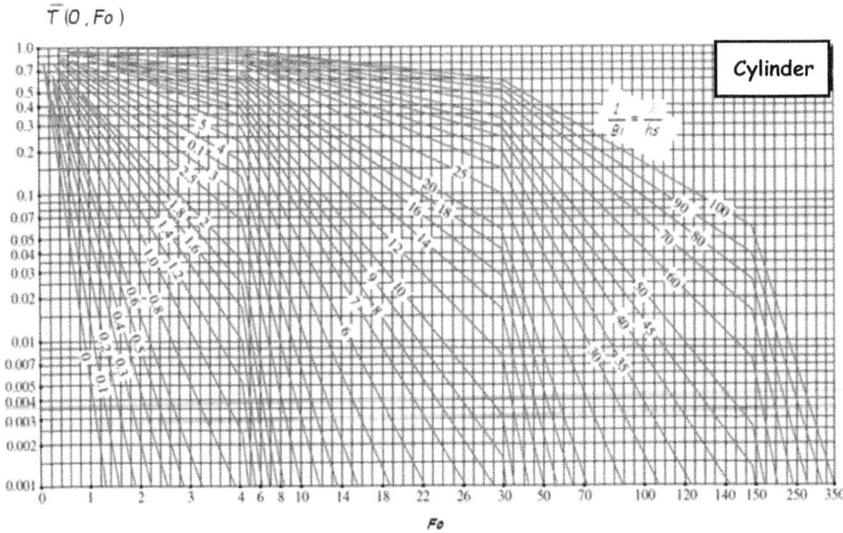

Figure 6.6. *The central temperatures as a function of* Fo *and of* $\frac{1}{Bi}$ *(cylindrical conductor:* Fo > *0.2). For a color version of this figure, see www.iste.co.uk/benallou/energy2.zip*

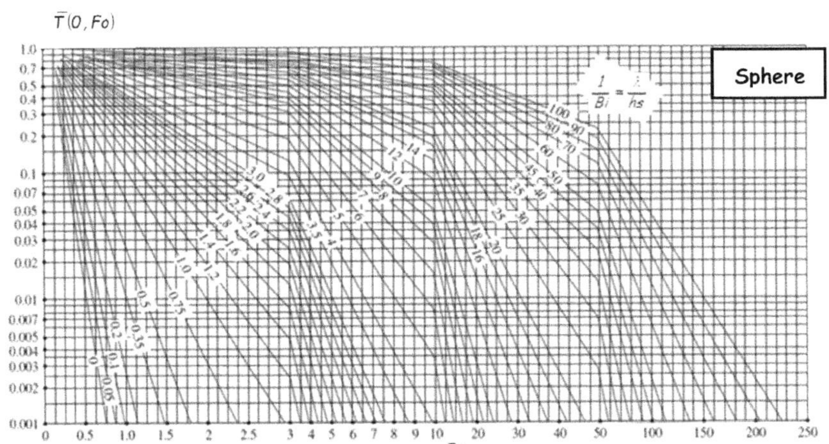

Figure 6.7. *The central temperatures as a function of* Fo *and of* $\frac{1}{Bi}$ *(spherical conductor:* Fo > *0.2). For a color version of this figure, see www.iste.co.uk/benallou/energy2.zip*

6.7.2. *Using charts to determine temperature profile at center of solid*

Figures 6.5–6.7 enable the temperature at the center of the solid to be calculated, determining the $\bar{T}(0, Fo)$ parameter as a function of the Biot and Fourier numbers: $\bar{T}(0, Fo) = f_1(Bi, Fo)$.

The value of $\theta_0(t)$ is thus obtained as follows:

– *Step 1*: determine which chart is to be used according to the nature of the solid concerned: plane (Figure 6.5), cylinder (Figure 6.6) or sphere (Figure 6.7).

– *Step 2*: start by calculating the Biot number, then deduce therefrom $\frac{1}{Bi}$, and therefore the curve that is to be employed.

– *Step 3*: for each instant, *t*, calculate $Fo = \frac{\alpha t}{s^2}$, then refer to the chart to read the value of $\bar{T}(0, Fo)$.

– *Step 4*: if $\bar{T}(0, Fo)$ is known, deduce therefrom $\theta_0 = \theta_e + (\theta_i - \theta_e)\bar{T}(0, Fo)$.

By carrying out these steps for the different instants of interest to us, determine the evolution of the temperature at the center of the conductor as a function of time.

Thus, the charts presented in Figures 6.5 to 6.7 enable the temperatures at the center of the solid considered (plan, cylinder or sphere) to be calculated at different instants, as a function of the Fourier number and the inverse of the Biot number.

What about the temperature distribution inside the conductor?

6.7.3. *Temperature distribution inside the solid*

This distribution is generated as a function of the temperature at the center, $\bar{T}(0, Fo)$, bearing in mind that the $\frac{\bar{T}(\bar{x}, Fo)}{\bar{T}(0, Fo)}$ or $\frac{\bar{T}(\bar{r}, Fo)}{\bar{T}(0, Fo)}$ ratio can be obtained by dividing the expressions of $\bar{T}(\bar{x}, Fo)$ given in Table 6.8 by those of $\bar{T}(0, Fo)$ indicated in Table 6.10.

Conductor	$\frac{\overline{T}(\overline{x},Fo)}{\overline{T}(0,Fo)}$ or $\frac{\overline{T}(\overline{r},Fo)}{\overline{T}(0,Fo)}$	Where ω_1 is the first solution of the equation
Plane: - of thickness 2s - of conductivity λ - the two faces of which are subject to a convection, *h*	$\frac{\overline{T}(\overline{x},Fo)}{\overline{T}(0,Fo)} \approx \cos(\omega_1 \overline{x})$	$\omega \, tg(\omega) = Bi$
Cylindrical: - of radius *r* - of conductivity λ subject to a convection, *h*	$\frac{\overline{T}(\overline{r},Fo)}{\overline{T}(0,Fo)} \approx J_0(\omega_1 \overline{r})$	$\omega \frac{J_1(\omega)}{J_0(\omega)} = Bi$
Spherical: - of radius *r* - of conductivity λ - subject to a convection, *h*	$\frac{\overline{T}(\overline{r},Fo)}{\overline{T}(0,Fo)} \approx \frac{\sin(\omega_1 \overline{r})}{\omega_1 \overline{r}}$	$1 - \omega\cot(\omega) = Bi$

Table 6.12. *Distribution $\overline{T}(\overline{x},Fo)$ or $\overline{T}(\overline{r},Fo)$ as a function of $\overline{T}(0,Fo)$ (case where Fo > 0.2)*

Figures 6.8 to 6.10 represent these ratios as a function of the reduced length, $\frac{x}{s}$ or $\frac{r}{s}$, and of the inverse of the Biot number.

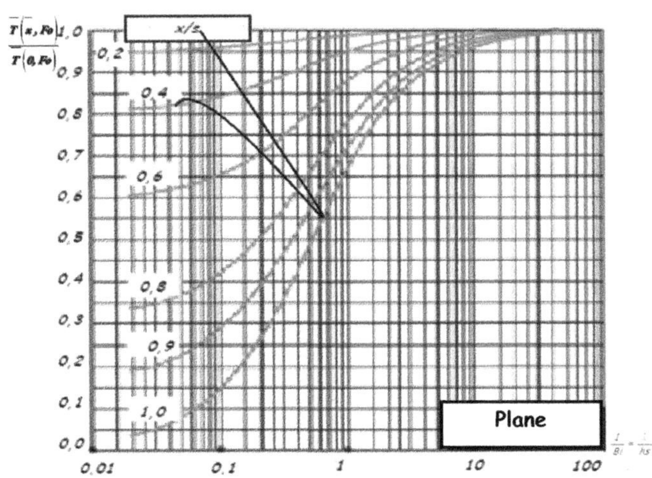

Figure 6.8. *Temperature distribution in a plane conductor (Fo > 0.2). For a color version of this figure, see www.iste.co.uk/benallou/energy2.zip*

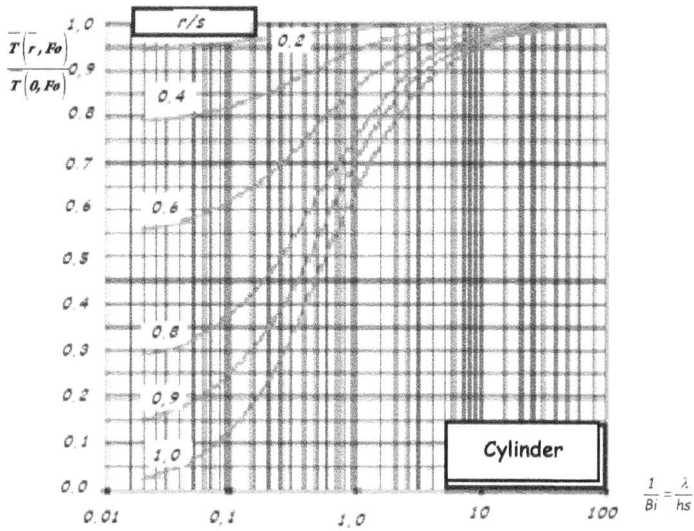

Figure 6.9. *Temperature distribution in a cylindrical conductor (*Fo > *0.2). For a color version of this figure, see www.iste.co.uk/benallou/energy2.zip*

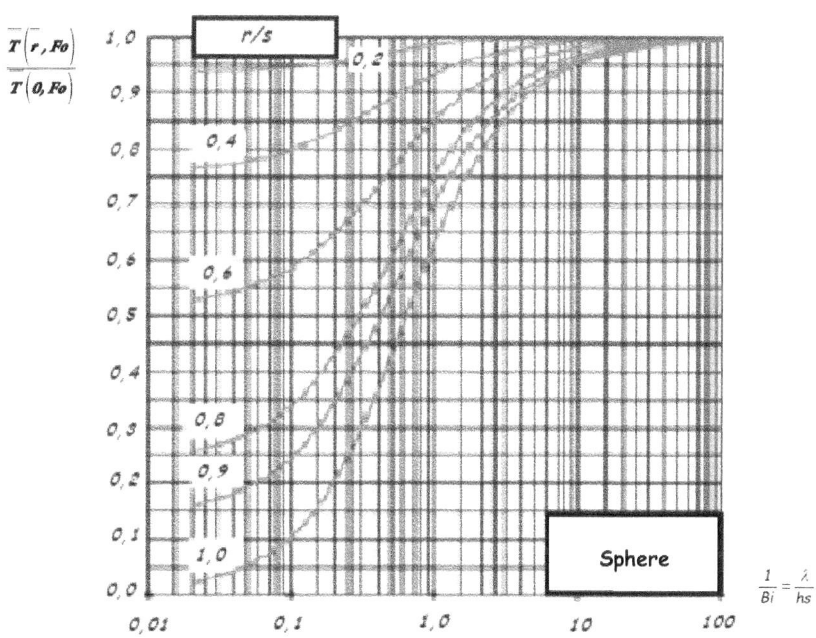

Figure 6.10. *Temperature distribution in a spherical conductor (*Fo > *0.2). For a color version of this figure, see www.iste.co.uk/benallou/energy2.zip*

6.7.4. *Using Figures 6.8 to 6.10 to determine the temperature distribution inside the solid*

With the temperature at the center at time *t* being known, we now need to determine the temperature at different lengths from the inside of the solid. To do this, firstly define a mesh on the solid considered, we propose to proceed as follows:

– *Step 1*: each point in the mesh has a corresponding length, *x* (or *r* in the case of a cylinder or a sphere).

– *Step 2*: for each value of *x* (or *r*, as applicable), calculate the corresponding $\frac{x}{s}$ or $\frac{r}{s}$.

– *Step 3*: for each value of $\frac{x}{s}$, determine which curve to use.

– *Step 4*: place $\frac{1}{\mathrm{Bi}}$ on the *x* axis, then read the value of $\frac{\overline{T}(\overleftarrow{x},\mathrm{Fo})}{\overline{T}(0,\mathrm{Fo})}$ on the *y* axis.

– *Step 5*: if $\frac{\overline{T}(\overleftarrow{x},\mathrm{Fo})}{\overline{T}(0,\mathrm{Fo})}$ and $\overline{T}(0,\mathrm{Fo})$ are known, it is then possible to deduce therefrom $\overline{T}(\overleftarrow{x},\mathrm{Fo})$, i.e. $\theta(x,t)$, for each value of *x* or *r*.

NOTE.– The figures presented may be used for Biot numbers above 0.01.

For Biot numbers below 0.01, the analysis applicable to thermally-thin systems needs to be used.

Thus, Figures 6.5 to 6.10 enable determination at different instants of the temperature at the center, as well as the temperature distribution inside the conductors. In most situations, calculating the fluxes exchanged between the conductor and the bath will also be worthy of interest.

6.7.5. *Calculating the fluxes exchanged*

Assuming that we wait for a sufficiently long time, the temperature of the solid would ultimately reach that of the bath, θ_e. In this case, the energy transferred between the solid and the bath would correspond to the maximum exchangeable, given by:

$$Q_{max} = \rho V C_p (\theta_i - \theta_e),$$

where V is the volume of the conductor, ρ is its density and C_p its specific heat.

At any finite instant, when the temperature of the solid passes from θ_i to θ, the energy exchanged between the latter and the bath is given by:

$$Q = \rho C_p \iiint_V (\theta_i - \theta)\, dV$$

$$\text{Hence: } = \frac{1}{V} \frac{\iiint_V (\theta_i - \theta)\, dV}{(\theta_i - \theta_e)}.$$

$(\theta_i - \theta_e)$ being a constant, this results in:

$$\frac{Q}{Q_{max}} = \frac{1}{V} \iiint_V \frac{(\theta_i - \theta)}{(\theta_i - \theta_e)} dV = \frac{1}{V} \iiint_V \left[1 - \frac{(\theta - \theta_e)}{(\theta_i - \theta_e)}\right] dV$$

Thus, $\left(d\overline{x} = \frac{dV}{V}\right)$, we have $\frac{Q}{Q_{max}} = 1 - \int \overline{T} d\overline{x}$.

Yet, for a plane conductor: $\overline{T}(\overline{x}, Fo) \approx \overline{T}(0, Fo)\ \cos(\omega_1 \overline{x})$.

$\overline{T}(0, Fo)$ not depending on $\overline{x}$, we have:

$$\left[\frac{Q}{Q_{max}}\right]_{Plan} \approx 1 - \overline{T}(0, Fo) \int_0^1 \cos(\omega_1 \overline{x})\ d\overline{x}\ .$$

$$\text{Hence: } \left[\frac{Q}{Q_{max}}\right]_{Plan} = 1 - \overline{T}(0, Fo) \left[\frac{\sin(\omega_1 \overline{x})}{\omega_1}\right]_0^1.$$

$$\text{i.e.: } \left[\frac{Q}{Q_{max}}\right]_{Plan} = 1 - \overline{T}(0, Fo)\ \frac{\sin(\omega_1)}{\omega_1}.$$

By proceeding in the same way for cylindrical or spherical solids, we obtain the expressions of $\left[\frac{Q}{Q_{max}}\right]$ given in Table 6.13.

Conductor	$\left[\dfrac{Q}{Q_{max}}\right]$	Where ω_l is the first solution to the equation
Plane: - of thickness 2s - of conductivity λ - the two faces of which are subject to a convection, h	$\left[\dfrac{Q}{Q_{max}}\right]_{Plan} = 1 - \overline{T}(0, Fo)\ \dfrac{\sin(\omega_1)}{\omega_1}$	$\omega\ tg(\omega) = Bi$
Cylindrical: - of radius r - of conductivity λ - subject to a convection, h	$\left[\dfrac{Q}{Q_{max}}\right]_{Cyl} = 1 - 2\overline{T}(0, Fo)\ \dfrac{J_1(\omega_1)}{\omega_1}$	$\omega \dfrac{J_1(\omega)}{J_0(\omega)} = Bi$
Spherical: - of radius r - of conductivity λ - subject to a convection, h	$\left[\dfrac{Q}{Q_{max}}\right]_{Sph} = 1 - 3\overline{T}(0, Fo)\ \dfrac{\sin(\omega_1) - \omega_1 \cos\omega_1}{\omega_1^3}$	$1 - \omega\cot(\omega) = Bi$

Table 6.13. $\left[\dfrac{Q}{Q_{max}}\right]$ *fractions of the energy transferred (case where* Fo *> 0.2)*

NOTE.– The expressions of $\left[\dfrac{Q}{Q_{max}}\right]$ given in Table 6.13 can be expressed as a function of ω_l and of Fo only. Such expressions are more practical if $\left[\dfrac{Q}{Q_{max}}\right]$ is calculated without needing to calculate $\overline{T}(0, Fo)$. If, however, $\overline{T}(0, Fo)$ needs to be calculated, it is preferable to use the expressions given in Table 6.12.

Conductor	$\left[\dfrac{Q}{Q_{max}}\right]$	Where ω_l is the first solution of the equation
Plane: - of thickness 2s - of conductivity λ - the two faces of which are subject to a convection, h	$\left[\dfrac{Q}{Q_{max}}\right]_{Plan} = 1 - \dfrac{2\sin^2(\omega_1)}{\omega_1\left[\omega_1 + \sin(\omega_1)\cos(\omega_1)\right]} e^{-\omega_1^2 Fo}$	$\omega\ tg(\omega) = Bi$

Cylindrical: - of radius r - of conductivity λ - subject to a convection, h	$\left[\frac{Q}{Q_{max}}\right]_{Cyl} = 1 - \frac{4J_1^2(\omega_1)}{\omega_1^2\left[J_0^2(\omega_1)+J_1^2(\omega_1)\right]} e^{-\omega_1^2 Fo}$	$\omega\frac{J_1(\omega)}{J_0(\omega)} = Bi$
Spherical: - of radius r - of conductivity λ - subject to a convection, h	$\left[\frac{Q}{Q_{max}}\right]_{Sph} = 1 - \frac{6\left[\sin(\omega_1)-\omega_1\cos(\omega_1)\right]^2}{\omega_1^3\left[\omega_1 - \sin(\omega_1)\cos(\omega_1)\right]} e^{-\omega_1^2 Fo}$	$1-\omega\cot(\omega) = Bi$

Table 6.14. $\left[\frac{Q}{Q_{max}}\right]$ *as a function of* Fo *and of* ω_1 *(case where* Fo > 0.2*)*

Figures 6.11 to 6.13 offer graphical representations of the expressions presented in Tables 6.12 and 6.13. They make it possible to graphically obtain the amounts of energy, Q, exchanged between the conductor and the bath between the initial instant and instant t, in relation to the maximum exchangeable energy, Q_{Max}, defined at the start of section 6.7.5.

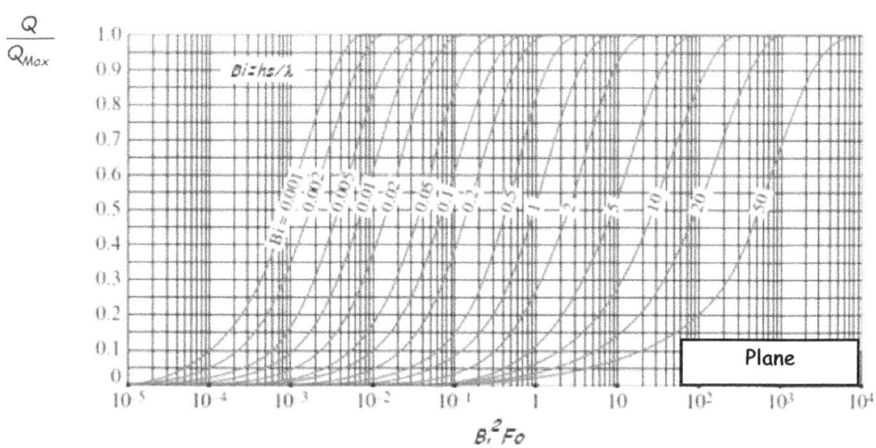

Figure 6.11. *Flux exchanged with the outside (plane conductor,* Fo > 0.2*). For a color version of this figure, see www.iste.co.uk/benallou/energy2.zip*

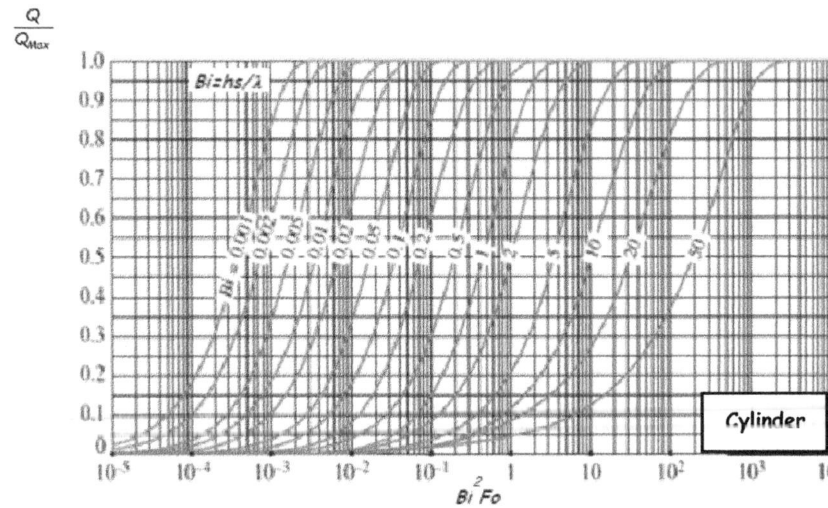

Figure 6.12. *Flux exchanged with the outside (cylindrical conductor,* Fo > *0.2). For a color version of this figure, see www.iste.co.uk/benallou/energy2.zip*

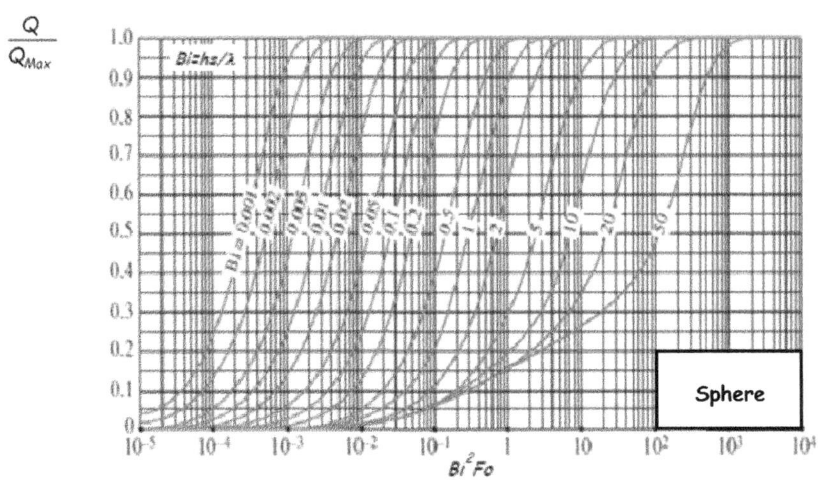

Figure 6.13. *Flux exchanged with the outside (spherical conductor,* Fo > *0.2). For a color version of this figure, see www.iste.co.uk/benallou/energy2.zip*

6.7.5.1. *General comment regarding the use of the graphical method*

The graphical method explored in the previous section makes it possible to fairly rapidly obtain orders of magnitude that are often sufficient to enable decisions to be made during the analysis of industrial processes. Nevertheless, it presents the following disadvantages:

– it is limited by the hypothesis that conduction is one-directional. As a result, it cannot be used for problems where heat visibly propagates in more than one direction;

– it is fairly approximate and offers limited accuracy;

– difficulties associated with graph reading and interpolations are reflected in the method's accuracy.

6.7.5.2. *Illustration: tempering of a cable*

A metal cable *50 cm* in diameter is initially at uniform temperature, $\theta_i = 760$ °C. The cable is tempered in a bath at $\theta_e = 38$ °C.

Questions

1) Determine graphically an estimation of the time needed for the temperature at the center of the cylinder to reach 108 °C.

2) Define at this instant the temperature profile inside the cable.

Data:

$h = 24$ kcal/m^2 °C

$C_p = 0.49$ J/kg °K

$\lambda = 50$ kcal/hm °C

$\rho = 7{,}850$ kg/m^3

Solutions

1) Calculating the time it takes for θ_0 to reach 108 °C

The initial excess temperature is given by:

$$T_i = \theta_i - \theta_e = 760 - 38 = 722 \text{ °C}$$

The Biot number is defined by: $\mathrm{Bi} = \dfrac{hs}{\lambda} = 0.12$

Hence: $\dfrac{1}{\mathrm{Bi}} = 8$.

The temperature sought at the center of the cylinder is:

$$T_0 = \theta_0 - \theta_e = 108 - 38 = 70\ °C$$

i.e.: $\frac{T_0}{T_i} = 0.1.$

By recording this data in Figure 6.6, we obtain:

Hence: $Fo = 10$.

Yet: $t = \frac{s^2 Fo}{\alpha}$ and: $\alpha = \frac{\lambda}{\rho C_p}$.

i.e.: $t = s^2 Fo \frac{\rho C_p}{\lambda}$.

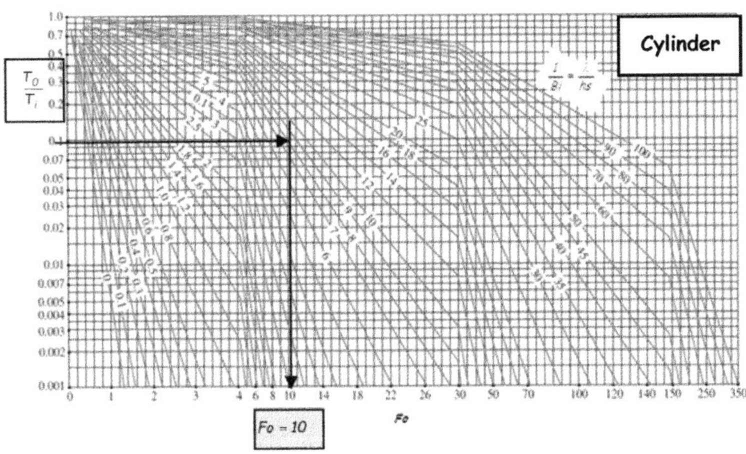

Figure 6.14. *Determination of* Fo *from* $\frac{T_0}{T_i}$. *For a color version of this figure, see www.iste.co.uk/benallou/energy2.zip*

NUMERICAL APPLICATION.–			
$s = 0.25$ m	$Fo = 10$	$\rho = 7{,}850$ kg/m^3	$C_p = 0.49$ J/kg °K
$\lambda = 50$ kcal/hm °C		$\alpha = 54.3$ m^2/hr	
$t = 41.4$ secs			

2) Temperature profile in the cable at t = 41.4 seconds

$$\frac{1}{\mathrm{Bi}} = 8 \text{ and } T_0 = 70\ °\mathrm{C}$$

By recording $\frac{1}{\mathrm{Bi}}$ on Figure 6.12, we determine $\frac{\mathrm{T}}{\mathrm{T_0}}$ for different values of $\frac{\mathrm{r}}{\mathrm{s}}$:

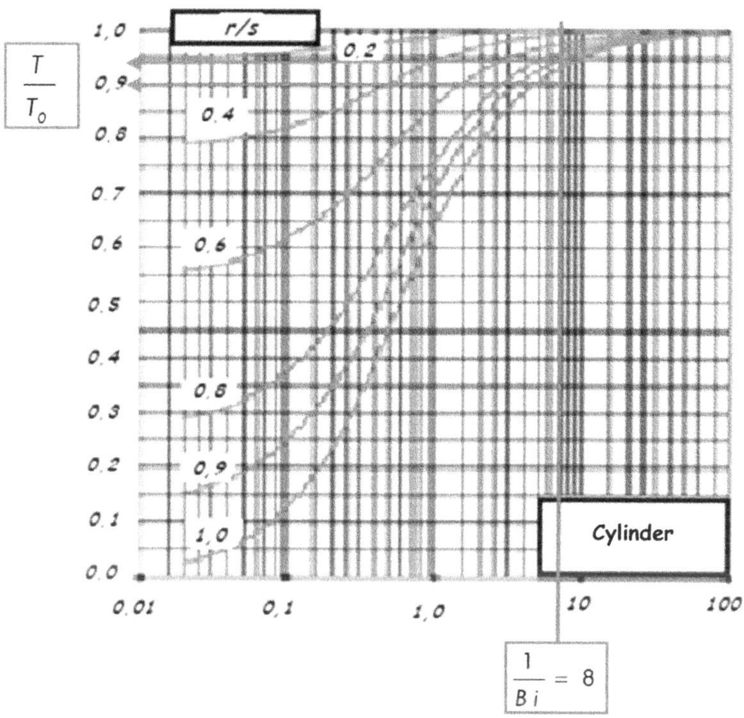

Figure 6.15. *Determination of $\frac{1}{Bi}$ from $\frac{T}{T_0}$. For a color version of this figure, see www.iste.co.uk/benallou/energy2.zip*

We deduce therefrom the values of θ for different values of $\frac{\mathrm{r}}{\mathrm{s}}$ from the following defining equation:

$$\theta = \theta_e + \left(\frac{\mathrm{T}}{\mathrm{T_0}}\right)(\theta_0 - \theta_e)$$

$\frac{r}{s}$	1.0	0.9	0.8	0.6	0.4	0.2
$\frac{T}{T_0}$	0.94	0.96	0.97	0.98	0.99	1
θ	103.8	105.2	105.9	106.6	107.3	108.0

Table 6.15. *Temperatures for different values of* $\frac{r}{s}$

Hence the profile:

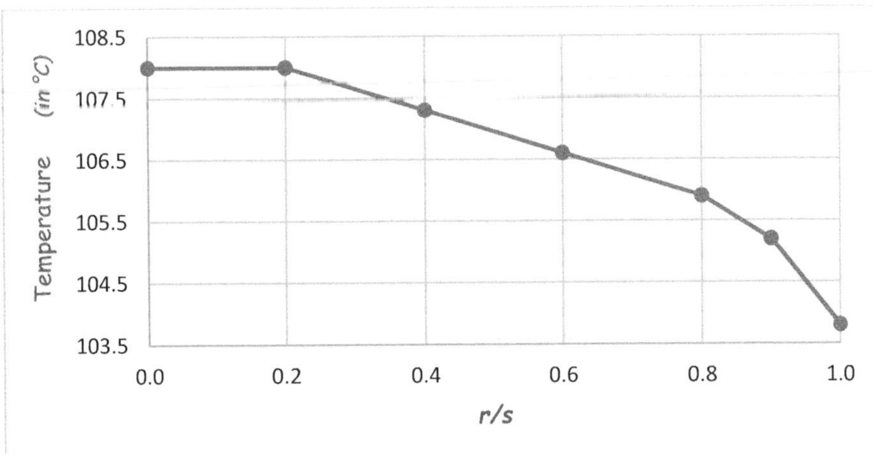

Figure 6.16. *Temperature profile at* t = *41.4 seconds*

6.7.5.3. *Illustration: tempering of a sphere*

A steel sphere with a radius of 20 cm is initially at uniform temperature, θ_i.

It is suddenly immersed in a liquid-metal bath, which is at uniform, constant temperature, θ_e.

Questions

1) Calculate the temperature at the center of the sphere at $t = 0.5$ hrs, $t = 1$ hr, $t = 2$ hrs and $t = 2.5$ hrs.

2) To cool a sphere that is *five* times' larger, the turbulence in the bath is increased such that the convection transfer coefficient, h, is multiplied by *four*.

a) How long will it take for the $\frac{T_0}{T_i}$ ratio to reach 50%, 10% and 1%?

b) Plot the temperature profiles inside the sphere at these instants.

Data:

$\theta_I = 350$ °C $\theta_e = 50$ °C $\eta = 32$ kcal/m^2°C

$C_{pSteel} = 0.11$ kcal/kg °C $\lambda = 64$ kcal/hm °C $\rho_{Steel} = 7{,}600$ kg/m^3

Solutions

1) Temperature at the center of the sphere at instants t = 0.5 hrs, t = 1 hr, t = 2 hrs and t = 2.5 hrs

The initial excess temperature is given by:

$$T_i = \theta_i - \theta_e = 350 - 50 = 300 \text{ °C}$$

The Biot number is given by: $\mathrm{Bi} = \dfrac{hs}{\lambda}$,

i.e.: $Bi = 0.1$.

Hence: $\dfrac{1}{\mathrm{Bi}} = 10$.

Moreover, we have: $\alpha = \dfrac{\lambda}{\rho C_p}$,

i.e.: $\alpha = 7.66\ 10^{-2}$m^2/hr.

And: $\mathrm{Fo} = \dfrac{\alpha t}{s^2} = \dfrac{(7{,}66)\ 10^{-2}}{(0{,}2)^2} t$,

i.e.: $Fo = 1.91\ t$, where t is in hours.

To calculate the evolution in the temperature at the center of the sphere with respect to time, the Fourier number will be evaluated at different instants, then indicated in Figure 6.7 (with $\frac{1}{\mathrm{Bi}} = 10$) to deduce the corresponding values of $\frac{T_0}{T_i}$ (see Figure 6.17).

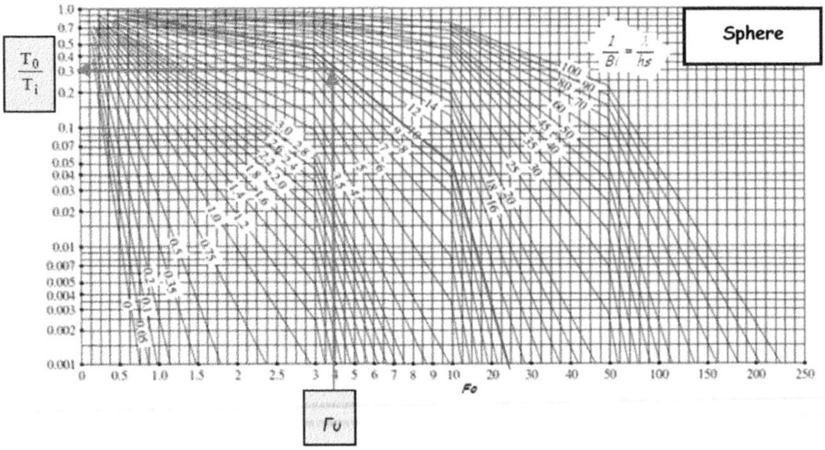

Figure 6.17. *Determination of* $\frac{T_o}{T_i}$ *from the Fourier number. For a color version of this figure, see www.iste.co.uk/benallou/energy2.zip*

The values of $\theta_0\,(t)$ are then obtained from the defining equation:

$$\theta_0 = \theta_e + \left(\frac{T_0}{T_i}\right)(\theta_i - \theta_e)$$

By reading the curve corresponding to $\frac{1}{Bi} = 0.5$, we obtain, at $t = 0.5$ hrs, $t = 1$ hr, $t = 2$ hrs and $t = 2.5$ hrs:

t (hours)	0.5	1.0	2.0	2.5
Fo	0.96	1.91	3.83	4.78
T_0/T_i	0.88	0.6	0.35	0.26
θ_0	314	230	155	128

Table 6.16. *Temperatures at the center at different instants*

2) Profile at t = 1 hr and an increased turbulence

We will have:

$s' = 5$ secs $= 1$ m

$h' = 4$ hrs $= 128$ kcal/hm^2 °C

$Bi' = 2$

Hence: $\frac{1}{\text{Bi}'} = 0{,}5$.

<u>*a) We firstly determine the Fourier numbers corresponding to the situations considered*</u>

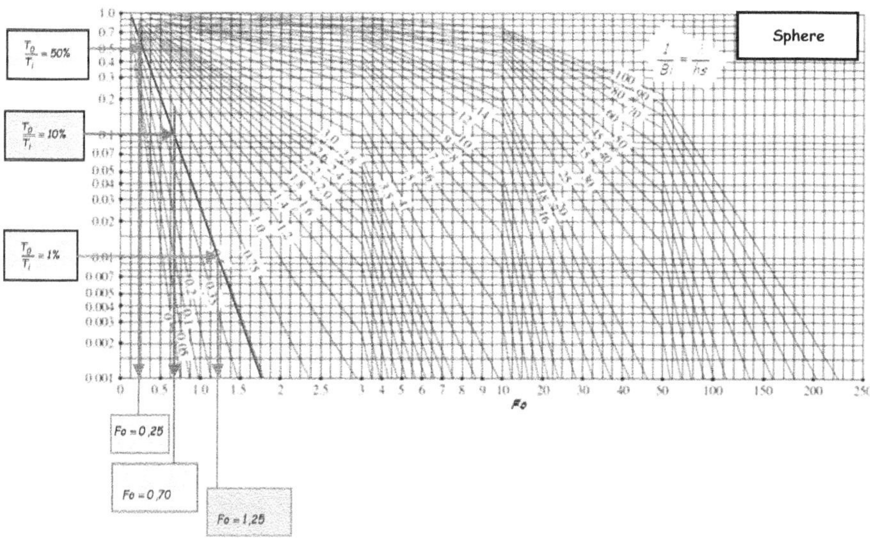

Figure 6.18. *Determination of Fourier numbers for different temperatures at the center. For a color version of this figure, see www.iste.co.uk/benallou/energy2.zip*

We deduce therefrom the values of *t*: $t = \frac{s^2 \text{Fo}}{\alpha}$.

T_0/T_i	50%	10%	1%
Fo	0.25	0.7	1.25
t (hours)	3.27	9.14	16.33

Table 6.17. *Calculation of times for different T_0/T_i ratios*

b) We define a meshing for the solid:

$\frac{r}{s} = 0.2; 0.4; 0.6; 0.8; 0.9$ and 1

For each $\frac{r}{s}$ value, we now use the 6.10 chart with $\frac{1}{Bi} = 0.5$.

The vertical line passing through $\frac{1}{Bi} = 0.5$ gives the values of $\frac{T}{T_0}$ for different values of $\frac{r}{s}$.

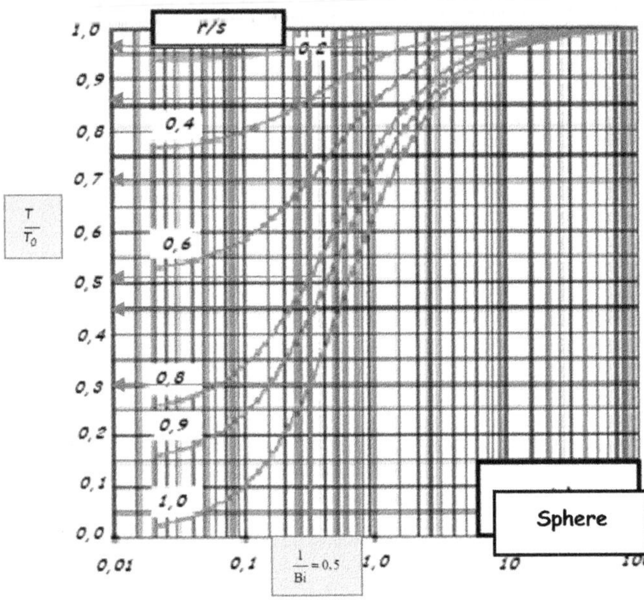

Figure 6.19. *Determination of $\frac{T}{T_o}$ as a function of $\frac{1}{Bi}$. For a color version of this figure, see www.iste.co.uk/benallou/energy2.zip*

$\frac{r}{s}$	0	0.2	0.4	0.6	0.8	0.9	1.0
$\frac{T}{T_0}$	1	0.92	0.86	0.70	0.50	0.43	0.30

Table 6.18. *Calculation of $\frac{T}{T_o}$ for different ratios of $\frac{r}{s}$*

We then calculate T for each value of T_0.

For each value of θ_0 and $\frac{T}{T_0}$, we determine the values of θ from the following relation:

$$\theta = \theta_e + \left(\frac{T}{T_0}\right)(\theta_0 - \theta_e)$$

The results obtained are gathered in Table 6.19.

$t =$	3,27	H		$\theta_0 =$	200	°C	
$\frac{r}{s}$	0	0,2	0,4	0,6	0,8	0,9	1
$\frac{T}{T_0}$	1	0,92	0,86	0,7	0,50	0,43	0,30
θ (°C)	200	188	179	155	125	114,5	95
$t =$	9,14	H		$\theta_0 =$	80	°C	
$\frac{r}{s}$	0	0,2	0,4	0,6	0,8	0,9	1
$\frac{T}{T_0}$	1	0,92	0,86	0,7	0,5	0,43	0,3
θ	80	77,6	75,8	71	65	62,9	59
$t =$	16,33	H		$\theta_0 =$	53	°C	
$\frac{r}{s}$	0	0,2	0,4	0,6	0,8	0,9	1
$\frac{T}{T_0}$	1	0,92	0,86	0,7	0,5	0,43	0,3
θ	53	52,76	52,58	52,1	51,5	51,29	50,9

Table 6.19. *Calculation of temperatures at different instants, as a function of the $\frac{r}{s}$ ratios*

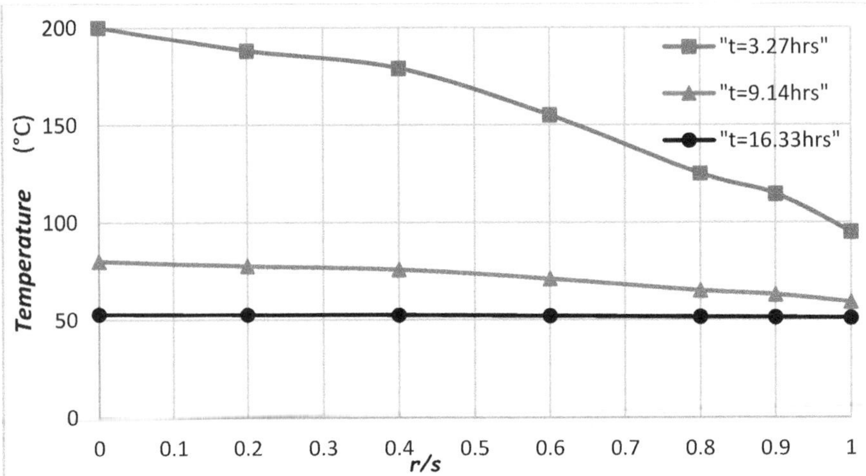

Figure 6.20. *Evolution of temperature profiles inside the solid as a function of time. For a color version of this figure, see www.iste.co.uk/benallou/energy2.zip*

6.8. Case study: comparison of graphical and numerical methods

Both the graphical method that we presented in section 6.7 and the numerical method explored in section 6.4 are approximate approaches to resolving the heat equation. The advantage of the numerical method is that the user can improve accuracy up to a certain limit by opting for increasingly smaller integration increments. However, as we have seen (section 4.2), this apparent freedom in the choice of integration increments is curtailed by the convergence constraints.

In this section, we propose conducting a comparison of these resolution methods by studying a specific case. To do this, let us take the example of a flat glass, 8 mm thick, which is initially at 350 °C and is suddenly immersed in a bath at 50 °C (see Figure 6.21). The transfer coefficient between the bath and the glass is $h = 325$ W/h m^2 °C.

For the first step, we will use the Heisler graphical method (figures in section 6.7) in order to determine the approximate temperature profile and its evolution over time. In this perspective, we will determine the temperature on the central plane of the solid at instants t = 15, 30, 45, 60, 75, 90 and 110 seconds. Then, for each of the instants t = 15, 45 and 90 seconds, the temperature profile will be defined by graphically evaluating the temperatures at the planes defined by x = 1, 2, 3 and 4 millimeters.

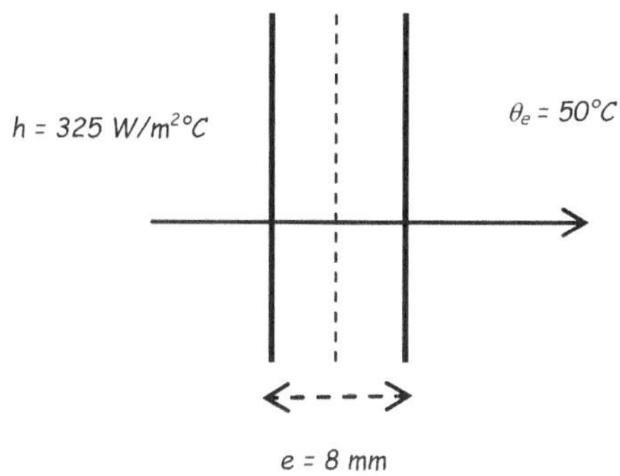

Figure 6.21. *Tempering 8 mm flat glass*

During the second step, the same temperature profiles are calculated using the numerical integration method presented in section 6.4.

Particular care is taken when choosing the Δx and Δt increments in order to guarantee integration stability. The same type of result is obtained using MAPLE (Gander and Hrebicek, 2004) or "PEDToolbox" in MATLAB (Merrien, 2007; Quarteroni, Sacco and Saleri, 2006).

Lastly, an analytical resolution is attempted in order to draw up comparisons between the different resolution methods. Indeed, in some straightforward cases, analytical solutions can be obtained using integration methods such as functional decompositions, or Laplace transforms.

6.8.1. *Resolution using the graphical method*

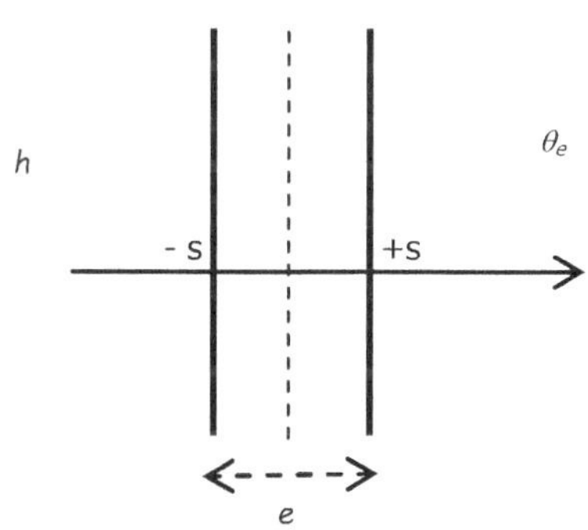

Figure 6.22. *Conduction in a plane solid of infinite thickness*

The example addressed here is straightforward enough to attempt to develop an analytical solution. Indeed, we are in the presence of one-dimensional conduction in a plane solid of finite thickness, without any heat generation.

The problem data are as follows:

$\rho = 2{,}300 \text{ kg/m}^3$

$\lambda = 1.05 \text{ W/m°C}$

$C_p = 837 \text{ J/kg °C}$

$\theta_i = 350\text{°C}$

The half-thickness of the planc is:

$$s = \frac{8}{2}$$

$$s = 4 \text{ mm}$$

The Biot number is given by:

$$\text{Bi} = \frac{h\,s}{\lambda}$$

$$\text{Bi} = \frac{325 * 4\ 10^{-3}}{1.05}$$

$$Bi = 1.24$$

Hence: $\frac{1}{\text{Bi}} = 0.8$.

The calculation of the temperature at the center is as follows:

$$\alpha = \frac{\lambda}{\rho C_p}$$

$$\alpha = \frac{1.05}{2{,}300(837)}$$

i.e.: $\alpha = 5.45\ 10^{-7} \text{m}^2/\text{s}$.

Hence the Fourier number: $Fo = \frac{\alpha\ t}{s^2}$, $Fo = \frac{5.45\ 10^{-7}}{16\ 10^{-6}}\ t$,

i.e.: $Fo = 0.0341\ t$, where t is in seconds.

Using Figure 6.5, we deduce $\frac{T_0}{T_i}$ for different values of Fo (see Figure 6.23). We then deduce the corresponding values of θ_0 from the equation defining $\frac{T_0}{T_i}$:

$$\theta_0 = \theta_e + \left(\frac{T_0}{T_i}\right)(\theta_i - \theta_e)$$

i.e.: $\theta_0 = 50 + 300\left(\frac{T_0}{T_i}\right)$.

T (seconds)	15	30	45	60	75	90	110
Fo	0.51	1.02	1.53	2.05	2.56	3.07	3.75
$\tau = \frac{s^2}{\alpha\ Bi}$	0.8	0.5	0.36	0.20	0.15	0.078	0.045
θ_0	290	200	158	110	95	73.4	63.5

Table 6.20. *Axial temperatures at different instants*

Temperature profiles at t = 15 seconds, 45 seconds and 90 seconds

Let us use Figure 6.8 with $\frac{1}{Bi} = 0,8$ (see Figure 6.24). Note that as the temperatures need to be assessed at the planes defined by x=1, 2, 3 and 4 millimeters ($\frac{x}{s} = 0$; 0.25; 0.5; 0.75 and 1), we will need to conduct interpolations for $\frac{x}{s} = 0.25$; 0.5 and 0.75.

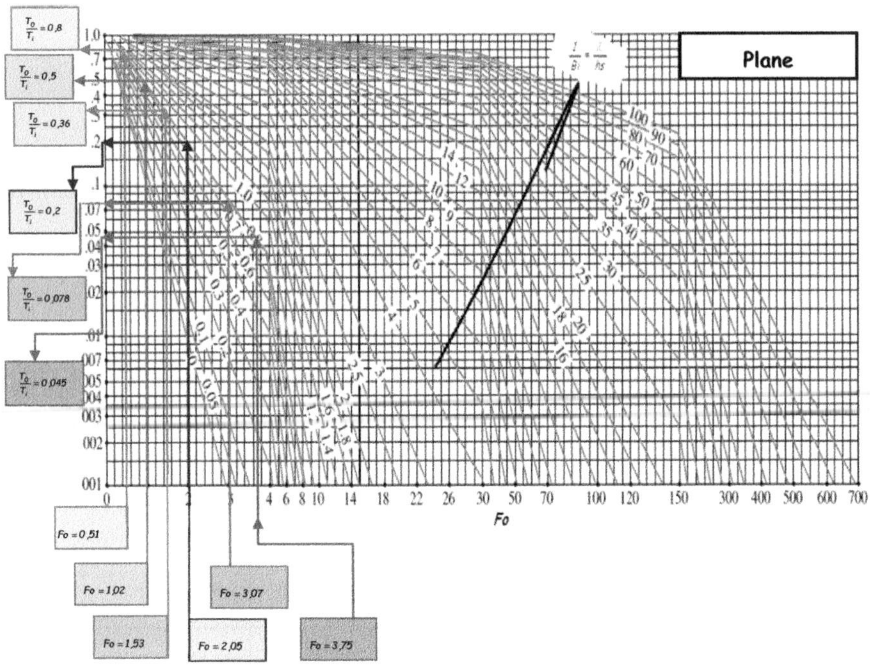

Figure 6.23. *Determination of temperatures at the center at different instants. For a color version of this figure, see www.iste.co.uk/benallou/energy2.zip*

We therefore obtain the following values of $\frac{T}{T_0}$ for $x = 0$, $x = 1$ mm, $x = 2$ mm, $x = 3$ mm and $x = 4$ mm:

x (mm)	0	1	2	3	4
x/s	0	0.25	0.5	0.75	1
T/T_0	1	0.96	0.88	0.76	0.58

Table 6.21. $\frac{T}{T_0}$ *ratios at different distances*

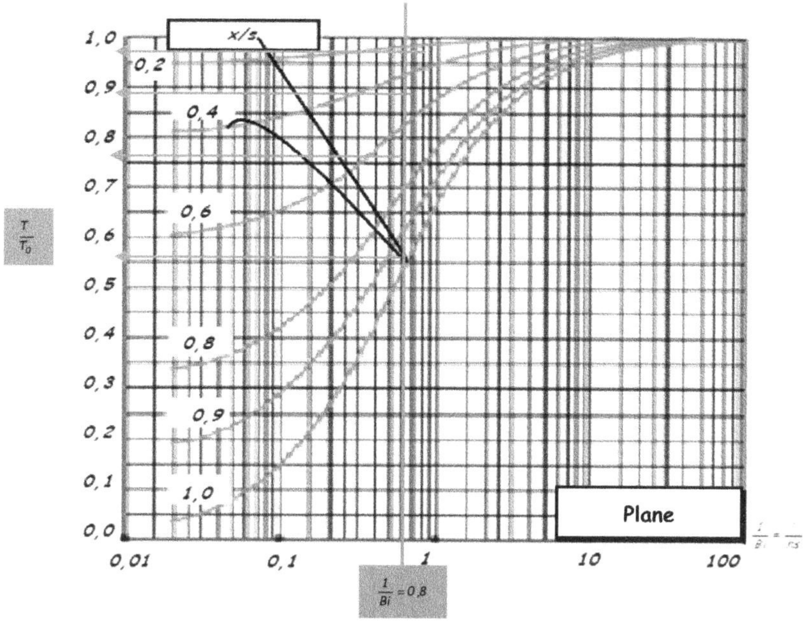

Figure 6.24. *Determination of temperatures at different distances from the center. For a color version of this figure, see www.iste.co.uk/benallou/energy2.zip*

We therefore deduce therefrom the values of θ as a function of x for the following instants:

	T		
	15	45	90
θ_0	290	158	73.4
θ_e	240	108	23.4
x (mm)	θ (°C)		
0	290	158	73.4
1	280.4	153.7	72.5
2	261.2	145.0	70.6
3	232.4	132.1	67.8
4	189.2	112.6	63.6

Table 6.22. *Calculations of the profiles at different instants*

6.8.2. *Resolution using the numerical method*

In this case, the discretized conduction equation is:

$$\theta_i^{n+1} = \theta_i^n + (Fo_\Delta)\left[\theta_{i+1}^n - 2\theta_i^n + \theta_{i-1}^n\right],$$

with: $Fo_\Delta = \alpha\frac{(\Delta t)}{(\Delta x)^2}$.

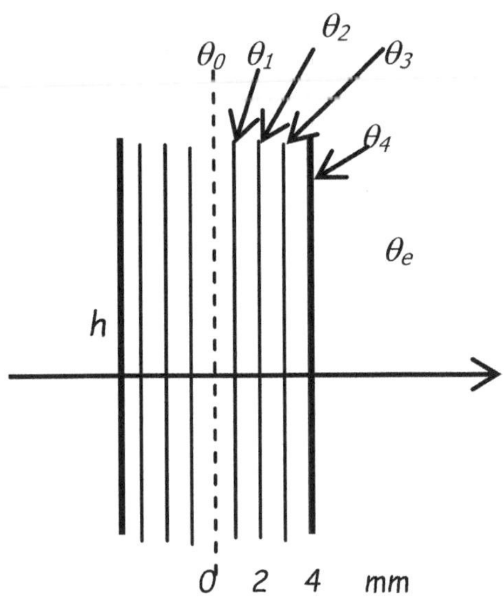

Figure 6.25. *Discretization of the plane considered*

The mesh defined in Figure 6.25 gives the value of Δx:$\Delta x = 1$ mm.

Choice of Δt:

To guarantee integration stability (see Exercise 29 in Chapter 7), Δt need to be chosen in order to have $Fo_\Delta \leq \frac{1}{2}$; or:

$$\Delta t \leq \frac{(\Delta x)^2}{2\alpha}$$

Knowing the value of $\alpha(\alpha$ = 5.45 10^{-7}m²/sec), we deduce therefrom the constraint on Δt for Δx, set at 1 mm:

$$\Delta t \leq \frac{\left(10^{-3}\right)^2}{2(5.45)10^{-7}}$$

i.e.: $\Delta t \leq 0.92$ seconds.

We will then take $\Delta t = 0.5$ seconds

As the value of α (α = 5.45 10^{-7}m²/sec) is known, we will deduce therefrom the discrete Fourier number:

$$Fo_\Delta = 5.45\,10^{-7} \frac{0.5}{(10^{-3})^2}$$

i.e.: $Fo_\Delta = 2.73\ 10^{-1}$.

Discretization of the boundary conditions:

$$t = 0:\ \forall x \quad : \quad \theta^0(x) = 350\ °C$$

i.e.: $\forall i \quad : \quad \theta_i^0 = 350\ °C$.

$$x = s:\ \forall j \geq 1 \ : \quad -\lambda \frac{\theta_4^j - \theta_3^j}{\Delta x} = h\left(\theta_4^j - \theta_e\right)$$

Or: $\forall j \geq 1 \ : \quad \theta_4^j = \theta_3^j - \dfrac{h(\Delta x)}{\lambda}\left(\theta_4^j - \theta_e\right)$.

If we use the definition of the discrete Biot number: $Bi_\Delta = \dfrac{h(\Delta x)}{\lambda}$, this secondary boundary condition becomes:

$$\forall j \geq 1 \quad : \quad \theta_4^j\left(1 + Bi_\Delta\right) = \theta_3^j + \left(Bi_\Delta\right)\ \theta_e$$

i.e.:. $\forall j \geq 1 \ : \quad \theta_4^j = \dfrac{\theta_3^j}{\left(1 + Bi_\Delta\right)} + \dfrac{Bi_\Delta}{\left(1 + Bi_\Delta\right)}\ \theta_e$

Implementation of calculations:

The equation system connecting temperatures θ_o, θ_1, θ_2, θ_3, and θ_4 at instants n and $n+1$ is given by:

$$\begin{cases} \theta_0^{n+1} = \theta_0^n + (Fo_\Delta)\left[\theta_1^n - 2\theta_0^n + \theta_1^n\right] \\ \theta_1^{n+1} = \theta_1^n + (Fo_\Delta)\left[\theta_2^n - 2\theta_1^n + \theta_0^n\right] \\ \theta_2^{n+1} = \theta_2^n + (Fo_\Delta)\left[\theta_3^n - 2\theta_2^n + \theta_1^n\right] \\ \theta_3^{n+1} = \theta_3^n + (Fo_\Delta)\left[\theta_4^n - 2\theta_3^n + \theta_2^n\right] \\ \theta_4^{n+1} = \dfrac{\theta_3^{n+1}}{(1+Bi_\Delta)} + \dfrac{Bi_\Delta}{(1+Bi_\Delta)}\,\theta_e \end{cases}$$

We have: $Bi_\Delta = \dfrac{h(\Delta x)}{\lambda}$ $Bi_\Delta = \dfrac{35(10^{-3})}{1{,}05}$,

i.e.: $Bi_\Delta = 0.310$.

Furthermore: $Fo_\Delta = 2.73\ 10^{-1}$.

To implement the resolution we start from the uniform initial profile ($n = 0$):

$$\left(\theta_0^0 \quad \theta_1^0 \quad \theta_2^0 \quad \theta_3^0 \quad \theta_4^0\right) = (350 \quad 350 \quad 350 \quad 350 \quad 350)$$

We then obtain, for each iteration, n, the values presented in Table 6.23 for the θ_i^n values.

Iteration	t (s)	$x = 0$: θ_0	$x = 1$mm: θ_1	$x = 2$mm: θ_2	$x = 3$mm: θ_3	$x = 4$mm: θ_4
0	0	350	350	350	350	350
2	1	350	350	350	331	264
4	2	350	349	339	308	247
10	5	336	329	308	272	219
20	10	297	290	270	237	192

30	15	262	256	238	209	172
40	20	231	226	210	186	154
50	25	205	200	187	166	139
60	30	182	178	167	149	126
70	35	163	160	150	135	115
80	40	146	144	136	123	105
90	45	132	130	123	112	97
100	50	120	118	112	103	90
120	60	101	100	96	89	80
140	70	88	86	83	78	72
160	80	77	77	74	71	66
180	90	70	69	68	65	62
200	100	65	64	63	61	58
240	120	58	58	57	56	54
360	180	51	51	51	51	51

Table 6.23. *Calculating the temperature profiles at different distances*

6.8.3. *Comparison of the numerical and graphical results*

The results obtained through the graphical method ("Gra") and through numerical integration ("Num") are grouped together in Table 6.24 for comparison purposes. Note that *the difference between the two results is below 20%*. This may seem a lot, but the graphical method presents the advantage of offering rapid estimations, which has a cost! Thus, the graphical method is always used to obtain estimations, which prove quick.

Comparison	θ_0			θ_1			θ_2			θ_3			θ_4		
t (s)	Gra	Num	Δ%	Gra	Num	Δ%	Gra	Num	Δ%	Gra	Num	Δ%	Gra	Num	Δ%
15	290.0	262	11%	280.4	256	10%	261.2	238	10%	232.4	209	11%	189.2	172	10%
45	158.0	132	19%	153.7	130	18%	145.0	123	18%	132.1	112	18%	112.6	97	16%
90	73.4	70	5%	72.5	69	4%	70.6	68	4%	67.8	65	4%	63.6	62	3%

Table 6.24. *Comparisons of the results obtained using the graphical and numerical methods*

We also note that estimations given by the graphical method barely differ from those of the numerical solution *towards the end of transient state* (significant lengths of time), *where the deviations are below 5%.* At the start of transient state, however, (t < 15 secs), the deviation is closer to 10%; see θ_0 and θ_4, in Figure 6.26, for example. This situation worsens in the middle of transient states where *the deviations are close to 20%.*

Moreover, whilst the gain in accuracy via the numerical method is proven, we do not really know to what degree the numerical method is accurate. In other words, we do not know which solution (graphical or numerical) is closest to the exact solution. To get an idea, since the example being explored is straightforward enough for us to seek an analytical solution, in the following section we will compare the numerical results with those of an analytical solution.

6.8.4. Comparison with the analytical solution

The precise analytical solution is of course an infinite sequence (see section 6.5). For a plane conductor, it is given by:

$$\overline{T} = \frac{\theta - \theta_e}{\theta_i - \theta_e} \quad \theta = \theta_e + (\theta_i - \theta_e)\overline{T}$$

$$\overline{T}(\overline{x}, Fo) = \sum_{j=1}^{\infty} \left[A_{\omega_j} e^{-\omega_j^2 Fo} \cos\left(\omega_j \overline{x}\right) \right],$$

with: $A_{\omega_j} = \dfrac{2\sin(\omega_j)}{\omega_j + \sin(\omega_j)\cos(\omega_j)}$ where, for a given Biot number, the ω_j values are the solutions of the equation: $\omega_j \, tg(\omega_j) = Bi$

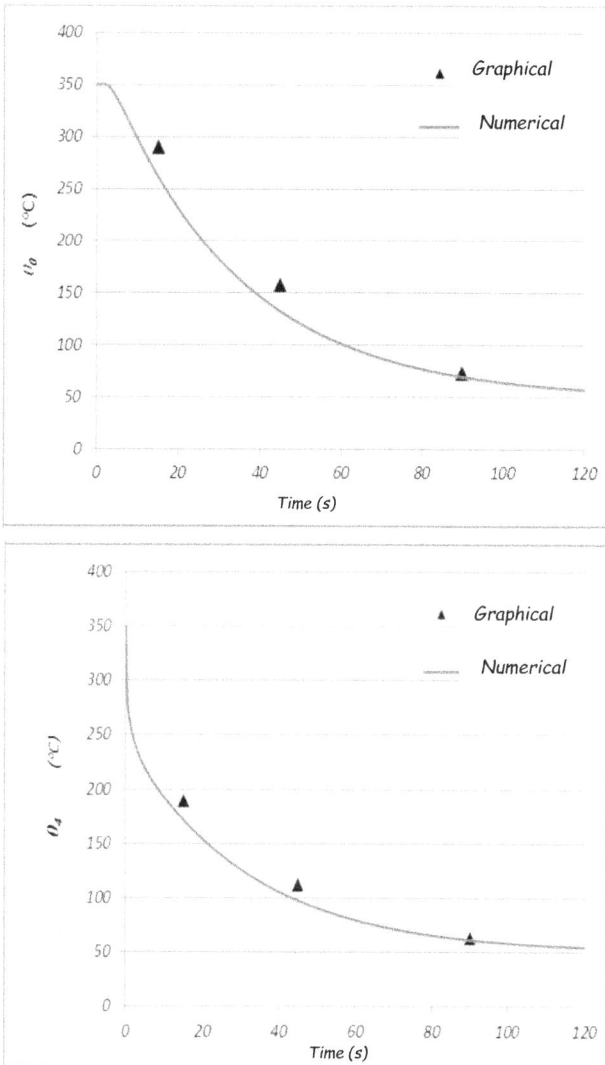

Figure 6.26. *Graphical representation of the numerical and graphical results. For a color version of this figure, see www.iste.co.uk/benallou/energy2.zip*

In practice we note that the first terms determine the numerical values of the solution with an accuracy greater than a threshold yet to be defined.

In the case of interest to us (*Bi* = 1.24), let us examine the contributions of the first five solutions to $\omega \operatorname{tg}(\omega) = \mathrm{Bi}$. These solutions are as follows:

$$\omega_1 = 0.9282;\ \omega_2 = 3.4835;\ \omega_3 = 6.472472;\ \omega_4 = 9.55385;\ \omega_5 = 12.663975$$

Table 6.25 presents, for each solution ω_j, the $A_{\omega j}$ coefficients and the $A_{\omega j}\, e^{-\omega_j^2}$ contributions involved in the calculation of $T(\overline{x}, Fo)$.

J	ω_j	$A_{\omega j}$	$A_{\omega j}\, e^{-\omega_j^2}$
1	0.9282	1.137178006	0.480462797
2	3.4835	-0.176494523	-9.47691E-07
3	6.472472	0.056527191	3.61747E-20
4	9.55385	-0.026589691	-6.08175E-42
5	12.663975	0.015273064	3.41507E-72

Table 6.25. *The parameters involved in the solution*

Thus, it is clear to see that whereas the first contribution is of the order of units, the second is already of the order of ten millionths (10^{-7}). Given that the different contributions still need to be multiplied by the exponential of $-\omega_j^2 Fo$ (that is: $A_{\omega j}\, e^{-\omega_j^2 Fo}$) and that for $j \geq 3$, $e^{-\omega_j^2 Fo} \leq e^{-41 Fo}$, we note that the other contributions quickly become negligible: orders of magnitude of 10^{-20} to 10^{-72}. As a result, in the expression of the exact solution we will only retain the terms corresponding to ω_1 and ω_2,

i.e.: $$\overline{T}(\overline{x}, Fo) \approx 1{,}137178\ e^{-(0{,}9282)^2 Fo} \cos(0{,}9282\overline{x}) - 0{,}176494523\ e^{-(3{,}4835)^2 Fo} \cos(3{,}4835\overline{x})$$

The profiles at 15, 45 and 90 seconds are calculated and compared to the results obtained previously using the numerical and graphical methods. Table 6.26 presents the deviations recorded between the analytical results, on the one hand, and the numerical and graphical results, on the other.

	t	15		45		90	
x (mm)	Method			$\theta(x,t)$			
0	Analytical	277	Δ (%)	144	Δ (%)	75	Δ (%)
	Numerical	262	-5%	144	0%	70.0	-7%
	Graphical	290	11%	158	10%	73.4	5%
1	Analytical	271	Δ (%)	141	Δ (%)	74	Δ (%)
	Numerical	256	-6%	130	-8%	69.0	-7%
	Graphical	280.4	10%	153.7	18%	72.5	5%
2	Analytical	253	Δ (%)	134	Δ (%)	72	Δ (%)
	Numerical	238	-6%	123	-8%	68.0	-6%
	Graphical	261.2	10%	145	18%	70.6	4%
3	Analytical	224	Δ (%)	122	Δ (%)	69	Δ (%)
	Numerical	209	-7%	112	-8%	65.0	-6%
	Graphical	232.4	11%	132.1	18%	67.8	4%
4	Analytical	186	Δ (%)	106	Δ (%)	65	Δ (%)
	Numerical	172	-8%	97	-8%	62.0	-5%
	Graphical	189.2	10%	112.6	16%	63.6	3%

Table 6.26. *Profiles at 15, 45 and 90 seconds*

These results are also presented in Figure 6.27, the curves illustrating the evolutions of θ_0 and θ_4 over time.

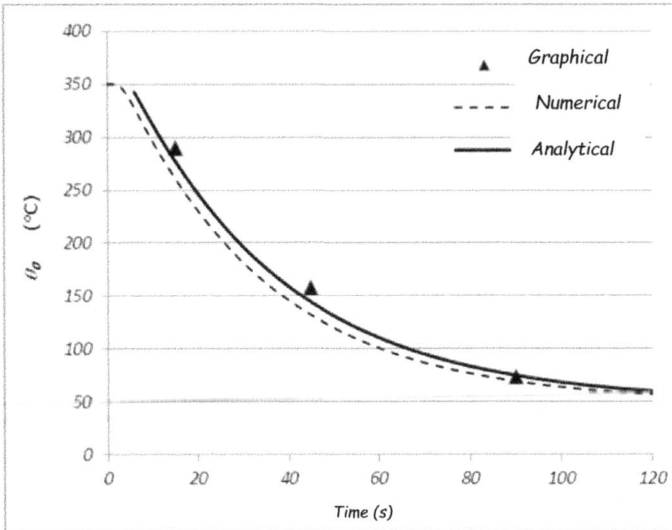

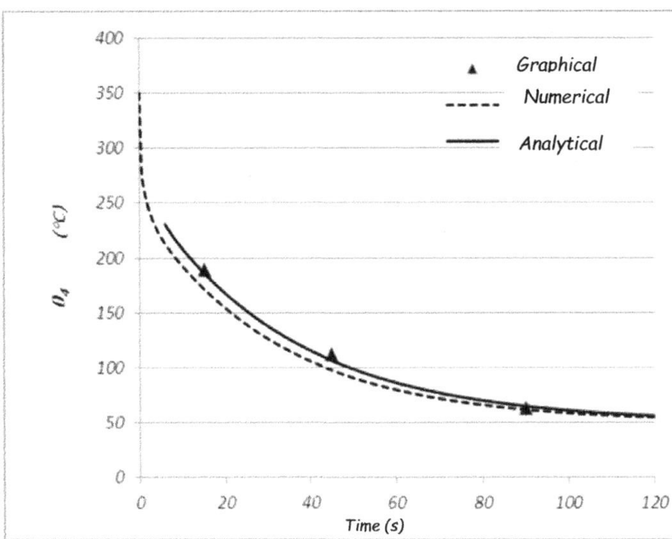

Figure 6.27. *Comparison of graphical, numerical and analytical solutions. For a color version of this figure, see www.iste.co.uk/benallou/energy2.zip*

This comparison enables us to make several essential remarks:

– although fairly approximative, the graphical method gives a result that is fairly close to the reality. Moreover, it is easy and quick to use;

– the numerical method generates a good representation of the analytical solution, both in space and time. It enables us to have deviations of less than 10%;

– the results given by the graphical method are not very accurate in the middle of transient states. Indeed, as shown in Table 6.26, the deviations reach 18% at $t = 45$ seconds;

– in engineering calculations, we often seek to increase accuracy. This is always associated with a calculation effort. In the most common scenarios featuring planes, cylinders or spheres, the effort needed in order to resolve the equation giving the values of ω_j is justified[2]. Indeed, once the first solutions to this equation are found, we have direct access to an applicable analytical solution, $\forall t$, for any point on the solid;

– analytical integration is often unattainable for more complex geometries, however. Therefore, the choice of method will depend on the intended use of the results sought:

- if the aim is to obtain orders of magnitude, the graphical method is sufficient,

- for more accurate results, numerical resolution is required, on the condition that particular care is taken when choosing the integration steps. In theory, the smaller the step, the greater the accuracy. Yet in practice, truncation errors generated by the multiple nature of the calculations resulting from very small steps may, in certain situations, lead to solution instabilities. Section 6.8.2 shows that these instabilities can be avoided if the integration steps, Δx and Δt, are chosen such that:

$$\alpha \frac{(\Delta t)}{(\Delta x)^2} \le \frac{1}{2} \text{ for one-dimensional cases}$$

$$\alpha \frac{(\Delta t)}{(\Delta x)^2} \le \frac{1}{4} \text{ for two-dimensional cases}$$

$$\alpha \frac{(\Delta t)}{(\Delta x)^2} \le \frac{1}{6} \text{ for three-dimensional cases}$$

2 Equation $\omega\,\mathrm{tg}(\omega) = \mathrm{Bi}$ for a plane; $\omega \frac{J_1(\omega)}{J_0(\omega)} = \mathrm{Bi}$ for a cylinder and $1 - \omega\cot(\omega) = \mathrm{Bi}$ for a sphere.

An illustration of the instability that can be generated through a poor choice of integration step is presented in Exercise 29 in Chapter 7.

It is important to note that the numerical integration method is applicable in cases where the heat propagates in more than one direction.

6.9. Reading: Jean-Baptiste Biot

Jean-Baptiste Biot was born in Paris on 21 April 1774. After attending Louis-le-Grand secondary school, he was admitted to the Ecole des Ponts et Chaussées engineering school in 1794. That same year, he would become one of the first students to join the newly inaugurated Ecole Polytechnique in France.

Figure 6.28. *Jean-Baptiste Biot (1774-1862)*

In 1797, at the age of just 23, he was appointed professor of mathematics at the Oise département school.

A highly-valued physicist, astronomer and mathematician, Jean-Baptiste Biot was particularly interested in studying the electromagnetic, optical and chemical characteristics of the atmosphere.

He became a Doctor of Science in August 1809, then holder of the astronomy chair at the faculty of sciences from 1809 to 1848, where he served as professor of astronomy.

His studies in the field of the analysis of the components of the atmosphere, magnetic field generation and the propagation of light may be considered as the basis of current developments in the same field, making it possible to comprehend the factors governing climate change.

Indeed, along with Gay-Lussac, Jean-Baptiste Biot was one of the pioneers initiating studies into characterization of the atmosphere.

Figure 6.29. *Biot and Gay-Lussac in hot-air balloon (1804)*

Moreover, it was within this context that in 1804 the two eminent physicists embarked on a hot-air balloon ride, reaching an altitude 4,000 meters, this being the altitude that they deemed necessary in order to be able to study the magnetic, electrical and chemical characteristics of the atmosphere.

Thus, the skills of the two scientists (Gay-Lussac providing teaching on heat, gas, hygrometry, electricity and galvanism) came to be complementary in addressing one of the problems that to this day continues to interest researchers, without them arriving at definite answers, namely, climate change and global warming.

This study and research work can be considered as the starting point of environmental and space analysis, to which we should add the significant contributions by Jean-Baptiste Biot to acoustics, electromagnetism and optics.

Between 1816 and 1826, he taught physics for applications in the study and propagation of sound, in magnetism and in optics.

His time spent teaching acoustics, magnetism and optics served as a prelude for Jean-Baptiste Biot to his studies and research into the analysis of the propagation of light through liquids, which he conducted in collaboration with Persoz. Within this same context, the laws of rotation of the plane of polarization of light passing through a liquid solution came to be established. The laws and developments subsequently made by Biot in this field favored the development of the saccharimeter, designed to determine the nature and quantity of sugars present in a solution.

In addition, the work conducted by Jean-Baptiste Biot in the field of electromagnetism allowed him to formulate, together with Félix Savart, Biot-Savart's law, which offers the possibility of obtaining the value of the magnetic field produced by an electric current at a point of space as a function of the distance from this point to the conductor crossed by the current.

Jean-Baptiste Biot was made Dean of the Paris Faculty of Sciences in 1840.

To immortalize this great scientist, the city of Paris decided to name one of its streets after Jean-Baptiste Biot. Rue Biot is located in the 17th arrondissement of Paris.

7

Exercises and Solutions

EXERCISE 1. Equation for conduction

In anisotropic media, the density of the energy flux transferred by conduction in the i direction ($i = x, y$ or z) is given by:

$$\phi_i = -\lambda_{ix}\left(\frac{\partial\theta}{\partial x}\right) - \lambda_{iy}\left(\frac{\partial\theta}{\partial y}\right) - \lambda_{iz}\left(\frac{\partial\theta}{\partial z}\right)$$

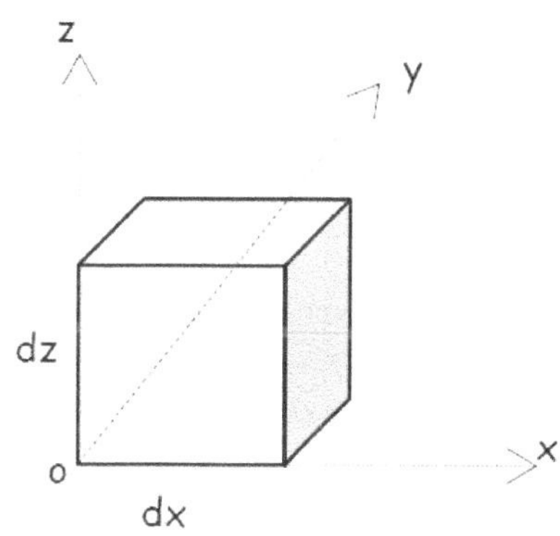

Questions

Consider a volume element of an anisotropic material:

1) Give the expression of the energy flux, ϕ_x, passing through the volume element in the ox direction, on the x-axis.

2) Express, as a function of ϕ_x, the flux, ϕ_{x+dx}, passing through the volume element in the x direction, on the $x+dx$ axis.

3) By writing $\phi_{(E-S)|x} = \phi_x - \phi_{x+dx}$, show that the energy balance on the x-axis can be put in the form

$$\phi_{(E-S)|x} = \frac{\partial}{\partial x}\left[\lambda_{xx}\left(\frac{\partial\theta}{\partial x}\right) + \lambda_{xy}\left(\frac{\partial\theta}{\partial y}\right) + \lambda_{iz}\left(\frac{\partial\theta}{\partial z}\right)\right] \text{dx dy dz}$$

4) Deduce therefrom equivalent expressions for $\phi_{(E-S)|y}$ and $\phi_{(E-S)|z}$.

5) Assuming that this volume element is the seat of energy generation, the generation power per unit volume is P and the density, ρ, and the heat capacity, C_P, are constant in this volume element, show that the general heat conduction equation in an anisotropic medium is given by:

$$\rho C_p \frac{\partial\theta}{\partial t} = \sum_{i=x,y,z}\left[\frac{\partial}{\partial i}\left(\sum_{j=x,y,z} \lambda_{ij} \frac{\partial\theta}{\partial j}\right)\right] + P$$

6) Taking $\lambda_{ij} = 0$ for $i \neq j$, and by writing $\lambda_{ii} = \lambda$ for $i = x, y$ or z, establish the conduction equation in the case of orthotropy.

7) Taking $\lambda_{ij} = 0$ for $i \neq j$, and by writing $\lambda_{ii} = \lambda$ for $i = x, y$ or z, establish the conduction equation in the case of isotropy.

Solutions

1) Expression of the energy flux, ϕ_x

The flux density on the x-axis is given by:

$$\phi_x = -\lambda_{xx}\left(\frac{\partial\theta}{\partial x}\right) - \lambda_{xy}\left(\frac{\partial\theta}{\partial y}\right) - \lambda_{xz}\left(\frac{\partial\theta}{\partial z}\right)$$

The flux on the x-axis is then obtained by multiplying the density by the transfer area, S_x:

$$\phi_x = \varphi_x S_x = \left[-\lambda_{xx}\left(\frac{\partial\theta}{\partial x}\right) - \lambda_{xy}\left(\frac{\partial\theta}{\partial y}\right) - \lambda_{xz}\left(\frac{\partial\theta}{\partial z}\right)\right] S_x$$

Yet, in the x direction, the transfer area is: $S_x = (dy)\ (dz)$.

Hence: $\phi_x = \varphi_x\ S_x = \left[-\lambda_{xx}\left(\frac{\partial\theta}{\partial x}\right)-\lambda_{xy}\left(\frac{\partial\theta}{\partial y}\right)-\lambda_{xz}\left(\frac{\partial\theta}{\partial z}\right)\right]\ (dy)(dz)$.

2) Expression of $\phi x + dx$

According to the Mean-Value Theorem, we have: $\phi_{x+dx} = \phi_x + \left(\frac{\partial\phi_x}{\partial x}\right)dx$.

I.e.: $\phi_{x+dx} = \phi_x + \frac{\partial}{\partial x}\left[-\lambda_{xx}\left(\frac{\partial\theta}{\partial x}\right)-\lambda_{xy}\left(\frac{\partial\theta}{\partial y}\right)-\lambda_{xz}\left(\frac{\partial\theta}{\partial z}\right)\right]\ (dx)(dy)(dz)$.

3) Energy balance on the x-axis

$$\phi_{(E-S)|x} = \phi_x - \phi_{x+dx}$$

Hence, by replacing $\phi_{(E-S)|x}$ with its expression:

$$\phi_{(E-S)|x} = \phi_x - \phi_x - \frac{\partial}{\partial x}\left[-\lambda_{xx}\left(\frac{\partial\theta}{\partial x}\right)-\lambda_{xy}\left(\frac{\partial\theta}{\partial y}\right)-\lambda_{xz}\left(\frac{\partial\theta}{\partial z}\right)\right]\ (dx)(dy)(dz)$$

I.e.: $\phi_{(E-S)|x} = \frac{\partial}{\partial x}\left[\lambda_{xx}\left(\frac{\partial\theta}{\partial x}\right)+\lambda_{xy}\left(\frac{\partial\theta}{\partial y}\right)+\lambda_{xz}\left(\frac{\partial\theta}{\partial z}\right)\right]\ (dx)(dy)(dz)$.

4) Equivalent expressions of $\phi_{(E-S)|y}$ and of $\phi_{(E-S)|y}$

Since the *x* direction presents no specificity, the result found in Question (3) is transposable in the *y* and *z* directions. We then have:

– in the *y* direction:

$$\phi_{(E-S)y} = \frac{\partial}{\partial y}\left[\lambda_{yx}\left(\frac{\partial\theta}{\partial x}\right)+\lambda_{yy}\left(\frac{\partial\theta}{\partial y}\right)+\lambda_{yz}\left(\frac{\partial\theta}{\partial z}\right)\right]\ (dx)(dy)(dz)$$

– in the z direction:

$$\phi_{(E-S)z} = \frac{\partial}{\partial z}\left[\lambda_{zx}\left(\frac{\partial \theta}{\partial x}\right) + \lambda_{zy}\left(\frac{\partial \theta}{\partial y}\right) + \lambda_{zz}\left(\frac{\partial \theta}{\partial z}\right)\right] (dx)(dy)(dz)$$

<u>*5) General heat conduction equation in an anisotropic medium*</u>

The general heat conduction equation translates the overall energy balance of the volume energy considered.

When the generation power per unit volume is P, the amount of energy generated, per unit time, in the volume element defined by $(dx)(dy)(dz)$ is given by:

$$G = P\ (dx)(dy)(dz)$$

If A designates the amount of energy accumulated, per unit time, in the volume element considered, the overall energy balance equation will be formulated as follows:

$$\phi_{(E-S)|x} + \phi_{(E-S)|y} + \phi_{(E-S)|z} + G = A$$

Yet: $A = mC_p\left(\frac{\partial \theta}{\partial t}\right)$ and $m = \rho(dV) = \rho\ dx\ dy\ dz$.

Hence: $A = \rho\ C_p\left(\frac{\partial \theta}{\partial t}\right) dx\ dy\ dz$.

The overall energy balance equation therefore becomes:

$$\phi_{(E-S)|x} + \phi_{(E-S)|y} + \phi_{(E-S)|z} + P\ dxdydz = \rho C_p\left(\frac{\partial \theta}{\partial t}\right) dxdydz$$

Or, by replacing $\phi_{(E-S)|x}$, $\phi_{(E-S)|x}$ and $\phi_{(E-S)|x}$ with their respective expressions and dividing by $dxdydz$:

$$\frac{\partial}{\partial x}\left[\lambda_{xx}\left(\frac{\partial \theta}{\partial x}\right) + \lambda_{xy}\left(\frac{\partial \theta}{\partial y}\right) + \lambda_{xz}\left(\frac{\partial \theta}{\partial z}\right)\right] + \frac{\partial}{\partial y}\left[\lambda_{yx}\left(\frac{\partial \theta}{\partial x}\right) + \lambda_{yy}\left(\frac{\partial \theta}{\partial y}\right) + \lambda_{yz}\left(\frac{\partial \theta}{\partial z}\right)\right]$$
$$+ \frac{\partial}{\partial z}\left[\lambda_{zx}\left(\frac{\partial \theta}{\partial x}\right) + \lambda_{zy}\left(\frac{\partial \theta}{\partial y}\right) + \lambda_{zz}\left(\frac{\partial \theta}{\partial z}\right)\right] + P = \rho C_p\left(\frac{\partial \theta}{\partial t}\right)$$

I.e.: $\rho C_p \dfrac{\partial\theta}{\partial t} = \sum_{i=x,y,z}\left[\dfrac{\partial}{\partial i}\left(\sum_{j=x,y,z}\lambda_{ij}\dfrac{\partial\theta}{\partial j}\right)\right] + P$.

6) Case of orthotropy

For an orthotropic material, we have: $\lambda_{ij} = 0$ for $i \neq j$, and $\lambda_{ii} = \lambda_i$ for $i = x, y$ or z.

The overall energy balance equation then becomes:

$$\frac{\partial}{\partial x}\left[\lambda_{xx}\left(\frac{\partial\theta}{\partial x}\right)\right]+\frac{\partial}{\partial y}\left[\lambda_{yy}\left(\frac{\partial\theta}{\partial y}\right)\right]+\frac{\partial}{\partial z}\left[\lambda_{zz}\left(\frac{\partial\theta}{\partial z}\right)\right]+P=\rho C_p\left(\frac{\partial\theta}{\partial t}\right)$$

7) Case of isotropy

For an isotropic material, we have: $\lambda_{ij} = 0$ for $i \neq j$, and $\lambda_{ii} = \lambda$ for $i = x, y$ or z.

The overall energy balance equation therefore becomes:

$$\frac{\partial}{\partial x}\left[\lambda\left(\frac{\partial\theta}{\partial x}\right)\right]+\frac{\partial}{\partial y}\left[\lambda\left(\frac{\partial\theta}{\partial y}\right)\right]+\frac{\partial}{\partial z}\left[\lambda\left(\frac{\partial\theta}{\partial z}\right)\right]+P=\rho C_p\left(\frac{\partial\theta}{\partial t}\right)$$

Hence: $\rho C_p\left(\dfrac{\partial\theta}{\partial t}\right)=\lambda\left(\dfrac{\partial^2\theta}{\partial x^2}\right)+\lambda\left(\dfrac{\partial^2\theta}{\partial y^2}\right)+\lambda\left(\dfrac{\partial^2\theta}{\partial z^2}\right)+P$.

I.e.: $\left(\dfrac{\rho C_p}{\lambda}\right)\left(\dfrac{\partial\theta}{\partial t}\right)=\nabla^2\theta+\dfrac{P}{\lambda}$.

Or: $\dfrac{1}{\alpha}\dfrac{\partial\theta}{\partial t}=\nabla^2\theta+\dfrac{P}{\lambda}$.

EXERCISE 2. Electrical analogy of a composite wall

The composite wall represented in the figure below comprises:

– a layer (A) of thickness eA, transfer area S and heat conductivity λA;

– a second layer composed of two superimposed materials, (*B*) and (*C*), of the same thickness (e_{BC}), of respective transfer areas S_B and S_C, such that $S_B + S_C = S$, and of heat conductivities λ_B and λ_C;

– a layer (*D*) of thickness e_D, transfer area *S* and heat conductivity λ_D.

Face 1 of this wall, of temperature T_1, is in contact with a medium at homogeneous, constant temperature, T_i; whereas Face 2, of temperature T_2, is in contact with a homogeneous medium, the constant temperature of which is equal to T_e.

Questions

1) Represent the electrical analogy of the system consisting of the internal environment, the wall and the external environment.

2) The contractor in charge of constructing this wall asks you to compare the heat losses in the two situations defined below and to recommend one of them to him, bearing in mind that their implementation costs are different.

a) calculate the thermal resistances of layers *A* and *D*: R_A and R_D, respectively;

b) calculate the thermal resistances of layers *B* and *C*: R_B and R_C, respectively, in these two situations;

c) calculate the overall thermal resistance, R_G (between T_i and T_e) in these two situations. What conclusions can you draw?

3) calculate the fluxes ϕ of heat losses in the two situations considered. What would you recommend to the contractor?

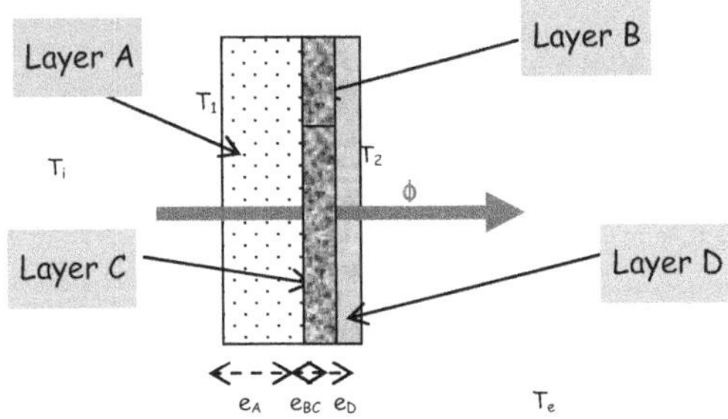

<u>*Data:*</u>

$T_i = 450$ °C	$T_e = 35$ °C	$e_A = 3$ cm
$\lambda_A = 150$ W/m °C	$\lambda_B = 50$ W/m °C	$\lambda_C = 5$ W/m °C
$e_D = 0.8$ cm	$\lambda_D = 0.4$ W/m °C	$S = 3.7$ m²
$h_i = 12$ W/m² °C	$h_e = 25$ W/m² °C	$e_{BC} = 1$ cm

Situation 1: $S_B = S_C$

Situation 2: $S_B = 0.6\ S_C$

Solutions

1) Electrical analogy of the system

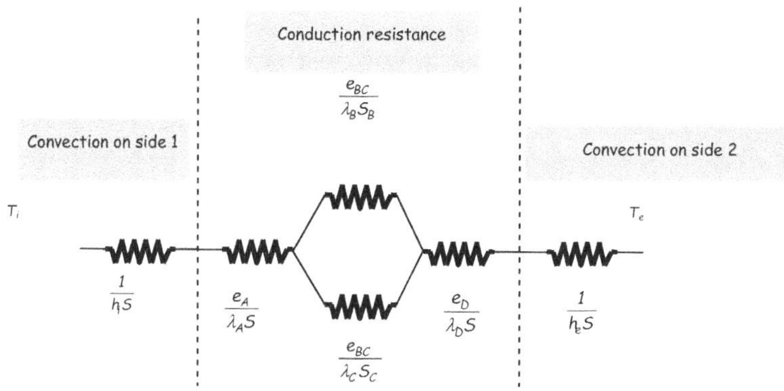

2) Calculating the thermal resistances

a) R_A and R_D

$$R_A = \frac{e_A}{\lambda_A S} \qquad R_D = \frac{e_D}{\lambda_D S}$$

NUMERICAL APPLICATION.–

$e_A = 0.03$ m $\qquad \lambda_A = 150$ W/m °C

$e_D = 0.008$ m $\qquad \lambda_D = 0.4$ W/m °C

$$R_A = \frac{0.03}{150(3.7)} \qquad R_D = \frac{0.008}{0.4(3.7)}$$

$R_A = 5.41\ 10^{-5}$ °C/W $\qquad R_D = 5.41\ 10^{-3}$ °C/W

b) R_B and R_C

$$R_B = \frac{e_{BC}}{\lambda_B S_B} \qquad R_C = \frac{e_{BC}}{\lambda_C S_C}$$

i) Situation 1: $S_B = S_C$ and $S_B + S_C = S$

Hence: $S_B = S_C = \frac{S}{2}$.

Thus: $R_B = \frac{2e_{BC}}{\lambda_B S}$ and $R_C = \frac{2e_{BC}}{\lambda_C S}$.

NUMERICAL APPLICATION.–

$e_{BC} = 1$ cm $\quad S = 3.7$ m²

$\lambda_B = 50$W/m °C $\quad \lambda_C = 5$W/m °C

$$R_B = \frac{2(0.01)}{50(3.7)} \qquad R_C = \frac{2(0.01)}{5(3.7)}$$

$R_B = 1.08\ 10^{-4}$ °C/W $\quad R_C = 1.08\ 10^{-3}$ °C/W

ii) Situation 2: $S_B = 0.6S_C$ and $1.6S_C = S$

$$S_C = \frac{S}{1,6} \text{ and } S_B = \frac{0,6}{1,6}S$$

Thus: $R_B = \frac{1,6\ e_{BC}}{0,6\ \lambda_B S}$ and $R_C = \frac{1,6\ e_{BC}}{\lambda_C S}$

$R_B = 1.44\ 10^{-4}$ °C/W and $R_C = 8.65\ 10^{-4}$ °C/W

c) Overall thermal resistances

The overall resistance of the circuit is noted R, such that:

$$R = \frac{1}{h_i S} + R_A + R_{BC} + R_D + \frac{1}{h_e S}$$

As resistances B and C are parallel, their equivalent resistance is R_{BC}, such that:

$$R_{BC} = \frac{R_B R_C}{R_B + R_C}$$

Hence: $R = \frac{1}{h_i S} + R_A + \frac{R_B R_C}{R_B + R_C} + R_D + \frac{1}{h_e S}$.

NUMERICAL APPLICATION.–

$h_i = 12$ W/m² °C

$h_e = 25$ W/m² °C

$R_A = 5.41\ 10^{-5}$ °C/W

$R_D = 5.41\ 10^{-3}$ °C/W

i) Situation 1: $R_B = R_C = 1.08\ 10^{-3}$ °C/W

$$T_{pe} = \frac{689.63}{2\pi * 0.13 * 10} + 20$$

$R = 4.07\ 10^{-2}$ °C/W

ii) Situation 2: $R_B = 1.44\ 10^{-4}$ °C/W and $R_C = 8.65\ 10^{-4}$ °C/W

$$R = \frac{1}{12(3.7)} + 5.41\ 10^{-5} + \frac{(1.44)(8.65)10^{-8}}{(1.44 + 8.65)10^{-4}} + 5.41\ 10^{-3} + \frac{1}{25(3.7)}$$

$R = 4.03\ 10^{-2}$ °C/W

The resistances are virtually identical. As a result, the heat losses are identical in the two situations. This is confirmed by the flux calculation below.

3) Calculating flux of heat losses in the two situations

$$\phi = \frac{T_i - T_e}{R}$$

i) Situation 1:

$R = 4.07\ 10^{-2}$ °C/W $\quad \phi = \frac{450 - 35}{4{,}07\ 10^{-2}} \quad \phi = 10.2$ kW

ii) Situation 2:

$R = 4.03\ 10^{-2}$ °C/W $\quad \phi = \frac{450 - 35}{4.03\ 10^{-2}} \quad \phi = 10.3$ kW

The contractor will therefore be able to choose either situation as the heat losses are virtually identical.

EXERCISE 3. Conduction through a plane wall with variable heat conductivity

Consider a plane wall of thickness, e, composed of a variable heat conductivity material with a temperature satisfying the following law:

$$\lambda(T) = \alpha + \beta T(x)$$

The inner face of this wall is in contact with an environment that is at uniform, constant temperature, T_i. The outer face, on the other hand, is protected by a metal sheath of thickness e_g and of constant heat conductivity λ_g.

Assume that the heat propagation is one-directional along x.

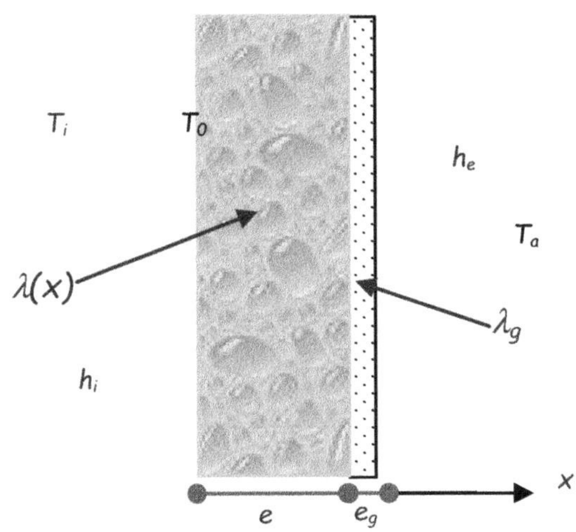

The convective heat transfer coefficients, internal and external, are designated h_i and h_e, respectively.

Questions

1) If we use the set of hypotheses corresponding to this case, what becomes of the equation governing conduction in transient state in the wall with variable heat conductivity?

2) Deduce therefrom the differential equations defining $T(x)$ in steady state for $0 \le x \le e + e_g$.

3) Give the boundary conditions necessary for resolving these equations.

4) Determine temperature $T(e+e_g)$ as a function of h_e, T_e, T_a, e_g and λ_g.

5) By integrating the differential equation relating to the variable heat conductivity material, show that, for $0 \leq x \leq e$, $T(x)$ must satisfy the following quadratic equation:

$$\alpha T(x)+\frac{\beta}{2}T^2(x)=K_1x+K_2,$$

where K_1 and K_2 are integration constants.

6) Using the boundary conditions established above, determine constants K_1 and K_2.

Solutions

1) Conduction equation in the variable heat conductivity wall

In this case, λ is variable as a function of x. The general heat conduction equation is therefore given by:

$$\rho C_p\left(\frac{\partial T}{\partial t}\right)=\frac{\partial}{\partial x}\left[\lambda\left(\frac{\partial T}{\partial x}\right)\right]+\frac{\partial}{\partial y}\left[\lambda\left(\frac{\partial T}{\partial y}\right)\right]+\frac{\partial}{\partial z}\left[\lambda\left(\frac{\partial T}{\partial z}\right)\right]+P$$

As there is no generation, $P = 0$.

Furthermore, with the the heat propagation being one-directional along the ox-axis, we have:

$$\frac{\partial T}{\partial y}=0 \text{ and } \frac{\partial T}{\partial z}=0$$

The conduction equation in the variable heat conductivity material is then:

$$\rho C_p\left(\frac{\partial T}{\partial t}\right)=\left[\alpha+\beta T(x)\right]\left(\frac{\partial^2 T}{\partial x^2}\right)+\beta\left(\frac{\partial T}{\partial x}\right)^2$$

2) Differential equations governing T(x) in steady state

The following two cases can be mentioned:

a) For $0 \leq x \leq e$

We have conduction in a variable heat conductivity material. The temperature therefore obeys the following differential equation:

$$\rho C_p\left(\frac{\partial T}{\partial t}\right)=\left[\alpha+\beta T(x)\right]\left(\frac{\partial^2 T}{\partial x^2}\right)+\beta\left(\frac{\partial T}{\partial x}\right)^2$$

In steady state: $\frac{\partial T}{\partial t}=0$.

The differential equation becomes: $\left[\alpha+\beta T(x)\right]\left(\frac{d^2 T}{dx^2}\right)+\beta\left(\frac{dT}{dx}\right)^2=0$.

Or: $\frac{d}{dx}\left[\lambda\left(\frac{dT}{dx}\right)\right]=0$.

b) For $e \leq x \leq eg$

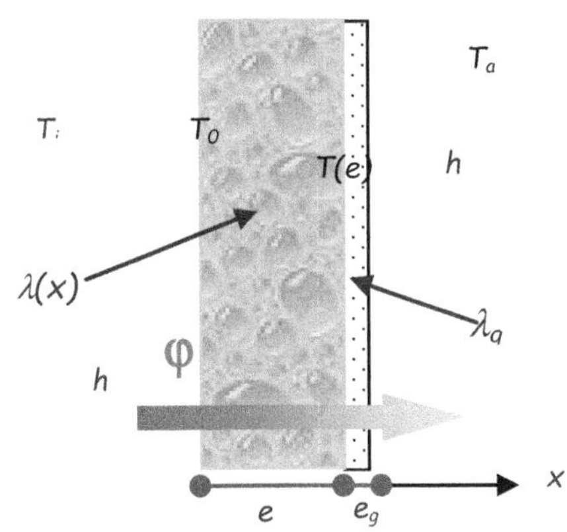

We will focus here on conduction in the metal sheath, with constant λ. The temperature variations are then governed by the conduction equation for isotropic materials, with constant λ:

$$\frac{1}{\alpha}\frac{\partial T}{\partial t}=\nabla^2 T+\frac{P}{\lambda}$$

I.e. in steady state, with no generation: $\nabla^2 T = 0$.

Considering that conduction is one-dimensional, the differential equation becomes:

$$\frac{\partial^2 T}{\partial x^2} = 0$$

Thus, the differential equations governing the variations of *T* are:

– for $0 \leq x \leq e$: $\dfrac{d}{dx}\left[\lambda\left(\dfrac{dT}{dx}\right)\right] = 0$;

– for $e \leq x \leq e_g$: $\dfrac{\partial^2 T}{\partial x^2} = 0$.

3) Boundary conditions necessary for resolving differential equations

a) For $0 \leq x \leq e$

CL_1: $x = 0$: $\phi = -\lambda\left(\dfrac{dT}{dx}\right)\Big|_{x=0} = h_i\left[T_i - T(0)\right]$.

CL_2: $x = e$: $\phi = \phi\left(e^-\right) = \phi\left(e^+\right)$.

Or: $-\lambda\left(\dfrac{dT}{dx}\right)\Big|_{x=e^-} = \dfrac{T(e) - T_a}{\dfrac{e_g}{\lambda_g} + \dfrac{1}{h_e}}$.

b) For $e \leq x \leq e_g$

CL_1: $x = e$: $\phi = \phi\left(e^-\right) = \phi\left(e^+\right)$.

Or: $-\lambda\left(\dfrac{dT}{dx}\right)\Big|_{x=e^-} = \dfrac{T(e) - T_a}{\dfrac{e_g}{\lambda_g} + \dfrac{1}{h_e}}$.

CL_2: $x = e + e_g$: $\phi = h_e\left[T\left(e + e_g\right) - T_a\right]$.

4) $T(e+e_g)$ as a function of h_e, T_e, T_a, e_g and λ_g

Given that conduction in the sheath occurs at constant λ, the standard expression giving the flux density as a function of $T(e)$ and T_a is applicable:

$$\phi = \frac{T(e) - T_a}{\frac{e_g}{\lambda_g} + \frac{1}{h_e}}$$

As the two flux expressions need to be equal, we have:

$$T(e+e_g) = T_a + \frac{T(e) - T_a}{\left(\frac{h_e e_g}{\lambda_g} + 1\right)}$$

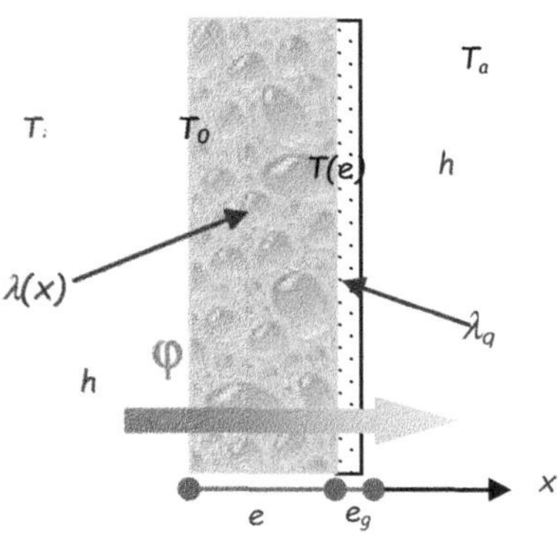

5) Quadratic equation needing to be satisfied by T(x) for $0 \leq x \leq e$

For $0 \leq x \leq e$, we have: $\frac{d}{dx}\left[\lambda\left(\frac{dT}{dx}\right)\right] = 0$.

The solution therefore takes the following form: $\lambda\left(\frac{dT}{dx}\right) = K_1$, where K_1 is a constant.

Yet: $\lambda(x) = \alpha + \beta T(x)$.

The solution therefore becomes: $[\alpha + \beta T(x)]\frac{dT}{dx} = K_1$.

Or: $\left[\alpha+\beta T(x)\right]dT = K_1 dx$.

I.e. after integration: $\alpha T(x)+\dfrac{\beta}{2}T^2(x)=K_1x+K_2$, K_2 being a constant.

6) Determining the integration constants

In this case, the boundary conditons are:

$$CL1: x=0:\ \phi=-\lambda\left(\frac{dT}{dx}\right)\bigg|_{x=0}=h_i\left[T_i-T(0)\right].$$

Or: $K_1=h_i\left[T(0)-T_i\right]$.

Hence: $T(0)=T_i+\dfrac{K_1}{h_i}$.

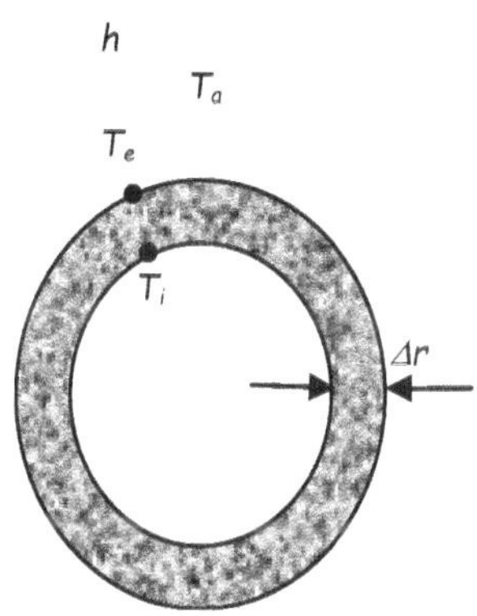

Yet: $\alpha T(0)+\dfrac{\beta}{2}T^2(0)=K_2$.

$$\text{Hence: } K_2=\alpha\left[T_i+\frac{K_1}{h_i}\right]+\frac{\beta}{2}\left[T_i+\frac{K_1}{h_i}\right]^2.$$

CL_2: $x = e$: $\phi = \dfrac{T(e) - T_a}{\dfrac{e_g}{\lambda_g} + \dfrac{1}{h_e}} = -\lambda \left(\dfrac{dT}{dx} \right)_{x=e}$.

Yet: $-\lambda \left(\dfrac{dT}{dx} \right)_{x=e} = -K_1$.

CL_2 thus gives: $K_1 = \dfrac{T_a - T(e)}{\dfrac{e_g}{\lambda_g} + \dfrac{1}{h_e}}$.

Hence the equation at the origin of $T(x)$:

$$\alpha T(x) + \frac{\beta}{2} T^2(x) = \left[\frac{T_a - T(e)}{\dfrac{e_g}{\lambda_g} + \dfrac{1}{h_e}} \right] x + \alpha \left[T_i + \frac{T_a - T(e)}{h_i \left(\dfrac{e_g}{\lambda_g} + \dfrac{1}{h_e} \right)} \right] + \frac{\beta}{2} \left[T_i + \frac{T_a - T(e)}{h_i \left(\dfrac{e_g}{\lambda_g} + \dfrac{1}{h_e} \right)} \right]^2$$

EXERCISE 4. Choosing heat lagging for a chimney

In this exercise, we wish to determine the heat losses towards the surrounding environment, originating from the wall of a cylindrical chimney of radius r.

This wall, of thickness Δr and heat conductivity λ, separates gases that are conveyed outside, on one side, and the surrounding environment, on the other side. The inner wall of the chimney is at temperature T_i, whereas the surrounding environment is at T_a.

Questions

In order to calculate the hourly heat losses per unit length of the chimney, ϕ_L, and to determine the temperature, T_e, of its outer wall:

1) Give the three expressions of ϕ_L: convective, conductive and overall.

2) Calculate the temperature, T_e.

3) Calculate the hourly heat losses for a chimney fifteen meters high.

4) What would the monetary value of these losses be (in Dh/year) if the energy cost was C_E?

5) What needs to be done in order for the losses to be halved?

6) If, for ergonomic reasons, your line manager requires you to use a lagging thickness that cannot exceed e, what would be the constraint on the heat conductivity of the lagging you choose?

7) Which lagging will you then recommend to your line manager, given that the chimney is subjected to an environment that may be humid at times?

The convective heat transfer coefficient between the the outer wall (at T_e) and the atmosphere will be noted h.

The chimney is in operation twenty-four hours a day, three hundred and fifty days a year.

Data:

$r = 25$ cm	$\Delta r = 5$ cm	$\lambda = 0.95$ kcal/h m °C
$h = 12$ kcal/h m^2°C	$T_i = 154$ °C	$T_a = 22$ °C
$C_E = 3.7$ Dh/kWh	$e = 3.5$ cm	

Solutions

1) Expressions of flux per unit length

– Convective expression: $\phi_L = 2\,\pi\,(r+\Delta r)\,h\,(Te - Ta)$.

– Conductive expression: $\phi_L = 2\pi \dfrac{T_i - T_e}{\dfrac{1}{\lambda}\text{Ln}\left(1+\dfrac{\Delta r}{r}\right)}$

– Overall expression: $\phi_L = 2\pi \dfrac{T_i - T_a}{\dfrac{1}{\lambda}\text{Ln}\left(1+\dfrac{\Delta r}{r}\right)+\dfrac{1}{(r+\Delta r)h}}$

2) Calculating T_e

The temperature, T_e, can be calculated from the convective expression of the flux:

$$T_e = T_a + \frac{\phi_L}{2\pi(r+\Delta r)h},$$

whereas ϕ_L can be calculated from the overall flux expression, with temperatures T_i and T_a being known.

NUMERICAL APPLICATION.–

$r = 0.25$ m $\Delta r = 0.05$ m $T_i = 154$ °C

$T_a = 22$ °C $h = 12$ kcal/h m^2°C $\lambda = 0.95$ kcal/h m °C

$$\varphi_L = 2\pi \frac{154-22}{\frac{1}{0.95}\text{Ln}\left(1+\frac{0.25}{0.05}\right)+\frac{1}{(0.3)12}}$$ $\phi_L = 1{,}765.8$ kcal/hm

$$T_e = 22 + \frac{1765.8}{2(3.14)(0.3)12}$$ $T_e = 100°C$

<u>*3) Hourly heat losses for a chimney fifteen meters high*</u>

$\phi = L\,\phi_L$ where L is the chimney height.

NUMERICAL APPLICATION.–

$\phi_L = 1{,}765.8$ kcal/hm

$L = 15$ m

$\phi = 15(1{,}765.8)$

ϕ = 26,487 kcal/hr, or 30.75 kW

<u>*4) Monetary value of these losses (in Dh/year) if CE was the energy cost*</u>

This value, V, is given by:

$V = \phi$ x (operating time) x C_E

NUMERICAL APPLICATION.–

$\phi = 30.75$ kW

Operating time = 350x24 hours

$C_E = 3.7$ Dh/kWh

Hence: $V = 30.75 \times (350 \times 24) \times 3.7$

V = 955,710 Dh/year

5) For the losses to be halved

Lagging needs to be added in order to halve the flux.

Therefore, if the flux per unit length, with heat lagging, is noted ϕ_c, we must have:

$$\phi_c = \frac{\phi_L}{2}$$

If the heat conductivity of the lagging is noted λ_c and the lagging's thickness used is designated e, then ϕ_c is given by:

$$\phi_c = 2\pi \frac{T_i - T_a}{\frac{1}{\lambda}\mathrm{Ln}\left(1+\frac{\Delta r}{r}\right)\frac{1}{\lambda_c}\mathrm{Ln}\left(1+\frac{e}{r+\Delta r}\right)+\frac{1}{(r+\Delta r+e)h}}$$

The condition $\phi_c = \frac{\phi_L}{2}$ becomes:

$$\frac{T_i - T_a}{\frac{1}{\lambda}\mathrm{Ln}\left(1+\frac{\Delta r}{r}\right)\frac{1}{\lambda_c}\mathrm{Ln}\left(1+\frac{e}{r+\Delta r}\right)+\frac{1}{(r+\Delta r+e)h}} = \frac{1}{2}\frac{T_i - T_a}{\frac{1}{\lambda}\mathrm{Ln}\left(1+\frac{\Delta r}{r}\right)+\frac{1}{(r+\Delta r)h}}$$

Hence:

$$\frac{1}{\lambda}\mathrm{Ln}\left(1+\frac{\Delta r}{r}\right)+\frac{2}{(r+\Delta r)h} = \frac{1}{\lambda_c}\mathrm{Ln}\left(1+\frac{e}{r+\Delta r}\right)+\frac{1}{(r+\Delta r+e)h}$$

6) Heat conductivity, λ_c, of the heat lagging to be used

The heat lagging to be recommended will need to have a heat conductivity, λ_c, such that:

$$\lambda_c = \frac{\mathrm{Ln}\left(1+\frac{e}{r+\Delta r}\right)+\frac{1}{(r+\Delta r+e)h}}{\frac{1}{\lambda}\mathrm{Ln}\left(1+\frac{\Delta r}{r}\right)+\frac{2}{(r+\Delta r)h}}$$

NUMERICAL APPLICATION.–

$r = 0.25$ m $\quad$ $\Delta r = 0.05$ m $\quad$ $e = 0.035$ m

$\lambda = 0.95$ kcal/h m °C $\quad$ $h = 12$ kcal/h m^2°C

$$e \le 0.035 \Rightarrow \lambda_c \le \frac{\mathrm{Ln}\left(1+\frac{0.035}{0.3}\right)+\frac{1}{(0.3+0.035)(12)}}{\frac{1}{0.95}\mathrm{Ln}\left(1+\frac{0.05}{0.25}\right)+\frac{2}{0.3(12)}}$$

$\lambda_c \le 0.48$ kcal/hm °C

7) Recommended lagging

The data in the appendix reveal two possibilities:

– plaster, with heat conductivity: $\lambda_c = 0.48$ W/m°C.

I.e.: $\lambda_c = 0.42$ kcal/m°C;

– polyethylene, with heat conductivity: $\lambda_c = 0.46$ W/m°C.

I.e.: $\lambda_c = 0.40$ kcal/hm°C.

Plaster has the lowest heat conductivity, meaning that it ought to be preferred, only it presents low resistance to moisture. As a result, we will recommend using polyethylene.

EXERCISE 5. Optimizing the lagging of a cylindrical pipe

The aim of this exercise is to determine the economic thermal insulation thickness, e^*, to be used in order to minimize heat loss from a cylindrical steel pipe transporting a hot fluid.

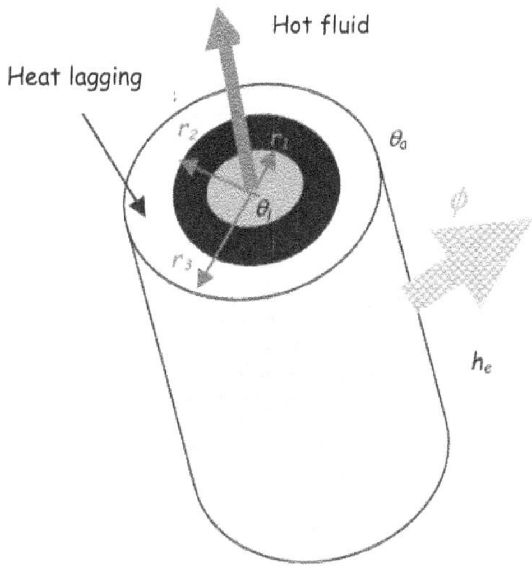

Questions

1) Calculate the flux of energy lost when the pipe has no lagging.

2) Determine the expression, $\phi(e)$, of the flux of energy lost, as a function of the thickness, e, of the lagging to be installed, and establish $\phi(0)$.

3) Show that the sign of $\frac{d\phi}{de}$ is the same as that of $\lambda_C - h_a r_3$.

4) Deduce therefrom that $\phi(e)$ admits a maximum and determine the location of this maximum (critical thickness, e_c).

5) Give the shapes of $\phi(e)$ according to the choice of lagging: $\lambda_C \leq r_2 h_a$ or $\lambda_C > r_2 h_a$.

6) Explain, first mathematically, then physically, why in the case where $\lambda_c \leq r_2 h_a$, the energy losses with heat lagging are greater than the losses without heat lagging, if the heat lagging presents low thicknesses: $\phi(e) > \phi(0)$.

7) Show that when $\lambda_C \leq r_2 h_a$, it is not enough to take $e > e_C$ in order to have $\phi(e) < \phi(0)$: it also necessary to ensure that the thickness of the lagging to be used is above e_{lim}, such that: $\phi(e_{lim}) = \phi(0)$.

8) Show that e_{lim} is the solution to the following equation:

$$\frac{r_2 h_a}{\lambda_C} Ln\left(1 + \frac{e_{lim}}{r_2}\right) + \frac{r_2}{(e_{lim} + r_2)} = 1$$

9) We choose to use polyurethane foam as the heat lagging. Which case will we be faced with and what will the critical thickness be?

10) Calculate the optimum thickness, e^*, to be used.

Data:

Pipe composition: ordinary steel

$d_1 = 12$ cm $\quad d_2 = 15$ cm $\quad L = 40$ m

$\theta_a = 25$ °C $\quad \theta_i = 160$°C (constant)

$\lambda_a = 60$ kcal/hm° C $\quad h_a = 25$ kcal/hm^2°C $\quad \lambda_c = 0.038$ kcal/hm°C

Cost of the heat lagging = 4.8 10^5 Dh/m^3

Lifetime of the heat lagging: 5 years

Energy cost: 3.2 Dh/kWh

Annual operating time = 8,760 hrs/year

Solutions

1) Flux of energy lost when the pipe has no lagging

$$\phi = \frac{2\,\pi\,L\,(\theta_i - \theta_a)}{\frac{1}{\lambda_a}\,\mathrm{Ln}\left(\frac{d_2}{d_1}\right) + \frac{2}{d_2 h_a}}$$

NUMERICAL APPLICATION.–

$L = 40$ m $\quad d_1 = 12$ cm $\quad d_2 = 15$ cm

$\theta_i = 160$ °C $\quad \theta_a = 25$ °C

λ_a=60 kcal/h m°C $\quad h_a$=25 kcal/hm^2°C

ϕ = 63,176.7 kcal/h

2) Expression of the flux of energy lost as a function of e

We have: $\phi(r_3)=\dfrac{2\,\pi\,L\,(\theta_i-\theta_a)}{\dfrac{1}{\lambda_a}Ln\left(\dfrac{r_2}{r_1}\right)+\dfrac{1}{\lambda_C}Ln\left(\dfrac{r_3}{r_2}\right)+\dfrac{1}{r_3h_a}}$, *e* being the thickness of the heat lagging: $r_3 = r_2 + e$

The flux expression can then be rewritten as a function of the thickness of the heat lagging material:

$$\phi(e)=\frac{2\,\pi\,L\,(\theta_i-\theta_a)}{\dfrac{1}{\lambda_a}Ln\left(\dfrac{r_2}{r_1}\right)+\dfrac{1}{\lambda_C}Ln\left(1+\dfrac{e}{r_2}\right)+\dfrac{1}{(e+r_2)h_a}}$$

The flux without heat lagging is calculated taking $e = 0$ in the previous expression:

$$\phi(0)=\frac{A}{\dfrac{1}{\lambda_a}Ln\left(\dfrac{r_2}{r_1}\right)+\dfrac{1}{r_2h_a}}$$

3) Sign of $\dfrac{d\varphi}{de}$

$$\phi(e)=\frac{2\,\pi\,L\,(\theta_i-\theta_a)}{\dfrac{1}{\lambda_a}Ln\left(\dfrac{r_2}{r_1}\right)+\dfrac{1}{\lambda_C}Ln\left(1+\dfrac{e}{r_2}\right)+\dfrac{1}{(e+r_2)h_a}}$$

Hence: $$\frac{d\phi}{de}=-2\,\pi\,L\,(\theta_i-\theta_a)\;\frac{\dfrac{1}{\lambda_C r_2}\dfrac{1}{1+\dfrac{e}{r_2}}-\dfrac{1}{h_a(e+r_2)^2}}{\left[\dfrac{1}{\lambda_a}Ln\left(\dfrac{r_2}{r_1}\right)+\dfrac{1}{\lambda_C}Ln\left(1+\dfrac{e}{r_2}\right)+\dfrac{1}{(e+r_2)h_a}\right]^2}.$$

Or: $$\frac{d\phi}{de} = 2\,\pi\,L\,(\theta_i - \theta_a)\,\frac{\dfrac{1}{(e+r_2)}\left(\dfrac{1}{h_a(e+r_2)} - \dfrac{1}{\lambda_C}\right)}{\left[\dfrac{1}{\lambda_a}\mathrm{Ln}\left(\dfrac{r_2}{r_1}\right) + \dfrac{1}{\lambda_C}\mathrm{Ln}\left(1+\dfrac{e}{r_2}\right) + \dfrac{1}{(e+r_2)h_a}\right]^2}.$$

For $\theta_i > \theta_a$, we have:

$$2\,\pi\,L\,(\theta_i - \theta_a)\,\frac{\dfrac{1}{(e+r_2)}}{\left[\dfrac{1}{\lambda_a}\mathrm{Ln}\left(\dfrac{r_2}{r_1}\right) + \dfrac{1}{\lambda_C}\mathrm{Ln}\left(1+\dfrac{e}{r_2}\right) + \dfrac{1}{(e+r_2)h_a}\right]^2} >$$

Consequently, the sign of $\frac{d\phi}{de}$ is that of $\left(\frac{1}{h_a(e+r_2)} - \frac{1}{\lambda_C}\right)$.

Or that of: $\lambda_C - h_a(e+r_2)$.

4) Maximum of $\phi(e)$

$$\frac{d\phi}{de} = 0$$

This results in: $\lambda_C - h_a(e_c + r_2) = 0$.

Hence: $e_c = \frac{\lambda_C}{h_a} - r_2$.

5) Shapes of $\phi(e)$ according to the choice of lagging

In general, two possible cases can occur depending on the value of e_c and the choice of lagging (λ_c):

– 1st case: $e_c > 0$, or $r_2h_a < \lambda_C$: ϕ is increasing over $[0; e_c[$ and decreasing over $[e_c; \infty[$;

– 2nd case: $e_c \leq 0$, or $r_2h_a \geq \lambda_C$: ϕ is decreasing $\forall\, e \in [0; \infty[$.

Thus, for the values of e of interest to us ($e > 0$), we obtain the following curves:

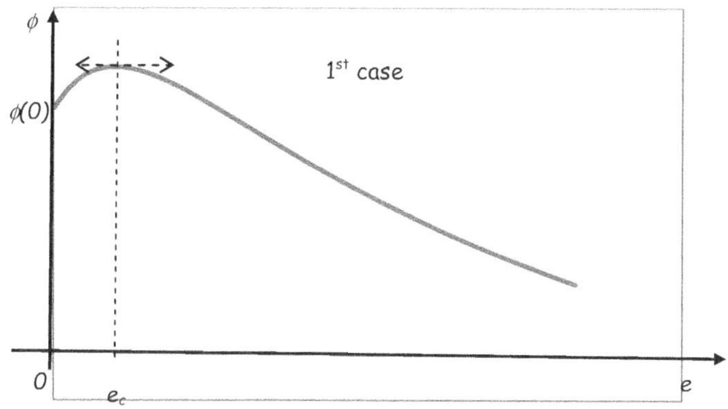

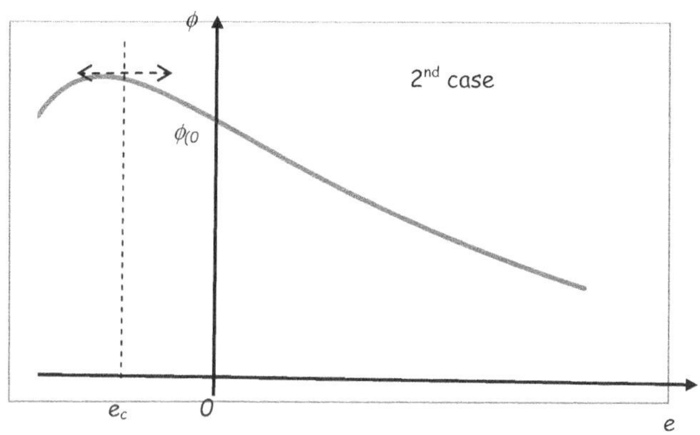

6) Mathematical and then physical explanations in the case where $\lambda_C \le r_2 h_a$

a) Mathematically

$\lambda_C > r_2 h_a$

Therefore:

– $\lim\limits_{e \to 0} \left[\lambda_C - h_a (e + r_2)\right] > 0$;

– The sign of $\dfrac{d\phi}{de}$ is positive;

– ϕ is therefore increasing when $e \to 0$.

– For low lagging thicknesses, the heat losses will be greater in this case than when the pipe has no lagging.

b) Physically

When $\lambda_C > r_2 h_a$ we will have: $\dfrac{r_2}{\lambda_C} < \dfrac{1}{h_a}$.

This would mean that the conduction resistance in the lagging is lower than the convection resistance.

Thus, by opting for a low thickness of lagging, the heat conductivity of which is such that $\lambda_C > r_2 h_a$, more energy will be lost than the when the pipe is left bare.

7) Condition on e in order to have $\phi(e) < \phi(0)$ in the case where $\lambda_C \le r_2 h_a$

$\phi(e) < \phi(0)$ for $e > e_{lim}$, where e_{lim} is such that $\phi(e_{lim}) = \phi(0)$.

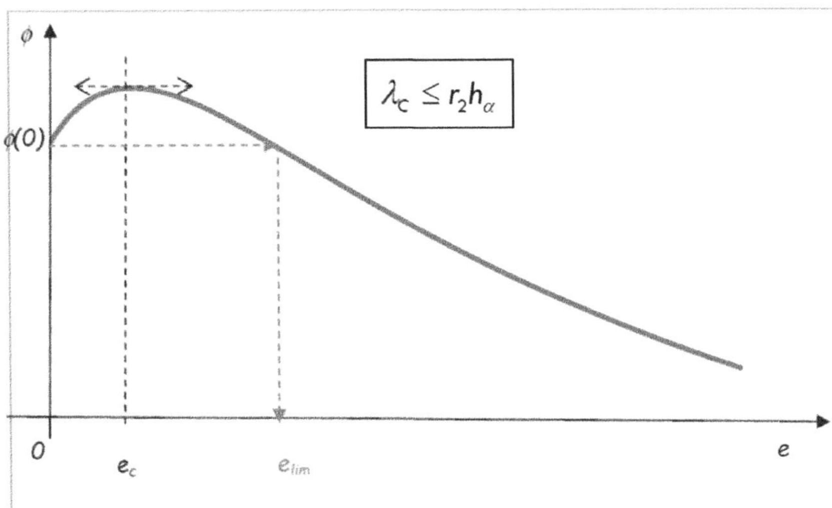

8) Expression of e_{lim}

$$\frac{2\pi L(\theta_i - \theta_a)}{\dfrac{1}{\lambda_a}\mathrm{Ln}\left(\dfrac{r_2}{r_1}\right) + \dfrac{1}{\lambda_C}\mathrm{Ln}\left(1 + \dfrac{e_{lim}}{r_2}\right) + \dfrac{1}{(e_{lim} + r_2)h_a}} = \phi(0)$$

$$\frac{1}{\lambda_a}\mathrm{Ln}\left(\frac{r_2}{r_1}\right) + \frac{1}{r_2 h_a} = \frac{1}{\lambda_a}\mathrm{Ln}\left(\frac{r_2}{r_1}\right) + \frac{1}{\lambda_C}\mathrm{Ln}\left(1 + \frac{e_{lim}}{r_2}\right) + \frac{1}{(e_{lim} + r_2)h_a}$$

e_{lim} is then the solution to the equation: $\frac{r_2 h_a}{\lambda_C} \mathrm{Ln}\left(1+\frac{e_{lim}}{r_2}\right)+\frac{r_2}{(e_{lim}+r_2)}=1$.

9) For polyurethane foam

We have: $e_c = \frac{\lambda_C}{h_a} - r_2$.

NUMERICAL APPLICATION.–

$r = 0.075$ m $\lambda_c = 0.038$ kcal/hm °C h_a=25 kcal/hm^2 °C

$e_c = -7.3$ cm

e_c is negative; we therefore find ourselves in the 2nd scenario.

10) Calculating the optimum thickness of the heat lagging

The function *total annual operating cost* is given, according to the lagging thickness, e, by:

$$C_T(e) = P(e) + C(e),$$

where $P(e)$ is the annual cost of the heat losses with heat lagging, given by: $P(e) = D_F P_E \phi(e)$, $C(e)$ being the annual lagging cost. In this case, it is given by:

$$C(e) = \frac{P_u \, V_c(e)}{D_v}$$

D_v: the lifetime of the heat lagging

P_u: the cost per m^3 of the lagging, which is constant in this case

$V_c(e)$: the necessary lagging volume: $V_c = 2\,\pi\, r_3 L\, e = 2\,\pi\, L\, e\,(r_2+e)$

D_F: the annual operating time

P_E: the unit price of electricity

Hence: $C_T(e) = D_F P_E\ \phi(e) + \frac{P_u \, V_c(e)}{D_v}$.

The optimum heat-lagging thickness, e^*, is that which cancels out the derivative:

$$\frac{dC_T(e)}{de} = 0$$

Yet: $\dfrac{dC_T(e)}{de} = D_F P_E \dfrac{d\phi(e)}{de} + \left(\dfrac{P_u}{D_v}\right) \dfrac{dV_c(e)}{de}$.

Hence: $\dfrac{dC_T(e)}{de} = D_F P_E \dfrac{d\phi(e)}{de} + \left(\dfrac{2\pi L P_u}{D_v}\right)(2e + r_2)$.

Or, by replacing $\dfrac{d\phi}{de}$ with its expression:

$$\frac{dC_T(e)}{de} = 2\,\pi\,L \left\{ \frac{\dfrac{D_F P_E(\theta_i - \theta_a)}{(e + r_2)}\left(\dfrac{1}{h_a(e + r_2)} - \dfrac{1}{\lambda_C}\right)}{\left[\dfrac{1}{\lambda_a}\text{Ln}\left(\dfrac{r_2}{r_1}\right) + \dfrac{1}{\lambda_C}\text{Ln}\left(1 + \dfrac{e}{r_2}\right) + \dfrac{1}{(e + r_2)h_a}\right]^2} + \frac{P_u}{D_v}(2e + r_2) \right\}$$

Thus, e^* is given by the solution to the following equation with e:

$$\frac{\dfrac{D_F P_E(\theta_i - \theta_a)}{(e + r_2)}\left(\dfrac{1}{h_a(e + r_2)} - \dfrac{1}{\lambda_C}\right)}{\left[\dfrac{1}{\lambda_a}\text{Ln}\left(\dfrac{r_2}{r_1}\right) + \dfrac{1}{\lambda_C}\text{Ln}\left(1 + \dfrac{e}{r_2}\right) + \dfrac{1}{(e + r_2)h_a}\right]^2} + \frac{P_u}{D_v}(2e + r_2) = 0$$

e^* is therefore the solution to the following equation:

$$\frac{D_F P_E(\theta_i - \theta_a)}{(e + r_2)(2e + r_2)}\left(\frac{1}{h_a(e + r_2)} - \frac{1}{\lambda_C}\right) + \frac{P_u}{D_v}\left[\frac{1}{\lambda_a}\text{Ln}\left(\frac{r_2}{r_1}\right) + \frac{1}{\lambda_C}\text{Ln}\left(1 + \frac{e}{r_2}\right) + \frac{1}{(e + r_2)h_a}\right]^2 = 0$$

$$\frac{D_F P_E(\theta_i - \theta_a)}{(e + r_2)(2e + r_2)}\left(\frac{1}{\lambda_C} - \frac{1}{h_a(e + r_2)}\right) = \frac{P_u}{D_v}\left[\frac{1}{\lambda_a}\text{Ln}\left(\frac{r_2}{r_1}\right) + \frac{1}{\lambda_C}\text{Ln}\left(1 + \frac{e}{r_2}\right) + \frac{1}{(e + r_2)h_a}\right]^2$$

We write: $f(e) = \dfrac{D_F P_E(\theta_i - \theta_a)}{(e + r_2)(2e + r_2)}\left(\dfrac{1}{\lambda_C} - \dfrac{1}{h_a(e + r_2)}\right)$.

And: $g(e)\frac{P_u}{D_v}\left[\frac{1}{\lambda_a}\text{Ln}\left(\frac{r_2}{r_1}\right)+\frac{1}{\lambda_C}\text{Ln}\left(1+\frac{e}{r_2}\right)+\frac{1}{(e+r_2)h_a}\right]^2$.

e^* is then the solution to the equation: $f(e) = g(e)$.

NUMERICAL APPLICATION.–		
$r_1 = 0.06$ m	$r_2 = 0.075$ m	$L = 40$ m
$\theta_i = 160$ °C	$\theta_a = 25$ °C	$D_F = 8{,}760$ hrs/year
$\lambda_a = 60$ kcal/hm °C	$h_a = 25$ kcal/hm^2 °C	$\lambda_c = 0.038$ kcal/hm °C
$P_u = 4.8\ 10^5$ Dh/m^3	$D_V = 5$ years	$P_E = 3.2$ Dh/kWh
$e^* = 0.0297726$ m	$f(e^*) = 8.09540E+06$	$g(e^*) = 8.09540E+06$
$e^* \approx$ **3 cm**		

EXERCISE 6. Lagging a steam pipe in a sugar-plant evaporator

Steam that is overheated (to T_v), necessary for the operation of a sugar-plant evaporator, is produced by a boiler that is at a distance, L, of the evaporator. This steam is then transported in a pipe of internal radius r_1 and external radius r_2. The temperature of the steam can reasonably be assumed to be constant and equal to T_v across its whole length, L. Likewise, the temperature of the surrounding environment is constant and equal to T_a.

Returning from a seminar on rational energy use, your supervisor wishes to conduct a comprehensive analysis of the heat losses generated by the transporting of steam between the boiler and the evaporator.

Questions

A) Of course, since the plant's construction *four years* ago, the pipe has been thermally insulated by a thickness, e, of thermal lagging with heat conductivity, λ_c. Despite this, your supervisor asks you to calculate the following parameters assuming that no lagging is used to protect the pipe:

1) the flux, ϕ_0, of the heat losses, in kW;

2) the cost of these energy losses (in Dh/year).

B) At the next meeting, when you present him with the results, your supervisor is proud to have fully recommended that this steam pipe be heat lagged at the time of the plant's installation. They then ask you to calculate:

1) the flux, ϕ_1, of the heat losses, in kW, for the pipe as it has been installed, i.e. lagged;

2) the cost of these energy losses (in Dh/year).

C) When you present your second series of results to your supervisor he is not happy. He asks you to check your calculations because your results appear to contradict those found in Part (A).

You are sure of your results and reassure him by explaining that such configurations are possible in some cases. You then offer to:

1) calculate the critical thickness of the heat lagging;

2) deduce therefrom an explanation of the apparent contradiction;

3) calculate the heat-lagging thickness to be exceeded so that $\phi(e) < \phi_0$;

4) calculate the optimum thickness of the heat lagging;

5) calculate, in monetary terms, the savings that would be made by putting in place the heat lagging with an optimum thickness.

Data:

$T_v = 160$ °C	$T_a = 20$ °C	
$r_2 = 6$ cm	$L = 65$ m	$r_1 = 50$ cm
$\lambda = 55$ W/m °C	$\lambda_c = 0.8$ W/m °C	$e = 2$ cm
$D_v = 5$ years	$P_u(e) = 850 + 2.1\ 10^5$ e	
$P_E = 2.8$ Dh/kWh	$D_F = 7{,}600$ hours/year	
$h_i = 250$ W/m^2 °C	$h_e = 17$ W/m^2 °C	

Solutions

A) With no heat lagging

1) Calculating the flux, ϕ_0, of the heat losses

The ϕ_0 flux is given by: $\theta_i = \theta_s - \dfrac{3\phi_{Ray}}{R\rho C_p}\sqrt{\dfrac{2H}{g}}$

NUMERICAL APPLICATION.–

$T_v = 160$ °C	$T_a = 20$ °C	
$r_1 = 2.5\ 10^{-2}$ m	$r_2 = 3\ 10^{-2}$ m	$L = 65$ m

$\lambda = 55$ W/m°C $h_i = 250$ W/m^2 °C $h_e = 20$ W/m^2 °C

$$\varphi_0 = \frac{2\pi L(65)(160-20)}{\frac{1}{2,5\ 10^{-2}(250)} + \frac{1}{55}\mathrm{Ln}\left(\frac{3}{2,5}\right) + \frac{1}{3\ 10^{-2}(20)}}$$

$\boldsymbol{\phi_0 = 31.2}$ **kW**

2) Cost of the energy losses generated by transporting the steam without heat lagging

This cost is given by: $P_0 = \phi_0 D_F P_E$.

NUMERICAL APPLICATION.–

$\phi_0 = 31.245$ kW $P_E = 3.8$ Dh/kWh $D_F = 7{,}600$ hrs/year

Hence: $P_0 = 31.245\ (7{,}600)\ (3.8)$

$\boldsymbol{P_0}$ **= 902,356 Dh/year**

B) Pipe installed

1) Flux of heat losses

The flux is given by:

$$\phi(e) = \frac{2\pi\ L\ (T_v - T_a)}{\frac{1}{r_1 h_i} + \frac{1}{\lambda}\mathrm{Ln}\left(\frac{r_2}{r_1}\right) + \frac{1}{\lambda_C}\mathrm{Ln}\left(1 + \frac{e}{r_2}\right) + \frac{1}{(r_2 + e)h_e}}$$

NUMERICAL APPLICATION.–

$T_v = 160$ °C $T_a = 20$ °C

$r_1 = 2.5\ 10^{-2}$ m $r_2 = 3\ 10^{-2}$ m

$L = 65$ m $\lambda = 55$ W/m °C

$h_i = 250$ W/m^2 °C $h_e = 20$ W/m^2 °C $e = 4$ cm

$$\varphi_1 = \frac{2\pi L(65)(160-20)}{\dfrac{1}{2{,}5\ 10^{-2}(250)} + \dfrac{1}{55}\mathrm{Ln}\left(\dfrac{3}{2{,}5}\right) + \dfrac{1}{0{,}8}\mathrm{Ln}\left(1+\dfrac{4}{3}\right) + \dfrac{1}{7\ 10^{-2}(20)}}$$

$\boldsymbol{\phi_I = 31.7}$ **kW**

2) Cost of the energy losses generated by transporting the steam with heat lagging

This cost is given by: $P_I = \phi_I D_F P_E$.

NUMERICAL APPLICATION.–

$\phi_I = 31.7$ kW $\quad$ $P_E = 3.8$ Dh/kWh $\quad$ $D_F = 7{,}600$ hrs/year

Hence: $P_I = 31.7\ (7{,}600)\ (3.8)$

$\boldsymbol{P_I = 915{,}496}$ **Dh/year**

C) Explanation of the apparent contradiction and optimization of the lagging

1) Critical heat-lagging thickness

The critical thickness, e_c, is given by: $e_c = \dfrac{\lambda_C}{h_e} - r_2$.

NUMERICAL APPLICATION.–

$r_2 = 3\ 10^{-2}$ m $\quad$ $\lambda_c = 0.8$ W/m °C $\quad$ $h_e = 20$ W/m^2 °C

$$e_c = \frac{0.8}{20} - 0.03$$

$\boldsymbol{e_c = 1}$ **cm**

2) Explanation of the apparent contradiction

The critical thickness being positive, the flux of losses will be greater than $\phi(0)$ if the lagging thickness used is less than the limit thickness.

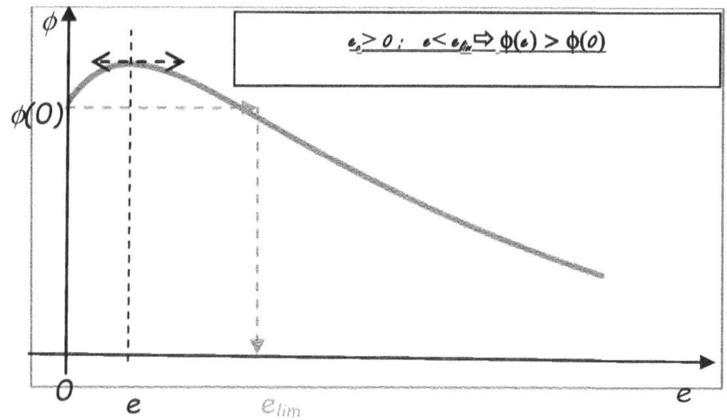

3) Calculating the limit thickness of the heat lagging, e_{lim}

e_{lim} is such that $\phi(e_{lim}) = \phi(0)$:

$$\frac{2\pi\, L\,(T_v - T_a)}{\dfrac{1}{r_1 h_i} + \dfrac{1}{\lambda}\mathrm{Ln}\left(\dfrac{r_2}{r_1}\right) + \dfrac{1}{\lambda_C}\mathrm{Ln}\left(1 + \dfrac{e_{lim}}{r_2}\right) + \dfrac{1}{(r_2 + e_{lim})\,h_e}} = \phi_0$$

e_{lim} must satisfy the following equation:

$$\frac{1}{\lambda_C}\mathrm{Ln}\left(1 + \frac{e_{lim}}{r_2}\right) + \frac{1}{(r_2 + e_{lim})\,h_e} = \frac{1}{r_2 h_e}$$

Or:

$$\frac{r_2 h_e}{\lambda_C}\mathrm{Ln}\left(1 + \frac{e_{lim}}{r_2}\right) + \frac{r_2}{(r_2 + e_{lim})} = 1$$

NUMERICAL APPLICATION.–

$T_v = 160$ °C $\quad T_a = 20$ °C $\quad r_1 = 2.5$ cm $\quad r_2 = 3$ cm

$L = 65$ m $\quad \lambda = 55$ W/m °C $\quad \lambda_c = 0.8$ W/m °C

$h_i = 250$ W/m^2 °C $\quad h_e = 20$ W/m^2 °C

$\phi_0 = 3.1245\ 10^4$W

e_{lilm} = 2.5 cm

Although it is greater than the critical thickness, the thickness of the lagging put in place during the plant's installation (e = 2 cm) is still below the limit thickness. As a result it is normal for the losses with lagging to be greater than the losses without lagging.

4) Calculating the optimum thickness of the heat lagging

The expression of the flux as a function of the thickness, e, of the lagging is given, in this case, by:

$$\phi(e) = \frac{A}{B + \frac{1}{\lambda_C}\mathrm{Ln}\left(1+\frac{e}{r_2}\right) + \frac{1}{(r_2+e)h_e}},$$

where: $A = 2\,\pi\,L\,(T_v - T_a)$ and $B = \frac{1}{r_1 h_i} + \frac{1}{\lambda}\mathrm{Ln}\left(\frac{r_2}{r_1}\right)$.

The optimum lagging thickness, e^*, is obtained by differentiating the expression of $\phi(e)$ and by equaling it to 0.

In this case, P_u is linear as a function of e:

$$P_u(e) = \alpha + \beta\, e$$

α = 850 Dh/m^2 and β = 2.1 10^5 Dh/m^3

e^* is then given by the solution to the following equation with e (see Chapter 3, section 3.3.2.2):

$$\frac{D_F P_E D_v (T_i - T_e)}{\alpha + \beta(2e + r_2)}\left(\frac{h_e(e+r_2) - \lambda_C}{\lambda_C h_e (e+r_2)^2}\right) = \left[\frac{1}{r_1 h_i} + \frac{1}{\lambda}\mathrm{Ln}\left(\frac{r_2}{r_1}\right) + \frac{1}{\lambda_C}\mathrm{Ln}\left(1+\frac{e}{r_2}\right) + \frac{1}{(e+r_2)h_e}\right]^2$$

NUMERICAL APPLICATION.–

T_v = 160 °C	T_a = 20 °C	r_1 = 2.5 cm	r_2 = 3 cm
L = 65 m	λ = 55 W/m°C	λ_c = 0.8 W/m °C	D_v = 5 years
α = 850 Dh/m^2	β = 2.1 10^{5}Dh/m^3		P_E = 0.0038 Dh/Wh
D_F = 7,600 hrs/year	h_i = 250 W/m^2 °C		h_e = 20 W/m^2 °C

The equation to be resolved can be presented in the following form:

$f(e) - g(e) = 0,$

where: $$f(e) = \frac{D_F P_E D_v (T_i - T_e)}{\alpha + \beta(2e + r_2)} \left(\frac{h_e(e + r_2) - \lambda_C}{\lambda_C h_e (e + r_2)^2} \right),$$

and: $$g(e) = \left[\frac{1}{r_1 h_i} + \frac{1}{\lambda} \mathrm{Ln}\left(\frac{r_2}{r_1} \right) + \frac{1}{\lambda_C} \mathrm{Ln}\left(1 + \frac{e}{r_2} \right) + \frac{1}{(e + r_2) h_e} \right]^2.$$

It admits two solutions: $e_1 = 1.29$ cm and $e_2 = 6.43$ cm.

as e_1 is less than $e_{lim,}$ we will take: $\boldsymbol{e^* = 6.4}$ **cm.**

5) Savings that can be made by optimizing the heat-lagging thickness

These savings can be calculated with respect to an operation without lagging *(situation 0)*, or to the plant's current operation, i.e. with 4 cm of lagging *(situation 1)*. The following table presents the savings that can be made in each of these situations.

With respect to situation 0	With respect to situation 1
$E_0 = P_0 - D_F P_E \phi(e^*)$	$E_1 = P_1 - D_F P_E \phi(e^*)$
E_0= 124,488 Dh/year	E_1=138,577 Dh/year

EXERCISE 7. Optimum heat-lagging thickness for a heat-transfer fluid pipe

At the Noor 1 Ouarzazate Solar Power Station, a cylindrical pipe of internal radius r_i and external radius r_e conveys, at high temperature (T_f), a heat-transfer fluid between a field of cylindro-parabolic solar energy collectors and the main boiler, over a length L equal to the sum of the three sections, L_1, L_2 and L_3 (see the following figure).

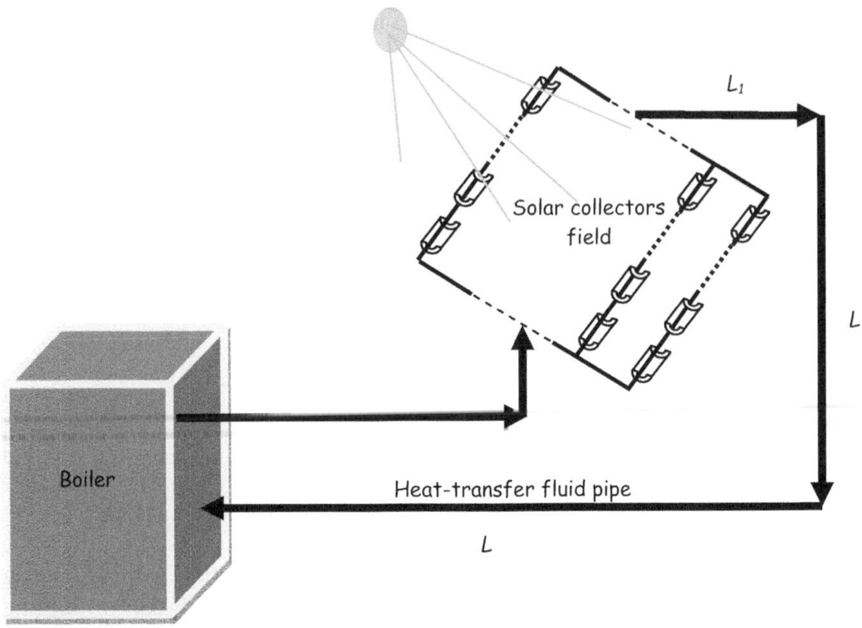

For the purposes of this analysis, the temperature of the heat-transfer fluid inside the pipe, T_f, can reasonably be assumed to be constant along its entire length, L.

Moreover, the temperature of the surrounding environment is considered constant and equal to T_j during the day and to T_n during the night. Of course, the daytime and nighttime temperatures depend on the month of the year. The following table gives the average daytime temperatures for each month of the year.

Solar energy production only takes place during hours of solar radiation: sunshine hours. Thus, the heat-transfer fluid is also only heated in the sensor field during periods when the sun is available. As a result, this fluid is only transported towards the boiler during sunshine hours.

Within the context of this problem, we wish to examine the opportunity for savings presented by heat lagging or the lack of heat lagging on this pipe. The choice will depend on the significance of any savings made by installing heat lagging, compared to the necessary investment. In this perspective, the project manager asks you to prepare the following work:

Questions

1) When the pipe is left bare, in contact with the surrounding environment, what would be the amount of energy, Q_0, lost, in MWh/year?

2) With a cost per kWh equal to C_E, deduce therefrom, in Dh/year, the cost generated by transporting the heat-transfer fluid between the solar field and the boiler.

3) Give the expression of the energy lost annually, $Q(e)$, as a function of the lagging thickness.

4) Calculate the critical radius of the lagging and, where applicable, the limit thickness.

5) Calculate the optimum thickness of the heat lagging.

6) Calculate, in monetary terms, the savings that can be made by putting in place the heat lagging with an optimum thickness.

7) Calculate the average payback period, defined by the time required for the expected saving from putting the lagging in place to equal the cost of the following installation: $\tau\, E_a = I_c$ where τ is the average payback period (in years), I_c is the investment needed in order to put the lagging in place and E_a is the annual saving made using the lagging (in Dh/year).

<u>*Data communicated by the project manager:*</u>

– Pipe:

$r_i = 10$ cm $\quad r_e = 12$ cm $\quad \lambda = 52$ W/m °C

$L_1 = 350$ m $\quad L_2 = 177$ m $\quad L_3 = 93$ m

– Convective transfer coefficients:

$h_i = 325$ W/m^2 °C $\quad h_e = 15$ W/m^2 °C

– Temperatures:

$T_f = 375$ °C

Average daily temperatures:

Month	1	2	3	4	5	6	7	8	9	10	11	12
Daytime temperatures (°C)	0	5	8	13	27	32	34	37	26	15	12	6

– Annual operating time:

$D_{Fm} = 12$ months/year

Average monthly sunshine duration per day

Month	1	2	3	4	5	6	7	8	9	10	11	12
Number of days	31	28	31	30	31	30	31	31	30	31	30	31
Sunshine duration (in hours per day)	6	5.5	6.5	9.5	10.5	12	13	14	11.5	9	7.5	6

– Heat lagging:

$\lambda_c = 0.035$ W/m°C

Cost per m² installed: $P_u(e) = \alpha + \beta e$

Lifetime: $D_v = 3$ years

$\alpha = 650$ Dh/m² $\beta = 3.5\ 10^5$ Dh/m³

– Energy: $C_E = 3.2$ Dh/kWh

Solutions

1) Amount of energy lost annually when the pipe is bare

Given that the outside temperature varies from one month to the next, so too will the flux. Thus, for month j, the flux, ϕ_{0j}, will be given by:

$$\phi_{0j} = \frac{2\ \pi\ (L_1 + L_2 + L_3)\ (T_f - T_j)}{\frac{1}{r_i h_i} + \frac{1}{\lambda} Ln\left(\frac{r_e}{r_i}\right) + \frac{1}{r_e h_e}},$$

where T_j is the average daytime temperature for month j.

The amount of energy lost during month j is then given by:

$$Q_{0j} = N_j\ D_j\ \varphi_{0j},$$

where N_j is the number of days in month j.

D_j is the average sunshine duration, u, of month j.

$$\text{I.e.: } Q_{0j} = \frac{2\ \pi\ (L_1 + L_2 + L_3)\ N_j\ D_j\ (T_f - T_j)}{\frac{1}{r_i h_i} + \frac{1}{\lambda} Ln\left(\frac{r_e}{r_i}\right) + \frac{1}{r_e h_e}}.$$

Hence the amount of energy lost during the course of one year:

$$Q_0 = \frac{2\pi\left(L_1 + L_2 + L_3\right)}{\frac{1}{r_i h_i} + \frac{1}{\lambda}\mathrm{Ln}\left(\frac{r_e}{r_i}\right) + \frac{1}{r_e h_e}} \sum_{j=1}^{12} N_j\, D_j\, \left(T_f - T_j\right)$$

NUMERICAL APPLICATION.–

$T_f = 375$ °C

$r_i = 10$ cm $\quad r_e = 12$ cm

$\lambda = 52$ W/m °C

$L_1 = 350$ m $\quad L_2 = 177$ m $\quad L_3 = 93$ m

$h_i = 325$ W/m^2 °C $\quad h_e = 15$ W/m^2 °C

$$\sum_{j=1}^{12} N_j D_j (T_f - T_j) = 1{,}195{,}873.5\ h\ °C$$

Q_0= 7,898.2 MWh/year

2) Cost of the energy losses generated by transporting the heat-transfer fluid

This cost is given by: $C_0 = Q_0\, C_E$.

NUMERICAL APPLICATION.–

Q_0 = 7,898.2 MWh/year

C_E = 3.2 Dh/kWh

Hence: C_0 = 7,898.2 (1,000)(3.2)

C_0 = 25.3 MDh/year

3) Expression of the energy lost annually as a function of the lagging thickness

With a lagging thickness, *e*, the amount of energy lost annually is given by:

$$Q(e) = \frac{2\pi\left(L_1 + L_2 + L_3\right)}{\frac{1}{r_i h_i} + \frac{1}{\lambda}\mathrm{Ln}\left(\frac{r_e}{r_i}\right) + \frac{1}{\lambda_c}\mathrm{Ln}\left(1 + \frac{e}{r_e}\right) + \frac{1}{\left(e + r_e\right)h_e}} \sum_{j=1}^{12} N_j\, D_j\, \left(T_f - T_j\right)$$

4) Critical radius of the heat lagging

The critical radius, r_c, is given by: $r_c = \dfrac{\lambda_C}{h_e}$.

> NUMERICAL APPLICATION.–
>
> $\lambda_c = 0.035$ W/m °C $\quad h_e = 15$ W/m² °C $\quad r_c = \dfrac{0.035}{15}$
>
> $\mathbf{r_c = 2.33}$ **mm**

Thus, the critical thickness ($e_c = r_c - r_e$) will be negative. As a result, it is not necessary to calculate the limit thickness.

5) Calculating the optimum thickness of the heat lagging

By adapting the function determined in section 3.3.2.2 (Chapter 3), the optimum lagging thickness, e^*, is given by the solution to the following equation with e:

$$\frac{C_E D_v \sum_{j=1}^{12} N_j D_j (T_f - T_j)}{\alpha + \beta(2e + r_e)} \left(\frac{h_e(e + r_e) - \lambda_C}{\lambda_C h_e (e + r_e)^2} \right) = \left[\frac{1}{r_i h_i} + \frac{1}{\lambda} \mathrm{Ln}\left(\frac{r_e}{r_i}\right) + \frac{1}{\lambda_C} \mathrm{Ln}\left(1 + \frac{e}{r_e}\right) + \frac{1}{(e + r_e) h_e} \right]^2$$

I.e.: $f(e) - g(e) = 0$,

where: $f(e) = \dfrac{C_E D_v \left[h_e(e + r_e) - \lambda_C \right]}{\lambda_C h_e \left[\alpha + \beta(2e + r_e) \right] (e + r_e)^2} \sum_{j=1}^{12} N_j D_j (T_f - T_j)$,

and: $g(e) = \left[\dfrac{1}{r_i h_i} + \dfrac{1}{\lambda} \mathrm{Ln}\left(\dfrac{r_e}{r_i}\right) + \dfrac{1}{\lambda_C} \mathrm{Ln}\left(1 + \dfrac{e}{r_e}\right) + \dfrac{1}{(e + r_e) h_e} \right]^2$.

> NUMERICAL APPLICATION.–
>
> $T_f = 375$ °C $\quad r_i = 10$ cm $\quad r_e = 12$ cm $\quad \lambda = 52$ W/m °C
>
> $h_i = 325$ W/m² °C $\quad h_e = 15$ W/m² °C $\quad \lambda_c = 0.035$ W/m °C
>
> $\sum_{j=1}^{12} N_j D_j (T_f - T_j) = 1{,}195{,}873.5$ h°C

The following figure shows curves *g(e)* and *f(e)*, as well as the curve representing the function, *f(e)-g(e)*.

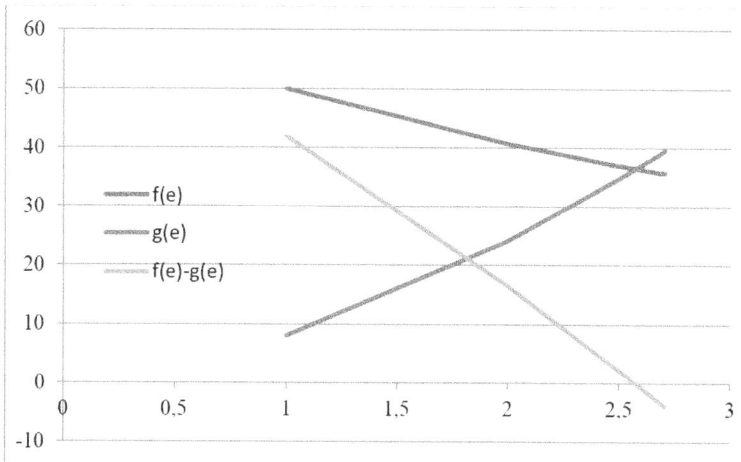

The solution to the equation is: $e^* = 2.57$ cm.

<u>*6) Savings*</u>

The savings made as a result of using heat lagging are given by:

$$E = [Q_0 - Q(e^*)]\, C_E,$$

where Q_0 corresponds to the energy loss where there is no lagging:

$$Q(e^*) = \frac{2\pi\,(L_1+L_2+L_3)}{\dfrac{1}{r_i h_i} + \dfrac{1}{\lambda}\mathrm{Ln}\left(\dfrac{r_e}{r_i}\right) + \dfrac{1}{\lambda_c}\mathrm{Ln}\left(1+\dfrac{e^*}{r_e}\right) + \dfrac{1}{(e^*+r_e)h_e}} \sum_{j=1}^{12} N_j\, D_j\,(T_f - T_j)$$

$$\text{I.e.: } E = C_0 - \frac{2\pi\, C_E\,(L_1+L_2+L_3)}{\dfrac{1}{r_i h_i} + \dfrac{1}{\lambda}\mathrm{Ln}\left(\dfrac{r_e}{r_i}\right) + \dfrac{1}{\lambda_c}\mathrm{Ln}\left(1+\dfrac{e^*}{r_e}\right) + \dfrac{1}{(e^*+r_e)h_e}} \sum_{j=1}^{12} N_j\, D_j\,(T_f - T_j).$$

NUMERICAL APPLICATION.–

$T_f = 375$ °C	$r_i = 10$ cm	$r_e = 12$ cm	$\lambda = 52$ W/m °C
$h_i = 325$ W/m^2 °C	$h_e = 15$ W/m^2 °C	$\lambda_c = 0.035$ W/m°C	$C_E = 0.0032$ Dh/W

$L_1 = 350$ m $\quad L_2 = 177$ m $\quad L_3 = 93$ m $\quad e^* = 0.0257$ m

$C_0 = 25.3$ MDh/year

$$\sum_{j=1}^{12} N_j D_j (T_f - T_j) = 1\,195\,873.5 \text{ h°C}$$

***E* = 22.8 MDh/year**

7) Average payback period

$$\tau = \frac{I_c}{E}, \text{ where } I_c = 2\pi (r_e + e^*)(L_1 + L_2 + L_3)[\alpha + \beta e^*].$$

NUMERICAL APPLICATION.–

$L_1 = 350$ m $\quad L_2 = 177$ m $\quad L_3 = 93$ m $\quad e^* = 0.0257$ m

$r_e = 12$ cm $\quad \alpha = 650$ Dh/m^2 $\quad \beta = 3.5\ 10^5$ Dh/m^3 $\quad E = 22.8\ 10^6$ Dh/year

$\tau = 2.88$ months.

EXERCISE 8. Thermal insulation of cylindrical storage tanks

An industrial cylindrical tank of height H, internal radius R_1 and external radius R_2 is used to store a fluid at high temperature T_0. The tank in question is made of special steel, of heat conductivity λ_S.

The average ambient temperatures vary between day and night and depending on the month in the year. The following table gives the average monthly daytime temperatures, Tj, and nighttime temperatures, T_n, along with the corresponding durations.

We will assume that the conduction, which is one-dimensional, follows direction *(r)*.

Questions

1) In the knowledge that the tank is not isolated, that $T(j,i)$ and $T(n,i)$ designate the monthly average daytime and nighttime temperatures for month i, and that $\delta(j,i)$ and $\delta(n,i)$ refer to the average day and night durations for month i, give, for month i, the expressions of the monthly daytime, $Q_0(j,i)$, nighttime, $Q_n(j,i)$, and total, $Q_0(T,i)$, heat losses to the outside.

2) Establish a spreadsheet application that provides the values of the daytime, nighttime and total heat losses (in kWh), as well as the cost of these losses (in Dh), for each month of the year. Deduce therefrom the annual cost of the total energy losses with no lagging.

3) We now wish to examine the opportunity of lagging the tank, and you are therefore asked to:

a) give the expression of the annual cost of the total energy losses as a function of the lagging thickness;

b) develop the expression of the lagging thickness in order to optimize the total annual energy losses;

c) ensure that you understand the term "optimize", used in the previous question;

d) find the optimum heat-lagging thickness in this case, and quantify the savings that can be made using this lagging.

Data:

$R_1 = 2.50$ m $R_2 = 2.52$ m $H = 5$ m

$\lambda_s = 65$ W/m °C $T_0 = 165$ °C $h_i = 145$ W/m^2 °C

$h_e = 15$ W/m^2 °C $P_E = 0.0028$ Dh/Wh

Annual operation: 365 days per year, 24 hours per day.

$\lambda_c = 0.042$ W/m °C $D_v = 5$ years

$P_u(e) = \alpha + \beta\, e$ $\alpha = 1{,}150$ Dh/m^2 $\beta = 5\ 10^4$ Dh/m^3

Average monthly daytime and nighttime temperatures and corresponding durations

Month	1	2	3	4	5	6	7	8	9	10	11	12
Average daytime temperatures, T_j (in °C)	15	14.8	15.8	17.5	22.5	26	29	30	26.5	23	17.5	15.5
Average day durations, δ_j (in hours)	8	9	10	11.5	13	14	15	15.5	14	11.5	10	8.5
Average nighttime temperatures, T_n (in °C)	9	8.5	9.8	11	16	18	20	22	20	17	11	9.5
Average night durations, δ_n (in hours)	16	15	14	12.5	11	10	9	8.5	10	12.5	14	15.5

Solutions

1) Expression of monthly energy losses to the outside

a) Daytime losses

The average daytime temperature, $T(j,i)$, and the average nighttime temperature, $T(n,i)$, are different for month i. As a result, the daytime flux, $\phi_0(j,i)$, will be different from the nighttime flux, $\phi_0(n,i)$.

The daytime flux, $\phi_0(j,i)$, of month i is given by:

$$\phi_0(j,i) = \frac{2\,\pi\,H\,\left[T_0 - T(j,i)\right]}{\dfrac{1}{R_1 h_i} + \dfrac{1}{\lambda_s} Ln\left(\dfrac{R_2}{R_1}\right) + \dfrac{1}{R_2 h_e}},$$

where $T(j,i)$ is the average daytime temperature for month i.

The daytime energy losses during month i are therefore given by:

$$Q_0(j,i) = N_i\,\delta(j,i)\,\varphi_0(j,i) = \frac{2\,\pi\,H\,N_i\,\delta(j,i)\,\left[T_0 - T(j,i)\right]}{\dfrac{1}{R_1 h_i} + \dfrac{1}{\lambda_s} Ln\left(\dfrac{R_2}{R_1}\right) + \dfrac{1}{R_2 h_e}},$$

where N_j is the number of days in month i and $\delta(j,i)$ is the average duration of daytime in month i.

b) Nighttime losses

Likewise, the nighttime flux, $\phi_0(n,i)$, for month i is given by:

$$\phi_0(n,i) = \frac{2\,\pi\,H\,\left[T_0 - T(n,i)\right]}{\dfrac{1}{R_1 h_i} + \dfrac{1}{\lambda_s} Ln\left(\dfrac{R_2}{R_1}\right) + \dfrac{1}{R_2 h_e}},$$

where $T(n,i)$ is the average nighttime temperature for month i.

The amount of energy lost, in the nighttime, during month i is then given by:

$$Q_0(n,i) = N_i\,\delta(n,i)\,\varphi_0(n,i) = \frac{2\,\pi\,H\,N_i\,\delta(n,i)\,\left[T_0 - T(n,i)\right]}{\dfrac{1}{R_1 h_i} + \dfrac{1}{\lambda_s} Ln\left(\dfrac{R_2}{R_1}\right) + \dfrac{1}{R_2 h_e}},$$

where N_j is the number of days in month i and $\delta(n,i)$ is the average duration of nighttime during month i.

c) Total monthly losses

The amount of energy lost, day and night, during month i is the sum of the daytime and nighttime losses:

$$Q_0(T,i) = Q_0(j,i) + Q_0(n,i)$$

$$Q_0(T,i) = \frac{2\ \pi\ H\ N_i \left\{T_0\left[\delta(j,i)+\delta(n,i)\right]-\delta(j,i)\ T(j,i)-\delta(n,i)\ T(n,i)\right\}}{\frac{1}{R_1 h_i}+\frac{1}{\lambda_s}\mathrm{Ln}\left(\frac{R_2}{R_1}\right)+\frac{1}{R_2 h_e}}$$

2) Monthly and annual cost of losses

The monthly cost of the losses is given by: $C_0(T,i) = P_E Q_0(T,i)$.

The annual cost is: $C_T = \sum_{i=1}^{12} C_0(T,i) = \sum_{i=1}^{12} P_E Q_0(T,i)$.

Month i	1	2	3	4	5	6	7	8	9	10	11	12
$T(j,i)$ (°C)	15	14.8	15.8	17.5	22.5	26	29	30	26.5	23	17.5	15.5
(j,i) (hrs/day)	8	9	10	11.5	13	14	15	15.5	14	11.5	10	8.5
$T(n,i)$ (°C)	9	8.5	9.8	11	16	18	20	22	20	17	11	9.5
$\delta(n,j)$ (hrs/day)	16	15	14	12.5	11	10	9	8.5	10	12.5	14	15.5
N_i	31	28	31	30	31	30	31	31	30	31	30	31
$Q_0(j,i)$ (MWh)	33.0	33.5	41.0	45.1	50.9	51.7	56.0	57.5	51.5	44.8	39.2	34.9
$Q_0(n,i)$ (MWh)	68.5	58.2	59.7	51.1	45.0	39.1	35.8	33.4	38.5	50.8	57.3	66.2
Total losses $Q_0(T,i)$ in MWh/month	101	91.8	100.6	96.2	95.9	90.8	91.9	90.8	90.1	95.6	96.5	101.1
Cost of total losses $C_0(T,i)$ in kDh/month	254	229	252	241	240	227	230	227	225	239	241	253

Hence the *cost of the annual losses* is the sum of the monthly costs:

$C_{OT} = 3.87$ MDh/year

3) *Study of the opportunity to heat-lag the tank*

a) Total annual cost of energy losses as a function of the thickness

With a heat-insulation thickness, *e*, the term defining the thermal resistance becomes:

$$\frac{1}{R_1 h_i} + \frac{1}{\lambda_s} \mathrm{Ln}\left(\frac{R_2}{R_1}\right) + \frac{1}{\lambda_c} \mathrm{Ln}\left(1 + \frac{e}{R_2}\right) + \frac{1}{(R_2 + e) h_e}$$

The total monthly losses are thus given by:

$$Q_T(e,i) = \frac{2\,\pi\,H\,N_i\left\{\left[\delta(j,i)+\delta(n,i)\right]\,T_0 - \delta(j,i)\,T(j,i) - \delta(n,i)\,T(n,i)\right\}}{\frac{1}{R_1 h_i} + \frac{1}{\lambda_s} \mathrm{Ln}\left(\frac{R_2}{R_1}\right) + \frac{1}{\lambda_c} \mathrm{Ln}\left(1 + \frac{e}{R_2}\right) + \frac{1}{(R_2 + e) h_e}}$$

The annual cost of the losses therefore corresponds to: $C_T(e) = \sum_{i=1}^{12} P_E Q_T(e,i)$.

The expression that gives this cost as a function of *e* takes the following form:

$$C_T(e) = \frac{2\,\pi\,H\,P_E}{\frac{1}{R_1 h_i} + \frac{1}{\lambda_s} \mathrm{Ln}\left(\frac{R_2}{R_1}\right) + \frac{1}{\lambda_c} \mathrm{Ln}\left(1 + \frac{e}{R_2}\right) + \frac{1}{(R_2 + e) h_e}}$$
$$\sum_{i=1}^{12} N_i\left\{\left[\delta(j,i)+\delta(n,i)\right]\,T_0 - \delta(j,i)\,T(j,i) - \delta(n,i)\,T(n,i)\right\}$$

b) Expression of the heat-lagging thickness that optimizes the energy losses

The total annual energy losses are given by:

$$Q_T(e) = \sum_{i=1}^{12} Q_0(T,i)$$

Hence:

$$Q_T(e) = \frac{2\,\pi\,H}{\dfrac{1}{R_1 h_i} + \dfrac{1}{\lambda_s}\mathrm{Ln}\left(\dfrac{R_2}{R_1}\right) + \dfrac{1}{\lambda_c}\mathrm{Ln}\left(1+\dfrac{e}{R_2}\right) + \dfrac{1}{(R_2+e)h_e}}$$

$$\sum_{i=1}^{12} N_i \left\{\left[\delta(j,i)+\delta(n,i)\right]\; T_0 - \delta(j,i)\; T(j,i) - \delta(n,i)\; T(n,i)\right\}$$

$Q_T(e)$ is therefore in the form: $Q_T(e) = \dfrac{N(e)}{D(e)}$.

The thickness that optimizes *$Q_T(e)$* is the value that optimizes *$D(e)$*, i.e.:

$$D\,(e) = \frac{1}{R_1 h_i} + \frac{1}{\lambda_s}\mathrm{Ln}\left(\frac{R_2}{R_1}\right) + \frac{1}{\lambda_c}\mathrm{Ln}\left(1+\frac{e}{R_2}\right) + \frac{1}{(R_2+e)h_e}$$

We therefore obtain e_{opt}, such that: $\dfrac{dD\,(e)}{de} = 0$:

$$\frac{dD\,(e)}{de} = \frac{1}{\lambda_c}\frac{1}{e+R_2} - \frac{1}{(R_2+e)^2 h_e}$$

I.e.: $\dfrac{dD\,(e)}{de} = \dfrac{1}{e+R_2}\left[\dfrac{1}{\lambda_c} - \dfrac{1}{(R_2+e)h_e}\right]$.

Hence: $\dfrac{1}{\lambda_c} - \dfrac{1}{\left(R_2+e_{opt}\right)h_e} = 0$.

Or: $e_{opt} = \dfrac{\lambda_c}{h_e} - R_2$.

c) Precautions

The thickness determined in the previous question is the critical thickness. It corresponds to the location of the maximum energy losses. The term "optimize" used in this question needs to be interpreted in the sense of "maximize" and not "minimize".

d) Optimum heat-lagging thickness and savings that can be made

In this case, $\dfrac{\lambda_C}{h_e} = 2{,}8\text{mm}$. The critical thickness will therefore be negative.

As a result, it is not necessary to calculate the limit thickness.

The cost of the total annual energy losses is given in this case by:

$$C_T(e) = \frac{2\ \pi\ H\ P_E\ \psi(D_F, \Delta T)}{\dfrac{1}{R_1 h_i} + \dfrac{1}{\lambda_s} \text{Ln}\left(\dfrac{R_2}{R_1}\right) + \dfrac{1}{\lambda_c} \text{Ln}\left(1 + \dfrac{e}{R_2}\right) + \dfrac{1}{(R_2 + e) h_e}},$$

where: $\psi(D_F, \Delta T) = \sum_{i=1}^{12} N_i \left\{ \left[\delta(j,i) + \delta(n,i)\right]\ T_0 - \delta(j,i)\ T(j,i) - \delta(n,i)\ T(n,i) \right\}$.

By adapting the function "total cost" (cost of total annual losses + heat-lagging cost) determined in section 3.3.2.2, the optimum heat-lagging thickness, e^*, is given by the solution to the following equation with e:

$$\frac{P_E D_V\ \psi(D_F, \Delta T)}{\alpha + \beta(2e + R_2)} \left(\frac{h_e(e + R_2)\ - \lambda_C}{\lambda_C h_e (e + R_2)^2} \right) = \left[\frac{1}{R_1 h_i} + \frac{1}{\lambda_s} \text{Ln}\left(\frac{R_2}{R_1}\right) + \frac{1}{\lambda_C} \text{Ln}\left(1 + \frac{e}{R_2}\right) + \frac{1}{(e + R_2) h_e} \right]^2$$

with: $\psi(D_F, \Delta T) = \sum_{i=1}^{12} N_i \left\{ \left[\delta(j,i) + \delta(n,i)\right]\ T_0 - \delta(j,i)\ T(j,i) - \delta(n,i)\ T(n,i) \right\}$.

I.e.: $f(e) - g(e) = 0$,

where: $$f(e) = \frac{P_E D_V \left[h_e(e + R_2)\ - \lambda_C \right]}{\lambda_C h_e \left[\alpha + \beta(2e + R_2) \right] (e + R_2)^2}\ \psi(D_F, \Delta T),$$

and: $$g(e) = \left[\frac{1}{R_1 h_i} + \frac{1}{\lambda_s} \text{Ln}\left(\frac{R_2}{R_1}\right) + \frac{1}{\lambda_C} \text{Ln}\left(1 + \frac{e}{R_2}\right) + \frac{1}{(e + R_2) h_e} \right]^2.$$

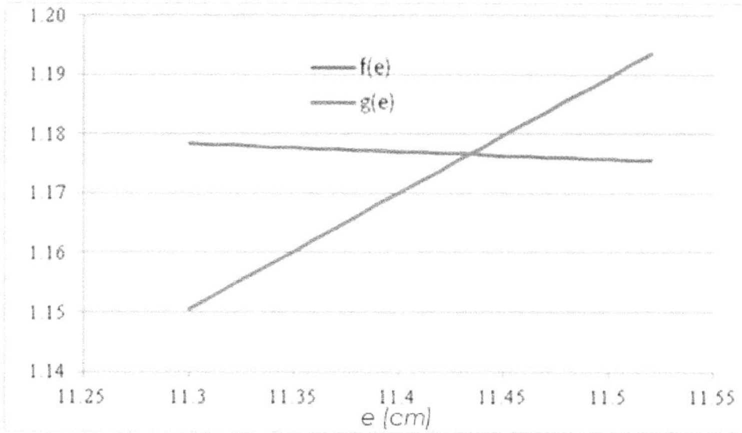

We deduce therefrom:

– the optimum thickness of the heat lagging: $e^* = 11.43$ cm;

– the cost of the annual energy losses with heat lagging: $C_{cT} = 0.10$ MDh/year.

Hence the saving that can be made using heat lagging:

$$E = C_{0T} - C_{cT} = 3.87\text{-}0.10 = 3.77 \text{ MDh/year}$$

EXERCISE 9. Storing a liquid reagent in a spherical tank

A liquid reagent needs to be stored under pressure in a spherical tank of internal radius R, the container wall of which is of thickness ΔR.

The temperature of the oxygen inside the storage sphere is assumed to be constant and equal to T_i.

As a first approximation, we will presume that the ambient temperature outside the storage sphere can reasonably be assumed to be constant and equal to T_a.

We will also suppose that the conduction is one-dimensional and follows the r direction.

Questions

1) In the specific case where the tank is not heat lagged, give the expression of the energy losses as a function of T_i and T_a.

2) Deduce therefrom the annual cost of the losses.

3) We now wish to examine the opportunity of lagging the tank, and you are therefore asked to:

a) give the expression of the annual cost of the energy losses as a function of the lagging thickness;

b) calculate the critical radius of the heat lagging;

c) calculate the optimum thickness of the heat lagging;

d) estimate, in this case, the new cost of the annual losses and the savings made by putting the heat lagging in place.

Data:

$R = 3$ m $\Delta R = 2$ cm $\lambda = 55$ W/m °C $T_i = 65$ °C

$T_a = 18$ °C $h_i = 11$ W/m^2 °C $h_e = 18$ W/m^2 °C $\lambda_c = 0.4$ W/m °C

$P_u(e) = 850 + 8\ 10^3 e$ Dh/m

Operating time: 365 days per year, 24 hours per day

Lifetime of the heat lagging: $D_v = 5$ years

$P_E = 2.5$ Dh/kWh

Solutions

1) Expression of energy losses without lagging

As conduction is one-dimensional, the heat flux passing through the container wall of the sphere will therefore be given by:

$$\varphi = \frac{4\ \pi\ (T_i, T_a)}{\dfrac{1}{h_i R^2} + \dfrac{\Delta R}{\lambda\ R\ (R + \Delta R)} + \dfrac{1}{h_e (R + \Delta R)^2}}$$

2) Annual cost of losses

$$C = D_F P_E \phi$$

$$\text{I.e.: } C = \frac{4\ \pi\ D_F P_E (T_i, T_a)}{\dfrac{1}{h_i R^2} + \dfrac{\Delta R}{\lambda\ R\ (R + \Delta R)} + \dfrac{1}{h_e (R + \Delta R)^2}}.$$

NUMERICAL APPLICATION.–

$R = 3$ m	$\Delta R = 2$ cm	$T_i = 65$ °C	$T_a = 18$ °C
$\lambda = 55$ W/m^2 °C	$h_i = 11$ W/m^2 °C		$h_e = 48$ W/m^2 °C
$D_F = 8{,}760$ hrs/year	$P_E = 0.0025$ Dh/Wh		

C = 1,040,977 Dh/year

3) Heat lagging the tank

a) Annual cost of energy losses as a function of the lagging thickness

This cost is given by: $C_c(e) = D_F P_E\ \phi_c(e)$; with (see Chapter 2, section 2.14):

$$\phi_c(e) = \frac{4\pi\ (T_i - T_e)}{\frac{1}{r_1^2 h_i} + \frac{1}{\lambda}\frac{r_2 - r_1}{r_2 r_1} + \frac{1}{\lambda_c}\frac{e}{r_2(r_2 + e)} + \frac{1}{(r_2 + e)^2 h_e}},$$

where $r_1 = R$ and $r_2 = R + \Delta R$.

$$\text{I.e.: } C_c(e) = \frac{4\pi\ D_F P_E (T_i - T_e)}{\frac{1}{r_1^2 h_i} + \frac{1}{\lambda}\frac{r_2 - r_1}{r_2 r_1} + \frac{1}{\lambda_c}\frac{e}{r_2(r_2 + e)} + \frac{1}{(r_2 + e)^2 h_e}}.$$

b) Critical radius of the heat lagging

The *critical thickness*, e_c, of the heat lagging is given by: $e_c = \frac{2\lambda_c}{h_c} - r_2$.

The critical radius then corresponds to: $r_c = \frac{2\lambda_c}{h_e}$.

NUMERICAL APPLICATION.–

$r_2 = 3.02$ m	$\lambda_c = 0.4$ W/m °C	$h_e = 48$ W/m^2 °C

Hence: **e_c = - 3 m** and **r_c = 1.6 cm**

As the critical thickness is negative, there is no need to establish the limit thickness.

c) Optimum heat-lagging thickness

The optimum thickness corresponds to the solution to the following equation with e (see Volume 1):

$$\left[\frac{1}{r_1^2 h_i}+\frac{1}{\lambda}\frac{r_2-r_1}{r_2 r_1}+\frac{1}{\lambda_c}\frac{e}{r_2(r_2+e)}+\frac{1}{(r_2+e)^2 h_e}\right]^2=\frac{(T_i-T_e)D_F P_E D_v}{(r_2+e)^4}\frac{\left[\frac{r_2+e}{\lambda_c}-\frac{2}{h_e}\right]}{\left[2P_u(e)+(r_2+e)\frac{dP_u(e)}{de}\right]}$$

where $r_1 = R$ and $r_2 = R+\Delta R$.

Here, $P_u(e)$ takes the following form: $P_u(e) = \alpha + \beta e$. The equation with e can then be rewritten as follows:

$$\frac{(r_2+e)^4(2\alpha+\beta r_2+3\beta e)}{(T_i-T_e)D_F P_E D_v}\left[\frac{1}{r_1^2 h_i}+\frac{1}{\lambda}\frac{r_2-r_1}{r_2 r_1}+\frac{1}{\lambda_c}\frac{e}{r_2(r_2+e)}+\frac{1}{(r_2+e)^2 h_e}\right]^2=\left[\frac{r_2+e}{\lambda_c}-\frac{2}{h_e}\right]$$

This expression can also be presented in the following form:

$f(e) = g(e)$,

where $f(e)$ and $g(e)$ are given by:

$$f(e)=\frac{(r_2+e)^4(2\alpha+\beta r_2+3\beta e)}{(T_i-T_e)D_F P_E D_v}\left[\frac{1}{r_1^2 h_i}+\frac{1}{\lambda}\frac{r_2-r_1}{r_2 r_1}+\frac{1}{\lambda_c}\frac{e}{r_2(r_2+e)}+\frac{1}{(r_2+e)^2 h_e}\right]^2$$

$$g(e)=\left[\frac{r_2+e}{\lambda_c}-\frac{2}{h_e}\right]$$

NUMERICAL APPLICATION.–

$R = 3$ m

$\Delta R = 2$ cm

$\lambda = 55$ W/m °C

$T_i = 65$ °C $\quad$ $T_a = 18$ °C

$h_i = 11$ W/m^2 °C $\quad$ $h_e = 48$ W/m^2 °C

$D_F = 8{,}760$ hrs/year $\quad$ $D_V = 5$ years

$P_E = 0.0025$ Dh/Wh

$\alpha = 850$ Dh/m^2 $\quad$ $\beta = 8\ 10^3$ Dh/m^3

The solution to the equation $f(e) = g(e)$ has been established on a spreadsheet.

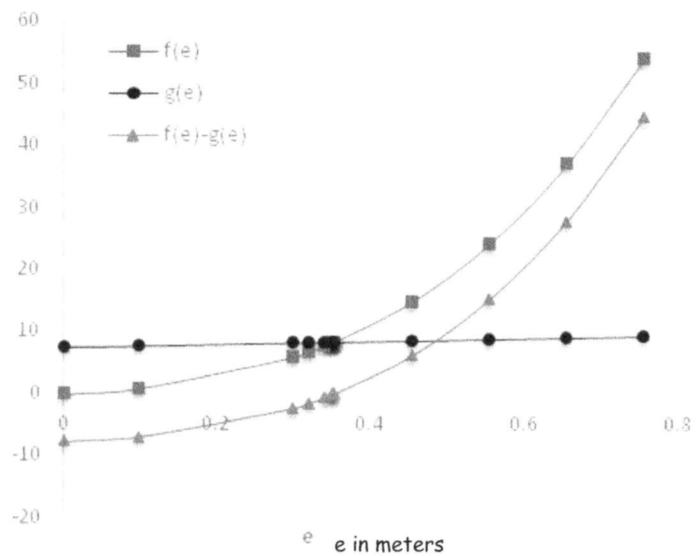

The optimum thickness of the heat lagging is: $e^* =$ **35.3 cm**.

d) Savings that can be made using the heat lagging

New cost of energy losses: $C_c = 52{,}469$ Dh/year.

Savings that can be made using the heat lagging: $E_c = 988{,}507$ Dh/year.

EXERCISE 10. Energy savings in the operation of an industrial furnace

After attending a conference on the potential of energy saving in the industry, you wish to examine the possibility of minimizing heat losses through the side walls of the large fusion furnace that you are responsible for at the plant.

The four side walls of the furnace are composed of refractory bricks and their external faces are in contact with the surrounding environment. The thickness Δx and the heat conductibility λ of the walls are constant. The internal face of the walls is at temperature T_i, the surrounding environment is at temperature T_a, and the convective heat transfer coefficient between the outer face of the walls (at T_e) and the atmosphere is noted h.

Questions

A) In this part, we wish to assess the annual cost of the heat losses in the event that no heat lagging is applied to the lateral surface of the furnace. In order to do this:

1) Calculate the temperature, T_e.

2) Calculate the heat losses.

3) Assuming that the furnace is supplied by electricity and that its annual operating time is D_F, estimate these losses in monetary terms.

B) We now wish to examine the possibility of reducing the heat losses using glass wool as heat lagging.

We apply a thickness, e, of this heat lagging onto the outer lateral faces of the wall. The heat conductibility of the heat lagging will be noted λ_c and its price per square meter will depend on the thickness according to the following function:

$$P_u(e) = \alpha + \beta e$$

1) Give, as a function of e, the expression of the flux of the heat losses with heat lagging.

2) Give, as a function of e, the expression of the overall annual operating cost.

3) What is the optimum thickness of the heat lagging?

4) What will be the savings made (in million Dh/year)?

We will represent the four walls of the furnace using a single wall of the same thickness, e, presenting an equivalent lateral area, S.

<u>*Data:*</u>

$T_i = 1{,}250$ °C	$T_a = 26$ °C	$\lambda = 29$ W/m °C	$h = 38$ W/m^2°C
$S = 22$ m^2	$\Delta x = 35$ cm	$\lambda_c = 0.042$ W/m°C	
$\alpha = 650$ Dh/m^2	$\beta = 4.05\ 10^4$ Dh/m^3		

Lifetime of the heat lagging: $D_v = 4.5$ years

Price of electricity: $P_E = 2.2$ Dh/kWh

Annual operating time: $D_F = 7{,}600$ hrs/year

Solutions

<u>*A) Without heat lagging*</u>

1) Calculating the temperature, T_e

The unknown temperature, T_e, can be calculated from the expressions of the flux. Indeed:

$$\varphi = \frac{T_i - T_a}{\frac{\Delta x}{\lambda} + \frac{1}{h}}$$

And: $\varphi = h\ (T_e - T_a)$.

Hence: $T_e = T_a + \frac{T_i - T_a}{\frac{h\ \Delta x}{\lambda} + 1}$.

NUMERICAL APPLICATION.–		
$\Delta x = 0.35$ m	$T_i = 1{,}250$ °C	$T_a = 26$ °C
$\lambda = 29$ W/m °C	$h = 38$ W/m^2°C	
$T_e = 25 + \frac{1250 - 26}{\frac{0.35 * 38}{29} + 1}$		
$\boldsymbol{T_e = 865}$ **°C**		

<u>*2) Calculating the heat losses*</u>

As T_e is now known, we will calculate ϕ from its convective expression, i.e.:

$$\phi = h\ S\ (T_e - T_a)$$

NUMERICAL APPLICATION.–			
$T_a = 26$ °C	$T_e = 865.15$ °C	$h = 38$ W/m^2 °C	$S = 22$ m^2
$\phi = 38\ (22)\ (865.15 - 26)$			
$\boldsymbol{\phi = 701.53}$ **kW**			

3) Annual losses in monetary terms

The annual cost, C, of the losses is given by: $C = P_E D_F \phi$.

NUMERICAL APPLICATION		
$\phi = 701.53$ kW	$P_E = 2.2$ Dh/kWh	$D_F = 7{,}600$ hrs/year
$C = 11.73$ MDh/year		

B) With a heat-lagging thickness, e_c

1) Expressing the flux of the heat losses

With heat lagging, the flux of the heat lost as a function of the thickness, e, of the heat lagging is $\phi_c(e)$, given by:

$$\phi_c(e) = \frac{S(T_i - T_a)}{\frac{\Delta x}{\lambda} + \frac{e}{\lambda_c} + \frac{1}{h}}$$

2) Overall annual cost as a function of e

The function "overall annual cost" is the following: $C_T(e) = C(e) + P_c(e)$, where $C(e)$ is the annual cost of the heat losses with heat lagging, given by: $C(e) = P_E D_F \phi_c(e)$ and $Pc(e)$ is the annual cost of the heat lagging, given by: $Pc(e) = \frac{P_u(e)\ S}{D_v}$.

Yet: $Pu(e) = \varepsilon$.

$$\text{Hence: } C_T(e) = \left[P_E D_F \frac{(T_i - T_a)}{\frac{\Delta x}{\lambda} + \frac{e}{\lambda_c} + \frac{1}{h}} + \frac{(\alpha + \beta e)}{D_v} \right] S.$$

3) Optimum heat-lagging thickness

This thickness, e^*, is obtained for: $\left. \frac{dC_T(e)}{de} \right|_{e=e^*} = 0$.

Yet: $$\frac{dC_T(e)}{de} = S\left[\frac{\beta}{D_v} - \frac{P_E D_F}{\lambda_c}\frac{(T_i - T_a)}{\left(\frac{\Delta x}{\lambda} + \frac{e}{\lambda_c} + \frac{1}{h}\right)^2}\right].$$

e^* is then given by the solution to the following equation:

$$\left(e^* + \frac{\lambda_c}{\lambda}\Delta x + \frac{\lambda_c}{h}\right)^2 = \frac{P_E D_F D_v \lambda_c}{\beta}(T_i - T_a)$$

Hence: $$e^* = \sqrt{\frac{P_E D_F D_v \lambda_c}{\beta}(T_i - T_a)} - \lambda_c\left(\frac{\Delta x}{\lambda} + \frac{1}{h}\right).$$

NUMERICAL APPLICATION.–

T_i = 1,250 °C T_a = 26°C λ = 29 W/m°C λ_c = 0.042 W/m°C

β = 4.05 10^4 Dh/m^3 h = 38 W/m^2°C

Thickness of the lateral walls of the furnace: Δx = 35 cm

Price of electricity: P_E = 2.2 10^{-3} Dh/Wh

Annual operating time: DF = 7,600 hrs/year

Lifetime of the heat insulation: D_v = 4.5 years

$$e_c^* = \sqrt{\frac{(2{,}2\,10^{-3})7600(4{,}5)0.042}{4.05\,10^4}(1250 - 26)} - 0.042\left(\frac{0.35}{29} + \frac{1}{38}\right)$$

e_c^* = 30 cm

4) Savings made

The annual cost of the heat losses with heat lagging is given by:

$$C_c(e^*) = P_E\ D_F\,\phi_c(e^*)$$

I.e.: $$C_c(e^*) = \frac{P_E D_F S\ (T_i - T_a)}{\frac{\Delta x}{\lambda} + \frac{e^*}{\lambda_c} + \frac{1}{h}}.$$

The annual savings made as a result of putting the heat lagging in place are then as follows:

$$E = S\ P_E D_F (T_i - T_a)\left[\frac{1}{\frac{\Delta x}{\lambda}+\frac{1}{h}} - \frac{1}{\frac{\Delta x}{\lambda}+\frac{e^*}{\lambda_c}+\frac{1}{h}}\right]$$

NUMERICAL APPLICATION.–

$T_i = 1{,}250$ °C $\quad T_a = 26$ °C $\quad \Delta x = 35$ cm

$P_E = 2.2\ 10^{-3}$ Dh/Wh $\quad D_F = 7{,}600$ hrs/year $\quad \lambda = 29$ W/m°C

$h = 38$ W/m²°C $\quad S = 22$ m² $\quad \lambda_c = 0.042$W/m °C

$e^* = 30.7$ cm

$$E = 22(0.0022)(7600)(1250-26)\left[\frac{1}{\frac{0.35}{29}+\frac{1}{38}} - \frac{1}{\frac{0.35}{29}+\frac{0.307}{0.042}+\frac{1}{38}}\right]$$

$E = 11.67$ MDh/year

EXERCISE 11. Energy savings on a sugar-plant evaporator

In a sugar plant, a steel cylindrical evaporator of internal radius r_1, external radius r_2 and height H is heat lagged with a thickness, e_1, of glass wool of heat conductivity λ_1.

To minimize the heat losses dissipating through the side walls of this evaporator, you conduct analyses demonstrating the possibility of making significant savings by changing the heat lagging and, above all, by optimizing its thickness. You thus decide to use polyurethane foam as the heat lagging.

Questions

1) Knowing that in this sugar plant, the heat supplied to the evaporator is produced from heavy fuel oil with an efficiency, η, show that the function "annual operating cost" ($C_T(e)$, in Dh/year) can be put in the following form, and give the literal expressions of parameters C_0 to C_6:

$$C_T(e) = C_0 e^2 + C_1 e + \frac{C_2}{C_4 + C_5 \text{Ln}\left(1 + \frac{e}{r_2}\right) + \frac{1}{eh_e + C_6}} + C_3$$

2) In what case does $C_T(e)$ admit an extremum?

3) Was e_1 optimum?

4) You are in charge of convincing the share holders to provide the investment necessary in order to change the heat lagging, by showing them that this operation will generate operating-cost savings.

a) What is the current annual operating cost?

b) What would this cost be after change and optimization of the heat lagging?

c) Calculate the investment needed in order to put the new heat lagging in place.

d) What saving would be made through this investment?

e) What is the necessary operating period for the saving made to cover the investment involved?

We will only take into consideration the heat losses through the side walls and we will consider conduction to be one-dimensional, in the r direction.

Data:

$r_1 = 0.75$ m $r_2 = 0.80$ m $e_1 = 25$ cm $H = 2.5$ m $T_i = 150$ °C

$T_a = 30$ °C $h_i = 30$kcal/hm² °C $h_e = 15$ kcal/hm² °C $\eta = 30\%$

$\lambda_{steel} = 75$ kcal/hm °C $\lambda_1 = 0.12$ kcal/hm °C $\lambda_2 = 0.03$ kcal/hm °C

Calorific value of the fuel oil: $P_c = 930$ kcal/kg

Fuel-oil price: $P_F = 12{,}000$ Dh/T

Operating time: $D_F = 8{,}760$ hrs/year

Cost per m² of heat laggings: $P_1(e) = 650 + 205\ 10^2 e$ $P_2(e) = 1{,}050 + 535\ 10^2 e$

Lifetimes of the laggings: $D_{v1} = 3.5$ years $D_{v2} = 5$ years

Solutions

1) Annual operating cost function

This function is given by: $C_T(e) = P(e) + C(e)$, where $P(e)$ is the annual cost of the heat losses with heat lagging and $C(e)$ is the annual cost of the heat lagging.

– If $\phi(e)$ represents the hourly flux of the heat losses, $P(e)$ is given by:

$$P(e) = \frac{P_F}{P_c} D_F \frac{\phi(e)}{\eta},$$

where: $$\phi(e) = \frac{2\pi H(T_i - T_a)}{\dfrac{1}{r_1 h_i} + \dfrac{1}{\lambda_{steel}} Ln\left(\dfrac{r_2}{r_1}\right) + \dfrac{1}{\lambda_c} Ln\left(1 + \dfrac{e}{r_2}\right) + \dfrac{1}{(e + r_2) h_e}}.$$

– $C(e)$ is given by:

$$C(e) = \frac{P_u(e)\ S}{D_v} = \frac{2\pi H}{D_v}(r_2 + e)(\alpha + \beta e)$$

Hence the "total cost" function:

$$C_T(e) = 2\pi H \left[\frac{P_F D_F (T_i - T_a)}{\eta P_c \left[\dfrac{1}{r_1 h_i} + \dfrac{1}{\lambda_{steel}} Ln\left(\dfrac{r_2}{r_1}\right) + \dfrac{1}{\lambda_C} Ln\left(1 + \dfrac{e}{r_2}\right) + \dfrac{1}{(e + r_2) h_e} \right]} + \frac{1}{D_v}(r_2 + e)(\alpha + \beta e) \right]$$

Or:

$$C_T(e) = \frac{2\pi H P_F D_F (T_i - T_a)}{\eta P_c \left[\dfrac{1}{r_1 h_i} + \dfrac{1}{\lambda_{steel}} Ln\left(\dfrac{r_2}{r_1}\right) + \dfrac{1}{\lambda_C} Ln\left(1 + \dfrac{e}{r_2}\right) + \dfrac{1}{(e + r_2) h_e} \right]} + \frac{2\pi H}{D_v}[(\beta e^2 + (r_2 \beta + \alpha) e + r_2\, \alpha]$$

$C_T(e)$ can be put in the following form:

$$C_T(e) = C_0 e^2 + C_1 e + \frac{C_2}{C_4 + C_5 \mathrm{Ln}\left(1 + \frac{e}{r_2}\right) + \frac{1}{eh_e + C_6}} + C_3,$$

where:

$$C_0 = \frac{2\beta\pi H}{D_v} \qquad C_1 = \frac{2\pi H(r_2\beta + \alpha)}{D_v}$$

$$C_2 = \frac{2\pi H P_F D_F (T_i - T_a)}{\eta\ P_c} \qquad C_3 = \frac{2\alpha\pi r_2 H}{D_v}$$

$$C_4 = \frac{1}{r_1 h_i} + \frac{1}{\lambda_{steel}} \mathrm{Ln}\left(\frac{r_2}{r_1}\right) \qquad C_5 = \frac{1}{\lambda_C}$$

$$C_6 = r_2 h_e$$

2) Extremum

$$C_T(e) = C_0 e^2 + C_1 e + \frac{C_2}{C_4 + C_5 \mathrm{Ln}\left(1 + \frac{e}{r_2}\right) + \frac{1}{eh_e + C_6}} + C_3$$

$C_T(e)$ admits an extremum for $\frac{dC_T(e)}{de} = 0$:

$$\frac{C_T(e)}{de} = 2C_0 e + C_1 - \frac{C_2\left[\frac{C_5}{r_2 + e} - \frac{h_e}{(eh_e + C_6)^2}\right]}{\left[C_4 + C_5 \mathrm{Ln}\left(1 + \frac{e}{r_2}\right) + \frac{1}{eh_e + C_6}\right]^2}$$

Therefore, $C_T(e)$ tolerates an extremum in cases where the following equation admits a solution such that:

$$C_2\left[\frac{C_5}{r_2 + e} - \frac{h_e}{(eh_e + C_6)^2}\right] = (2C_0 e + C_1)\left[C_4 + C_5 \mathrm{Ln}\left(1 + \frac{e}{r_2}\right) + \frac{1}{eh_e + C_6}\right]^2$$

3) Optimality of e_l

In this exercise, we need to determine the optimum thickness, e_l*, for the heat lagging of heat conductivity, λ_l. This thickness optimizes $C_T(e)$.

Let us recall that, amongst the parameters involved in defining the function total annual operating cost, C_0, C_1, C_3 and C_5 depend on the heat lagging used. Indeed:

$$C_0 = \frac{2\beta\pi H}{D_v} \quad C_1 = \frac{2\pi H(r_2\beta + \alpha)}{D_v} \quad C_3 = \frac{2\alpha\pi r_2 H}{D_v} \quad C_5 = \frac{1}{\lambda_C}$$

We will thus conduct a parameterized spreadsheet resolution of the optimality equation, which can be presented in the following form:

$f(e) = g(e)$,

where f and g are given by:

$$f(e) = C_2\left[\frac{C_5}{r_2 + e} - \frac{h_e}{(eh_e + C_6)^2}\right]$$

$$g(e) = (2C_0 e + C_1)\left[C_4 + C_5 \mathrm{Ln}\left(1 + \frac{e}{r_2}\right) + \frac{1}{eh_e + C_6}\right]^2$$

For $\lambda_c = \lambda_l$, putting these equations in a spreadsheet enables us to determine the solution as follows:

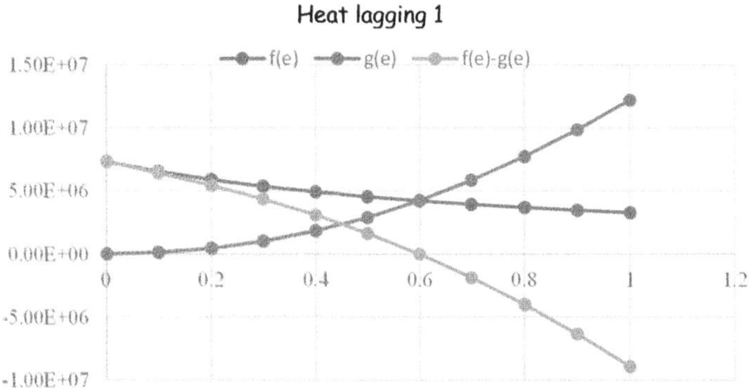

$e^* = 60$ cm

e_l was therefore *not optimal*.

4) Change of heat lagging

a) Annual operating cost in the current configuration

The heat lagging used is *1*, where λ_1 = 0.12 kcal/hm °C and e_1 = 25 cm. The annual operating cost is given in this case by:

$$C_T(e_1) = C_0 e_1^2 + C_1 e_1 + \frac{C_2}{C_4 + C_5 Ln\left(1+\frac{e_1}{r_2}\right) + \frac{1}{e_1 h_e + C_6}} + C_3,$$

where:

$$C_0 = \frac{2\beta_1 \pi H}{D_{v_1}} \qquad C_1 = \frac{2\pi H(r_2\beta_1 + \alpha_1)}{D_{v_1}}$$

$$C_2 = \frac{2\pi H P_F D_F (T_i - T_a)}{\eta \ P_e} \qquad C_3 = \frac{2\alpha_1 \pi r_2 H}{D_{v_1}}$$

$$C_4 = \frac{1}{r_1 h_i} + \frac{1}{\lambda_{steel}} Ln\left(\frac{r_2}{r_1}\right) \qquad C_5 = \frac{1}{\lambda_1} \qquad C_6 = r_2 h_e$$

I.e.: $C_T(e_1)$ = 326,258 Dh/year.

b) Cost after change and optimization of the heat lagging

This cost is given by:

$$C_T(e^*) = C_{0_2}\ e^{*2} + C_{1_2} e^* + \frac{C_2}{C_4 + C_{5_2} Ln\left(1+\frac{e^*}{r_2}\right) + \frac{1}{e^* h_e + C_6}} + C_{3_2},$$

where:

$$C_{0_2} = \frac{2\beta_2 \pi H}{D_{v_2}} \qquad C_{1_2} = \frac{2\pi H(r_2\beta_2 + \alpha_2)}{D_{v_2}}$$

$$C_2 = \frac{2\pi H P_F D_F (T_i - T_a)}{\eta \ P_e} \qquad C_{3_2} = \frac{2\alpha_2 \pi r_2 H}{D_{v_2}}$$

$$C_4 = \frac{1}{r_1 h_i} + \frac{1}{\lambda_{steel}} Ln\left(\frac{r_2}{r_1}\right) \qquad C_{5_2} = \frac{1}{\lambda_2} \qquad C_6 = r_2 h_e$$

Thus, in order to determine $C_T(e^*)$ it is necessary to know e^*. The spreadsheet solution of the optimality equation for $\lambda_c = \lambda_2$ gives $e^* = 27$ cm (see the following figure).

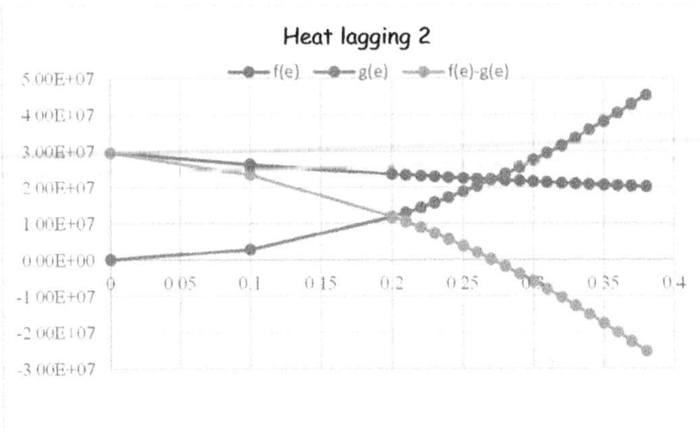

Hence: $C_T(e^*) = 124{,}393$ Dh/year.

c) Investment in order to put the new heat lagging in place

The necessary lagging surface, with the surface equal to e^*, is:

$S(e^*) = 2\pi r_2 + e^*)H$

Hence the installation cost: $I(e^*) = S(e^*)\, P_2(e^*)$.

Or: $I(e^*) = 2\pi\, H\, (r_2 + e^*)\ (\alpha_2 + \beta_2 e^*)$.

I.e.: $I(e^*) = 260{,}432$ Dh/year.

d) Estimated saving for this investment

$E(e^*) = C_T(e_1) - C_T(e^*)$

I.e.: $E(e^*) = 326{,}258 - 124{,}393 = 201{,}864$ Dh/year.

e) Time needed to recover the investment

$$\tau = \frac{I(e^*)}{E(e^*)}$$

I.e.: $\tau = 15.5$ months.

EXERCISE 12. Generation in a plane conductor

A plane conductor of thickness e and heat conductivity λ is found to be the seat of a heat generation with a power density, G, that is unknown but constant.

The faces of the conductor are in contact with a fluid, the temperature of which can be assumed to be constant and equal to T_a.

The heat transfer coefficient between the conductor and the fluid is noted h_a.

Questions

1) Considering that conduction occurs in the x direction only, determine the conduction equation and the associated boundary conditions.

2) Deduce therefrom the temperature profile, $T(x)$, in the interval $\left[-\frac{e}{2}, \frac{e}{2}\right]$, as a function of the wall temperature, T*, G, e and λ.

3) Calculate the flux density at the wall.

4) Give the expression of T^* as a function of G, e and λ.

5) Show that $T(x)$ passes via a maximum and give the expression of T_{Max} as a function of T^*, G, e and λ.

6) Using the information established previously, verify that the following condition is satisfied:

$$T_{Max} > T^* > T_a$$

7) What must the value of h_a be so that $T_{Max} < 650$ °C?

8) Calculate T^* for $h_a = 15$ W/m^2 °C.

Data:

$e = 20$ cm　　$G = 75$ kW/m^3

$\lambda = 25$ W/m °C　　$T_a = 40$ °C

Solutions

1) Equation of conduction and associated boundary conditions

With a single propagation direction and a generation power density, *G*, the equation of conduction is given by:

$$\frac{d^2T}{dx^2} + \frac{G}{\lambda} = 0,$$

with the following boundary conditions:

$$CL_1: \ x = \frac{e}{2} \Rightarrow T = T^*$$

$$CL_2: \ x = -\frac{e}{2} \Rightarrow T = T^*$$

2) Temperature profile

Integrating the equation of conduction gives:

$$T(x) = -\left(\frac{G}{2\lambda}\right) x^2 + \alpha\, x + \beta$$

Using the boundary conditions we obtain:

$$CL_1: \ x = \frac{e}{2} \Rightarrow T^* = -\left(\frac{G}{8\lambda}\right) e^2 + \frac{\alpha}{2} e + \beta$$

$$CL_2: \ x = -\frac{e}{2} \Rightarrow T^* = -\left(\frac{G}{8\lambda}\right) e^2 - \frac{\alpha}{2} e + \beta$$

By subtracting one member after another we arrive at: $0 = \alpha e \Rightarrow \alpha = 0$

The equation giving the temperature profile then becomes:

$$T(x) = -\left(\frac{G}{2\lambda}\right) x^2 + \beta$$

Using boundary condition CL_2, we obtain: $T^* = -\left(\frac{G}{8\lambda}\right) e^2 + \beta$.

Hence: $\beta = T^* + \left(\frac{G}{8\lambda}\right) e^2$.

We therefore have the following temperature profile, as a function of T^*, G, e and λ:

$$T(x) = T^* + \left(\frac{G}{2\lambda}\right)\left[\left(\frac{e}{2}\right)^2 - x^2\right]$$

3) Calculating the flux density at the wall

At the wall, the flux density is given by: $\varphi_{x=\frac{e}{2}} = -\lambda\left(\frac{dT}{dx}\right)_{x=\frac{e}{2}}$.

Yet: $\frac{dT}{dx} = -\frac{G}{\lambda}x$.

Hence: $\varphi_{x=\frac{e}{2}} = -\lambda\left(-\frac{G}{\lambda}\right)\left(\frac{e}{2}\right)$.

I.e.: $\varphi_{x=\frac{e}{2}} = \frac{G}{2}e$.

NUMERICAL APPLICATION.–

$e = 20$ cm $\qquad G = 75$ kW/m^3

$\phi_{x=\frac{e}{2}} = \frac{75}{2}(0,2) \qquad \phi_{x=\frac{e}{2}} = 7.5\ \text{kW}/\text{m}^2$

4) Expressing T as a function of G, e, T_a and h_a*

To determine T^*, we draw on the fact that there is equality of the conductive and convective fluxes at the wall. Hence:

$$-\lambda\left(\frac{dT}{dx}\right)_{x=\frac{e}{2}} = h_a\left(T^* - T_a\right)$$

Or: $-\lambda\left(-\frac{G}{2\lambda}\,e\right) = h_a\left(T^* - T_a\right) \Rightarrow \frac{G}{2}\,e = h_a\left(T^* - T_a\right)$.

I.e.: $T^* = T_a + \frac{G\,e}{2h_a}$.

5) Expressing T_{Max} as a function of T, G, e and λ*

The profile of *T(x)* is parabolic, in other words: $\frac{d^2T}{dx^2} = -\frac{G}{\lambda}$,

where *G* and λ are positive, $\frac{d^2T}{dx^2} < 0$.

T(x) therefore admits a maximum, T_{Max}.

Moreover, $\frac{dT}{dx} = -\frac{G}{\lambda}x \Rightarrow \frac{dT}{dx} = 0$ for $x = 0$.

T_{Max} is therefore obtained for $x = 0$.

Consequently: $T_{Max} = T(0) = T^* + \frac{G\,e^2}{8\lambda}$.

6) Verifying the condition $T_{Max} > T^* > T_a$

$$\forall x \in \left[-\frac{e}{2}, \frac{e}{2}\right] \text{ we have: } T(x) - T^* = \left(\frac{G}{2\lambda}\right)\left[\left(\frac{e}{2}\right)^2 - x^2\right]$$

$$\forall x \in \left[-\frac{e}{2}, \frac{e}{2}\right] : \frac{e}{2} > x \Rightarrow \left(\frac{e}{2}\right)^2 - x^2 > 0$$

Therefore: $\forall x \in \left[-\frac{e}{2}, \frac{e}{2}\right]$: $T(x) - T^* > 0 \Rightarrow T(x) > T^*$.

Yet: $T^* = T_a + \frac{Ge}{2h_a} \Rightarrow T^* > T_a \Rightarrow T(x) > T^* > T_a$.

Moreover: $\forall x \in \left[-\frac{e}{2}, \frac{e}{2}\right]$: $T_{Max} > T(x)$.

I.e.: $\forall x \in \left[-\frac{e}{2}, \frac{e}{2}\right]$: $T_{Maxx} > T^* > T_a$.

7) Value of h_a so that $T_{Max} < 650$ °C

We have: $T_{Max} = T^* + \frac{G\,e^2}{8\lambda}$ and $T^* = T_a + \frac{G\,e}{2h_a}$.

Hence: $T_{Max} = T_a + \frac{G\,e}{2h_a} + \frac{G\,e^2}{8\lambda} \Rightarrow T_a + \frac{G\,e}{2h_a} + \frac{G\,e^2}{8\lambda} > 650$.

Hence: $h_a > \frac{1}{2}\frac{G\,e}{650 - T_a - \frac{G\,e^2}{8\lambda}}$.

NUMERICAL APPLICATION.–

$e = 20$ cm $\quad G = 75$ kW/m^3 $\quad \lambda = 25$ W/m°C $\quad T_a = 40$ °C

$$h_a > \frac{1}{2}\frac{75{,}000\ (0.2)}{650 - 40 - \frac{75{,}000\ (0.2)^2}{8(25)}}$$

$h_a > 12.61$ W/m^2 °C

8) Calculating T^* for $h_a = 15$ W/m^2 °C

$$T^* = T_a + \frac{G\,e}{2h_a}$$

NUMERICAL APPLICATION.–

$e = 20$ cm $\quad G = 75$ kW/m^3 $\quad T_a = 40$ °C $\quad h_a = 15$ W/m^2°C

$$T^* = 40 + \frac{75{,}000\,(0.2)}{2(15)} \qquad T^* = 540\text{ °C}$$

EXERCISE 13. Thermal design of large buildings

Your consulting company is contacted by an architect who is currently involved in the design phase for the first two skyscrapers to be built in Casablanca Finance City. He asks you to perform thermal calculations for the two buildings in the following configurations:

– *Configuration 0:*

- wall construction: double-walled, according to the detailed diagram in Figure 1;

- windows and glazed partitions: 6 mm single glazing;

– *Configuration 1:*

- wall construction: double-walled, according to the detailed diagram in Figure 2;

- windows and bay windows: 6 mm double glazing, with 6 mm air gap;

– *Configuration 2:*

- wall construction: double-walled, according to the detailed diagram in Figure 2;

- windows and bay windows: 6 mm special double glazing, with 12 mm air gap;

– *Configuration 3:*

- wall construction: double-walled, according to the detailed diagram in Figure 2;

- windows and bay windows: 6 mm special double glazing, with 12 mm Argon gap.

Along with the plans of the two towers, the architect also provides you with the deductions and dimensions of the windows and bay windows looking out onto balconies, set out in the following table.

Tower	Frame type	Dimensions	Number
A	Windows	1.20 m x 0.80 m	7,650
	Windows	3.20 m x 0.80 m	3,240
	Bay windows	4.80 m x 2 m	2,860
B	Windows	1.20 m x 0.80 m	7,480
	Windows	3.20 m x 0.80 m	3,370
	Bay windows	4.80 m x 2 m	2,420

He also provides you with the following table listing the lateral faces of the walls in contact with the outside.

Tower	Lateral faces (m^2)
A	215,673
B	223,125

Questions

The architect asks you to prepare a technical file comprising the following elements, for each tower:

1) The total flux of the energy losses via the glazed walls and the side walls, expressed in kWh/year, for the different configurations considered.

2) The total annual costs of these energy losses, expressed in millions of Dh/year, for both towers and for each configuration considered, as well as the savings made with configurations 1, 2 and 3, compared to configuration 0.

3) The investment costs needed for configurations 1, 2 and 3 (we will only present the additional costs of these configurations with respect to the cost of configuration 0).

4) An investment profitability analysis by calculating, for each of the configurations 1, 2 and 3, the *average payback period,* representing the time needed for the saving made by using a given configuration to equal to the additional investment cost:

$$\tau = \frac{I_{Ad}}{E_a}$$

where τ is the average payback period (in years), I_{Ad} is the *additional* investment needed for the configuration considered and E_a is the annual saving made using this configuration;

5) The configuration that you recommend.

Dossier of data and detailed plans communicated by the architect:

Glazing type		U (W/m^2 °C)	Price (Dh/m^2)
Single	6 mm	5.5	1,100
Double	6 mm glass + 6 mm air + 6 mm glass	2.8	2,500
Double	6 mm special glass + 12 mm air + 6 mm special glass	1.6	3,900
Double	6 mm special glass + 12 mm Argon + 6 mm special glass	1.3	4,200

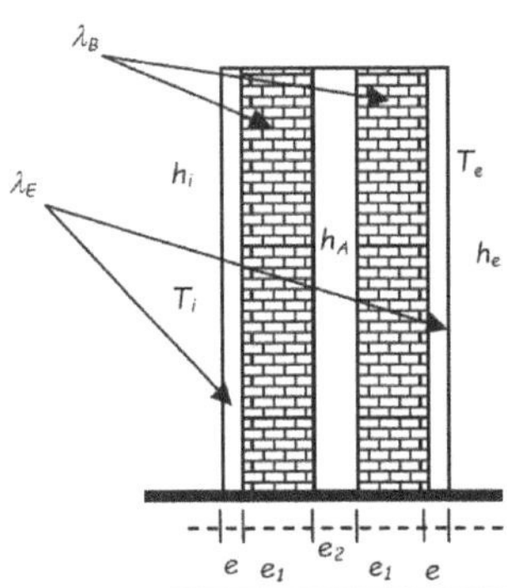

Figure ❶

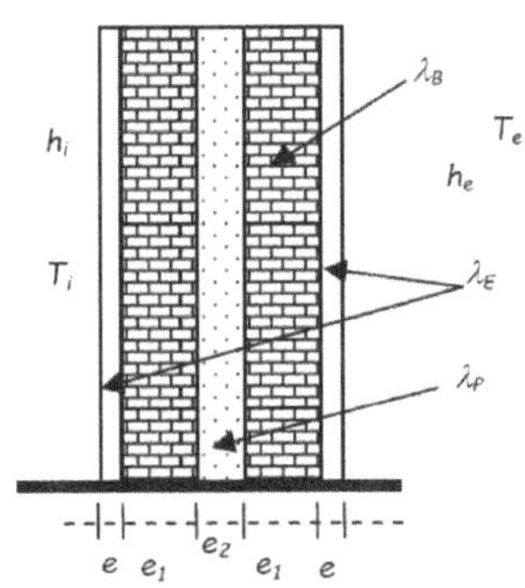

Figure ❷

$e = 20$ mm $\quad e_1 = 600$ mm $\quad e_2 = 650$ mm

$\lambda_B = 1.2$ W/m °C $\quad \lambda_E = 1.6$ W/m°C $\quad \lambda_P = 0.043$ W/m °C

$h_i = 8$ W/m^2 °C $\quad h_e = 8$ W/m^2 °C $\quad h_e = 23$ W/m^2 °C

$h_A > 15$ W/m^2 °C $\quad \Delta T = 15$ °C $\quad D_F = 8{,}760$ hrs/year

Electricity: $P_E = 2.75$ Dh/kWh $\quad$ Polystyrene: 300 Dh/m^3

Solutions

1) Flux of energy losses in the different configurations

For configuration *j*, the total flux of the heat losses is given by:

$$\phi_j = \phi_{M_j} + \phi_{Vj}$$

where ϕ_{M_j} is the flux through the walls and ϕ_{Vj} the flux through the panes.

ϕ_{M_j} is given by $\phi_{Mj} = \dfrac{(T_i - T_e)}{R_{M_j}}$ where R_{M_j} is the heat resistance of the wall in configuration *j*. If S_{Mj} is the surface of the walls in configuration *j*, then R_{M_j} is given by:

Configuration	Details	R_{Mj}
0	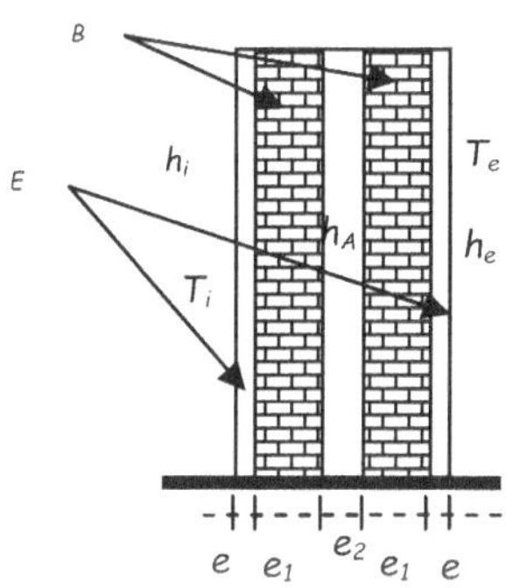	$R_{M_{ij}} = \dfrac{1}{S_{M_j}}\left(\dfrac{1}{h_i} + \dfrac{2e}{\lambda_E} + \dfrac{2e_1}{\lambda_B} + \dfrac{2}{h_A} + \dfrac{1}{h_e}\right)$
1, 2 and 3	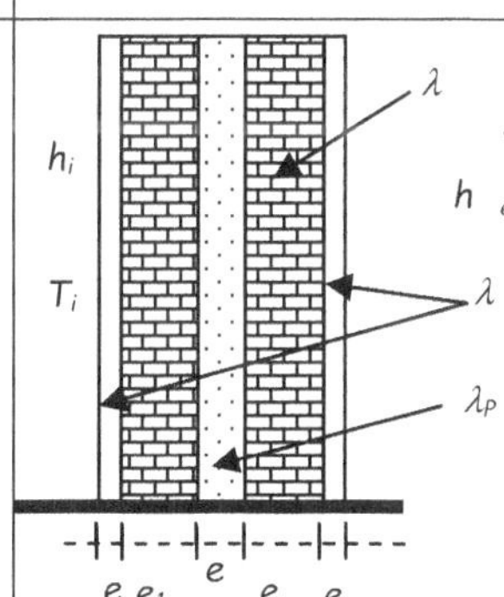	$R_{M_{ij}} = \dfrac{1}{S_{M_j}}\left(\dfrac{1}{h_i} + \dfrac{2e}{\lambda_E} + \dfrac{2e_1}{\lambda_B} + \dfrac{e_2}{\lambda_P} + \dfrac{1}{h_e}\right)$

ϕ_{Vj} is given by:

$$\phi_{Vj} = U_{V_j} S_{Vj} \left(T_i - T_e \right),$$

where S_{Vj} is the surface of the panes for configuration *j*.

The fluxes of the heat losses are indicated in the following table, for each configuration considered:

Configuration	R_{Mj}
0	$\phi_0 = \left[\dfrac{S_{M_0}}{\dfrac{1}{h_i} + \dfrac{2e}{\lambda_E} + \dfrac{2e_1}{\lambda_B} + \dfrac{2}{h_A} + \dfrac{1}{h_e}} + U_{V_0} S_{V0} \right] \left(T_i - T_e \right)$
1	$\phi_1 = \left[\dfrac{S_{M_1}}{\dfrac{1}{h_i} + \dfrac{2e}{\lambda_E} + \dfrac{2e_1}{\lambda_B} + \dfrac{e_2}{\lambda_P} + \dfrac{1}{h_e}} + U_{V_1} S_{V_1} \right] \left(T_i - T_e \right)$
2	$\phi_2 = \left[\dfrac{S_{M_2}}{\dfrac{1}{h_i} + \dfrac{2e}{\lambda_E} + \dfrac{2e_1}{\lambda_B} + \dfrac{e_2}{\lambda_P} + \dfrac{1}{h_e}} + U_{V_2} S_{V_2} \right] \left(T_i - T_e \right)$
3	$\phi_3 = \left[\dfrac{S_{M_3}}{\dfrac{1}{h_i} + \dfrac{2e}{\lambda_E} + \dfrac{2e_1}{\lambda_B} + \dfrac{e_2}{\lambda_P} + \dfrac{1}{h_e}} + U_{V_3} S_{V_3} \right] \left(T_i - T_e \right)$

It should be noted that configuration S_{Vj} is equal to the sum of the surface areas of the windows and bay windows of Tower A, or that of Tower B.

These last two sums are indicated in the table below.

	Dimensions	Number	S
Tower A			
Windows	1.20 m x 0.80 m	7,650	7,344
Windows	3.20 m x 0.80 m	3,240	8,294.4
Bay windows	4.80 m x 2 m	2,860	27,456
		Tower A total	43,094.4
Tower B			
Windows	1.20 m x 0.80 m	7,480	7,180.8
Windows	3.20 m x 0.80 m	3,370	8,627.2
Bay windows	4.80 m x 2 m	2,420	23,232
		Tower B total	39,040

Moreover, U_{V_j} is the global exchange coefficient of the glazing, *j*. The table below presents the U_{V_j} values and the S_{V_j} values to be used for each configuration.

<table>
<tr><th>Configuration</th><th>Glazing type</th><th>Details</th><th>U_{V_j} (W/m^2°C)</th><th>S_{Vj} (m^2)</th></tr>
<tr><td>0</td><td>Single</td><td>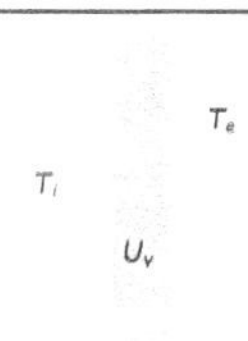
</td><td>5.5</td><td>Tower A: 43,094.4 m^2
Tower B: 39,040 m^2</td></tr>
<tr><td>1</td><td>Double air</td><td>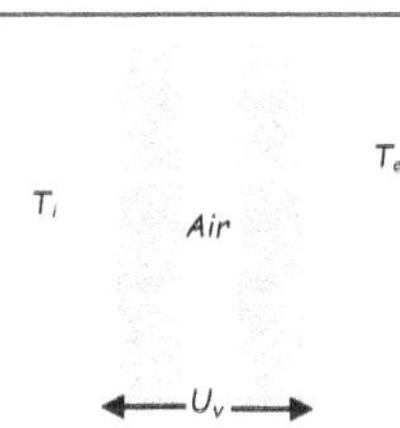
</td><td>2.8</td><td>Tower A: 43,094.4 m^2
Tower B: 39,040 m^2</td></tr>
</table>

<table>
<tr><td>2</td><td>Double special air</td><td>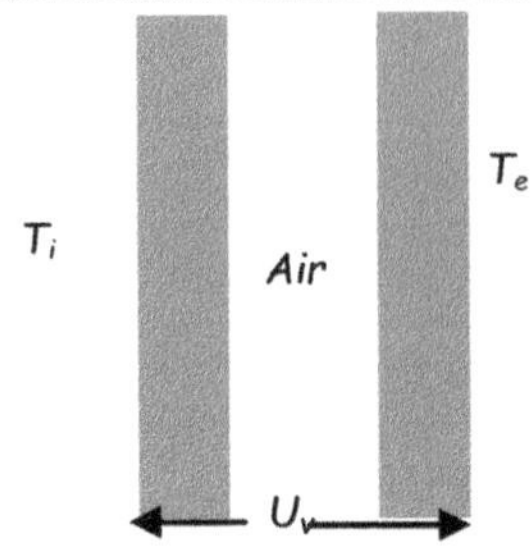</td><td>1.6</td><td>Tower A: 43,094.4 m²
Tower B: 39,040 m²</td></tr>
<tr><td>3</td><td>Double special argon</td><td>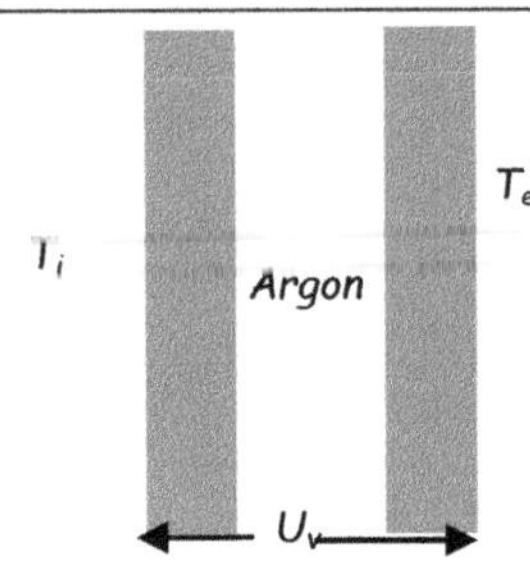</td><td>1.3</td><td>Tower A: 43,094.4 m²
Tower B: 39,040 m²</td></tr>
</table>

Hence the flux values for each tower and each configuration considered:

Tower	Surfaces	Config.	U_{vj} (W/m² °C)	Flux (kWh/year)
A	S_{Mj} (m²)	0	5.5	5.25E+07
	215,673	1	2.8	1.76E+07
	S_{Vj} (m²)	2	1.6	1.08E+07
	43,094.4	3	1.3	9.10E+06
B	S_{Mj} (m²)	0	5.5	5.03E+07
	223,125	1	2.8	1.62E+07
	S_{Vj} (m²)	2	1.6	1.00E+07
	39,040	3	1.3	8.47E+06

2) Total costs of energy losses for the two towers

The total of the costs is given by: $C_T = C_{T1} + C_{T2}$, where C_{T1} and C_{T2} are the costs of the losses of Towers A and B respectively. They correspond to:

$$C_{T1} = \phi_{T1} P_E \text{ and } C_{T2} = \phi_{T2} P_E$$

I.e.: $C_T = (\phi_{T1} + \phi_{T2}) P_E$.

The costs and the savings generated for each configuration are presented in the following table:

Towers	Configuration	Losses (kWh/year)	Costs (MDh/year)	Savings (%)
A and B	0	1.03E+08	283	0
	1	3.38E+07	93	67
	2	2.08E+07	57	80
	3	1.76E+07	48	83

3) *Costs of additional investments four configurations 1, 2 and 3*

If S_i and P_i designate, respectively, the surface and the unit price of the glazing in configuration i, the corresponding investment is given by: $I_i = S_iP_i$.

Hence the additional investment for configuration i: $I_{Ad} = I_i - I_0$.

The following table presents the results obtained for the additional investment of each configuration considered.

Towers	Config.	Glazed surface (m²)	Price (Dh/m²)	Investment (in MDh)	Additional inv. (in MDh)
A and B	0	82,134.4	1,100	90.35	0
	1		2,500	205.34	114.99
	2		3,900	320.32	229.98
	3		4,200	344.96	254.62

4) *Profitability analysis*

The "average payback period" in configuration j is given by: $\tau_j = \dfrac{I_{Ad}^j}{E_a^j}$.

The following table presents the payback periods obtained for each configuration.

Towers	Config.	I_{Ad} (in MDh)	Savings (MDh/year)	Payback period (in months)
A and B	1	114.99	190	7.27
	2	229.98	226	12.24
	3	254.62	234	13.03

5) Recommended configuration

This is the configuration that gives the shortest payback period, τ. We will therefore recommend configuration 1, which corresponds to double partition walls (see detailed diagram in Figure 2), and windows and bay windows with 6 mm double glazing featuring a 6 mm air gap.

Indeed, the investment corresponding to this solution will be recovered in under seven and a half months.

EXERCISE 14. Storing solar energy in zeolite walls

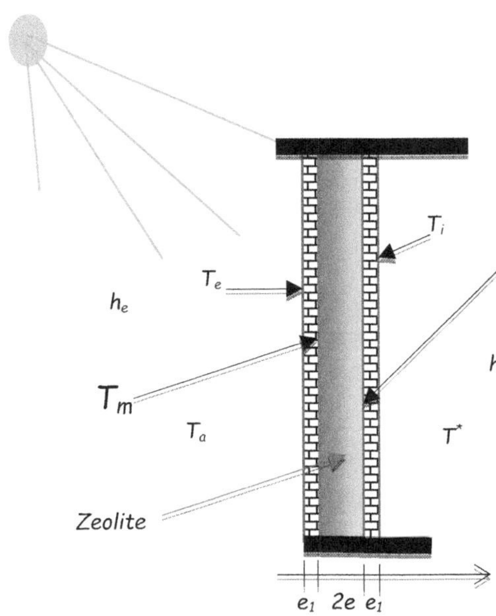

In certain solar-energy applications, the walls receiving the radiation contain a material that evaporates thanks to the solar contribution; this allows the energy supplied by radiation during the day to not cross the wall and favors the evaporation of a solution.

This fulfills a double purpose:

1) protection against the infiltration of energy through the wall during the day;

2) energy storage in the wall for later use.

The energy stored during the day in the form of latent heat of vaporization is released during the night when the outer wall becomes colder. In fact, when the outer partitions of the wall cool down, the zeolite vapors tend to condense, releasing the energy accumulated during the day.

In both phases of the energy storage–release cycle in zeolite, the generation power is given by: $P(x) = P_1\{1 + P_2[T(x) - T_m]\}$.

Since the inside of the wall is almost constantly subject to phase changes, it will be assumed that its two inner faces are at the same constant temperature, T_m.

Questions

1) Find the equation of the temperature distribution in zeolite.

2) Express temperatures T_e and T_i as a function of the outside ambient temperature, T_a, the inside temperature, T^*, and the convection transfer coefficients, h_e and h_i.

Solutions

<u>*1) Temperature profile in zeolite*</u>

In steady state: $\frac{d^2T}{dx^2} + \frac{P}{\lambda} = 0$,

where: $P(x) = P_1\{1 + P_2[T(x) - T_m]\}$.

Hence: $\frac{d^2T}{dx^2} + \frac{P_1P_2}{\lambda}T = \frac{P_1}{\lambda}(P_2T_m - 1)$,

which is of the form: $\frac{d^2T}{dx^2} + \omega^2T = K$,

where: $\omega^2 = \frac{P_1 P_2}{\lambda}$ and $K = \frac{P_1}{\lambda}(P_2 T_m - 1)$.

This equation has the following homogeneous solution:

$$T = u\cos(\omega x) + v\sin(\omega x)$$

And the following specific solution:

$$T_0 = \frac{K}{\omega^2} = T_m - \frac{1}{P_2}$$

Hence the general solution.

$$T = u\cos(\omega x) + v\sin(\omega x) + T_m - \frac{1}{P_2}$$

presenting the following boundary conditions:

$$T(e) = T(-e) = T_m$$

Hence: $v = 0$ and $u = \frac{1}{P_2 \cos(\omega e)}$.

Ultimately: $T - T_m = \frac{1}{P_2}\left(\frac{\cos(\omega x)}{\cos(\omega e)} - 1\right)$.

2) Temperatures T_e and T_i

T_e and T_i are linked to T_m by conduction through the wall: $h_e(T_a - T_e) = \frac{T_e - T_m}{\frac{e_1}{\lambda}}$

Hence: $T_e = \frac{T_a \frac{e_1}{\lambda} + \frac{1}{h_e} T_m}{\left(\frac{e_1}{\lambda} + \frac{1}{h_e}\right)}$

$$h_i\left(T_i - T^*\right) = \frac{T_m - T_i}{\frac{e_1}{\lambda}}$$

$$\text{Hence: } T_i = \frac{T^* \frac{e_1}{\lambda} + \frac{1}{h_i} T_m}{\left(\frac{e_1}{\lambda} + \frac{1}{h_i}\right)}$$

EXERCISE 15. Cooling a nuclear fuel rod

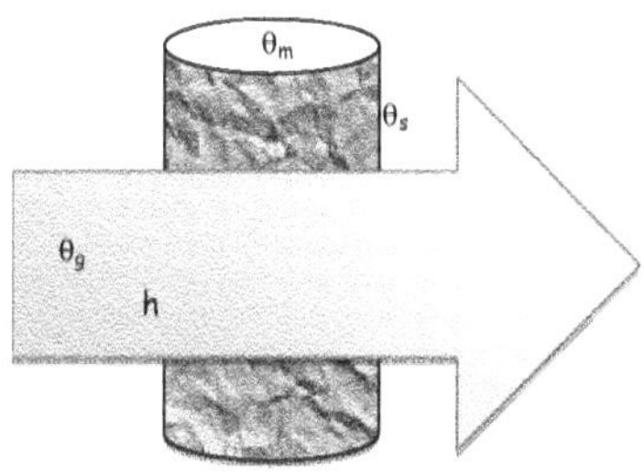

A cylindrical nuclear-fuel rod of radius R is flushed by a gas stream, at temperature θ_g, aimed at cooling it. This rod is the seat of a uniform heat release with power density, P.

We will assume that conduction in the rod is one-dimensional along r and we will use r to designate the convection heat transfer coefficient between the rod surface and the gas.

Questions

1) Express the temperature of the rod surface, θ_s, as a function of θ_g, of P and of h.

2) What is the temperature distribution inside the rod?

3) Calculate θ_s.

4) Calculate the temperature of the θ_m axis.

Data:

$R = 5$ cm $\quad h = 2{,}300$ kcal/h-m^2-degree $\quad P = 3.85\ 106$ kcal/hour-m^3

$\theta_g = 22$ °C $\quad \lambda = 43$ kcal/Hm °C

Solutions

1) Temperature of the rod surface, θ_s

The temperature, θ_S, is calculated from the convective expression of the flux density, i.e.:

$$\phi = h\,S\,(\theta_S - \theta_g),$$

where: $\phi = \pi R^2 L P$ and $S = 2\pi R L$.

Hence: $\theta_s = \theta_g + \frac{RP}{2h}$.

2) Temperature distribution inside the rod

As conduction is one-dimensional, the temperature in the rod is governed by:

$$\frac{d^2\theta}{dr^2} + \frac{1}{r}\frac{d\theta}{dr} + \frac{P}{\lambda} = 0,$$

with the following boundary conditions:

$r = R;\ \theta = \theta_s$

$r = 0$; finite

Through integration, we obtain: $\theta(r) = \theta_s + \frac{P\,R^2}{\lambda}\left[1 - \left(\frac{r}{R}\right)^2\right]$

i.e.: $\theta(r) = \theta_g + \frac{PR}{2h} + \frac{P\,R^2}{\lambda}\left[1 - \left(\frac{r}{R}\right)^2\right]$.

3) Calculating θ_s

$$\theta_s = \theta_g + \frac{RP}{2h}$$

NUMERICAL APPLICATION.–	
$R = 5$ cm	$h = 2{,}300$ kcal/h-m^2-degree
$P = 3.85\ 10^6$ kcal/HOUR-m^3	$\theta_g = 22$ °C
$\theta_s = 64$ °C	

4) Calculating θ_m

For $r = 0$: $\theta_m = \theta(0) = \theta_g + PR\left(\frac{1}{2h} + \frac{R}{\lambda}\right)$.

Note that $\theta(0)$ corresponds to the maximum of $\theta(r)$.

NUMERICAL APPLICATION.–

$R = 5$ cm	$h = 2{,}300$ kcal/h-m^2-degree	
$P = 3.85\ 10^6$ kcal/hour-m^3	$\lambda = 43$ kcal/h-m-degree	$\theta_g = 22°C$

$\boldsymbol{\theta_m} = \mathbf{288\ °C}$

EXERCISE 16. Protecting a nuclear reactor

With a view to protecting against ionizing radiation that can emanate from nuclear reactors in operation, the latter are generally confined (physically isolated from the outside) in enclosed spaces by means of walls that absorb radiation. These walls are often built from reinforced concrete, with lead or a combination of materials that neutralize radiation.

In this exercise, we are interested in the vertical confinement wall of a nuclear reactor. This wall, of thickness L, receives a flux of gamma rays, which generates heat. The power density thus produced is given by the following equation:

$$P = P_0 e^{-\alpha x},$$

where P_0 is the power density generated on the inner face of the wall, i.e. the face exposed to gamma rays, and α is a constant.

Questions

1) Establish an expression describing the temperature distribution in the wall, the inner and outer faces of which are maintained at temperatures T_i and T_e, respectively.

2) Show that the temperature of the vertical confinement wall admits a maximum, T^*.

3) Determine the x*-axis, where this maximum is reached.

4) Give the expression of T^*.

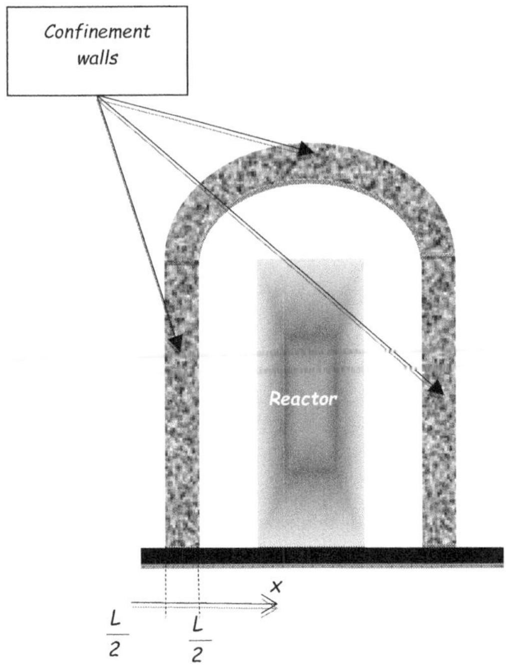

Solutions

1) Temperature distribution

Considering a one-dimensional conduction and knowing that heat conductivity does not vary with temperature, the differential equation governing the transfer will be given by:

$$\frac{d^2T}{dx^2} + \frac{P}{\lambda} = 0,$$

where: $P = P_0 e^{-\alpha x}$, with the following boundary conditions:

$$x = -\frac{L}{2},\ T = T_e$$

$$x = \frac{L}{2},\ T = T_i$$

If we replace P with its expression, the differential equation thus becomes:

$$\frac{d^2T}{dx^2}+\frac{P_0}{\lambda}e^{-\alpha x}=0$$

This equation admits a general solution taking the following form:

$$T(x)=-\frac{P_0}{\lambda\alpha^2}e^{-\alpha x}+K_1x+K_2$$

Using the boundary conditions enables the constants to be determined as follows:

$$T_e=-\frac{P_0}{\lambda\alpha^2}e^{\frac{\alpha L}{2}}-K_1\frac{L}{2}+K_2$$

$$T_i=-\frac{P_0}{\lambda\alpha^2}e^{-\frac{\alpha L}{2}}+K_1\frac{L}{2}+K_2$$

Hence:

$$K_2=\frac{T_e+T_i}{2}+\frac{P_0}{\lambda\alpha^2}\text{ch}\left(\alpha\frac{L}{2}\right)$$

$$K_1=\frac{T_i-T_e}{L}-\frac{P_0}{\lambda\alpha^2L}\text{sh}\left(\alpha\frac{L}{2}\right)$$

Hence the temperature distribution:

$$T(x)=-\frac{P_0}{\lambda\alpha^2}e^{-\alpha x}+\left[\frac{T_i-T_e}{L}-\frac{P_0}{\lambda\alpha^2L}\text{sh}\left(\alpha\frac{L}{2}\right)\right]x+\frac{T_e+T_i}{2}+\frac{P_0}{\lambda\alpha^2}\text{ch}\left(\alpha\frac{L}{2}\right)$$

I.e.: $$T(x)=\frac{P_0}{\lambda\alpha^2}\left[\text{ch}\left(\alpha\frac{L}{2}\right)-\left(\frac{x}{L}\right)\text{sh}\left(\alpha\frac{L}{2}\right)-e^{-\alpha x}\right]+T_i\left(\frac{1}{2}+\frac{x}{L}\right)+T_e\left(\frac{1}{2}-\frac{x}{L}\right).$$

2) Does the temperature admit a maximum?

The temperature admits a maximum if its derivative is canceled out and if its second derivative is negative:

$$\frac{dT}{dx} = 0 \text{ and } \frac{d^2T}{dx^2} = 0$$

To simplify presentation, let us consider this distribution in the following form:

$$T(x) = -\frac{P_0}{\lambda\alpha^2} e^{-\alpha x} + K_1 x + K_2$$

Hence: $\dfrac{dT}{dx} = \dfrac{P_0}{\lambda\alpha} e^{-\alpha x} + K_1$.

The equation $\dfrac{P_0}{\lambda\alpha} e^{-\alpha x} + K_1 = 0$ does indeed admit a solution on *R*.

Moreover, $\dfrac{d^2T}{dx^2} = -\dfrac{P_0}{\lambda} e^{-\alpha x}$ is negative as P_0 and λ are positive.

Thus, *T(x)* admits a maximum.

3) Location of maximum temperature

This is the solution to the equation:

$$\frac{P_0}{\lambda\alpha} e^{-\alpha x} + K_1 = 0 \Rightarrow e^{-\alpha x^*} = -\frac{K_1 \lambda\alpha}{P_0}$$

i.e.: $x^* = -\dfrac{1}{\alpha} \mathrm{Ln}\left[-\dfrac{\alpha\lambda}{P_0}\dfrac{T_i - T_e}{L} + \dfrac{1}{\alpha L}\mathrm{sh}\left(\alpha\dfrac{L}{2}\right)\right]$.

*4) Expression of T**

The maximum temperature is given by:

$$T^* = -\frac{P_0}{\lambda\alpha^2} e^{-\alpha x^*} + K_1 x^* + K_2$$

Yet: $e^{-\alpha x^*} = -\dfrac{K_1 \lambda\alpha}{P_0}$.

Hence: $T^* = K_1\left(x^* + \frac{1}{\alpha}\right) + K_2$.

i.e.:

$$T^* = \frac{1}{\alpha}\left[\frac{T_i - T_e}{L} - \frac{P_0}{\lambda\alpha^2 L}\mathrm{sh}\left(\alpha\frac{L}{2}\right)\right]\left\{1 - \mathrm{Ln}\left[-\frac{\alpha\lambda}{P_0}\frac{T_i - T_e}{L} + \frac{1}{\alpha L}\mathrm{sh}\left(\alpha\frac{L}{2}\right)\right]\right\}$$
$$+\frac{T_e + T_i}{2} + \frac{P_0}{\lambda\alpha^2}\mathrm{ch}\left(\alpha\frac{L}{2}\right)$$

EXERCISE 17. Transporting radioactive sources

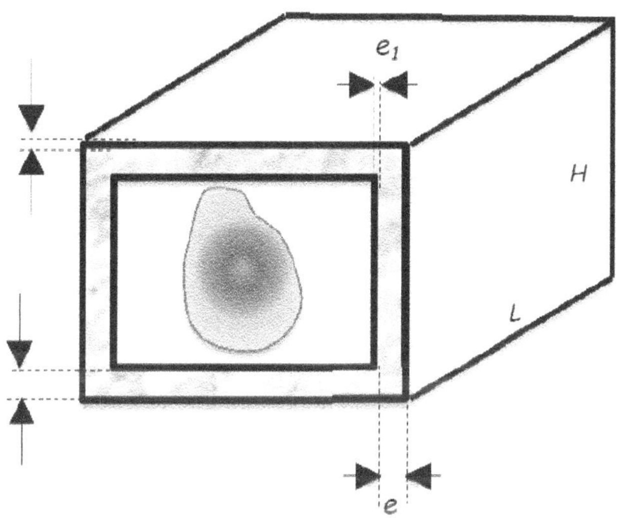

Radioactive sources are usually transported in "lead castles", or protective boxes, made of materials (usually lead) that absorb radioactive radiation and thus protect the environment during transport.

In the present case, we consider a parallelepiped-shaped lead castle, with internal dimensions L x W x H.

Each wall of the castle consists of a thickness, *e*, of lead, protected internally and externally by a thickness, e_1, of stainless steel composed of 25% chromium and 20% nickel.

The source essentially emits a nuclear radiation, which is completely absorbed into the lead layers of the castle walls.

The energy density thus neutralized is transformed into heat according to the following distribution:

$$\mathrm{Ln}(E) = \varepsilon_1 - \varepsilon_2 x,$$

where:

E is the density of the energy transformed into heat on the x-axis, expressed in W/m^3

ε_1 and ε_2 are the two constants given below.

The heat is then conveyed by conduction, then by convection to the outside. The ambient temperature on the outside is equal to T_a, whereas the inner temperature is assumed to be equal to T_i.

To simplify the problem, we will consider that it is equivalent to a conduction through a composite wall of thickness δ, the transfer area, Σ, of which is equal to the sum of the six internal surfaces of the lead castle (see figure below). δ and Σ are such that: $\delta = e + 2e_1$ and $\Sigma = 2LH + 2LW + 2HW$.

We will take the origin of the axes as indicated in the figure below.

Questions

1) Give the equations that govern the conduction in steel, then in lead.

2) Give, for each equation, all of the boundary conditions (on the temperatures and on the fluxes) to be satisfied.

Assuming that all of the temperatures are known, determine the temperature profiles in each thickness of the solid.

3) Does the temperature profile in lead admit a maximum?

4) Calculate, in MW/cm^3, the maximum power of the source admissible in this lead castle, in the knowledge that the internal temperature must not exceed T_{max}.

5) Using the same units, what would the flux be if the same castle was subject to the same temperature difference ($T_{max} - T_a$), but with no radioactive source?

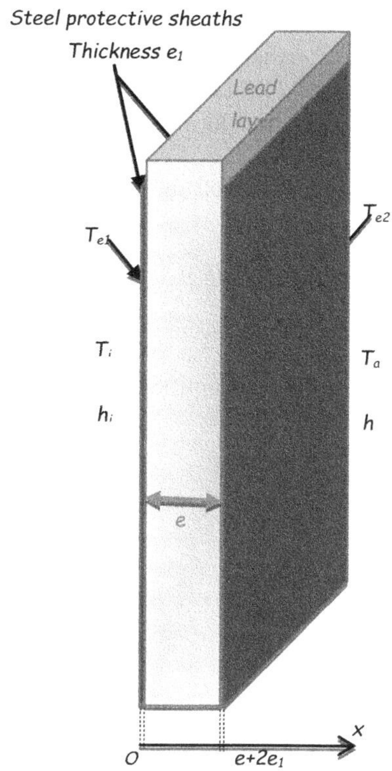

We will assume that:

– the radioactive radiation is absorbed in the same way in all three directions;

– the steel does not block or absorb the radiation;

– the convective transfer coefficient between the walls and the surrounding environment is the same for each of the six walls;

– the walls are in perfect contact.

Data:

$T_{max} = 55$ °C	$T_a = 22$ °C	$L = 10$ cm
$l = 8$ cm	$H = 4$ cm	$e = 6$ cm
$e_1 = 3$ mm	$\varepsilon_1 = 27.5$	$\varepsilon_2 = 5\ 10^{-8}\ m^{-1}$
$h_i = 10.25\ W/m^2\ °C$	$h_e = 18.50\ W/m^2°C$	
$\lambda_{lead} = 35\ W/m\ °C$	$\lambda_{steel\ (25\%Cr,\ 20\%\ Ni)} = 13\ W/m°C$	

Solutions

The absorption of nuclear radiation takes place only in the lead layer, of thickness e. Thus, conduction in this thickness will occur *with generation*, given that the absorbed nuclear radiation will transform into heat in this layer.

Conduction in steel thicknesses e_1, on the other hand, will take place without generation.

1) Conduction equations in lead and in steel

In steel thicknesses e_1, conduction occurs without generation. In addition, the equation of conduction in steel can be simplified as follows: $\frac{d^2T}{dx^2} = 0$.

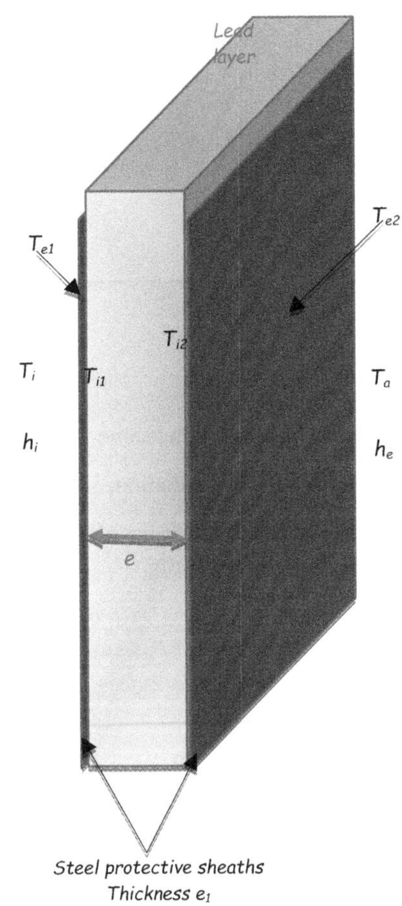

Conduction in lead layer *e*, on the other hand, occurs with generation.

Hence the following equation of conduction, in lead: $\frac{d^2T}{dx^2}+\frac{E}{\lambda}=0$ where: $\mathrm{Ln}(E)=\varepsilon_1-\varepsilon_2 x$ or: $E=e^{\varepsilon_1-\varepsilon_2 x}$.

<u>*2) Boundary conditions*</u>

a) Conditions on temperatures

– For the first steel layer:

$x=0$, $T=T_{e1}$

$x=e_1$, $T=T_{i1}$

– For the second steel layer:

$x=e_1+e$, $T=T_{i2}$

$x=e+2e_1$, $T=T_{e2}$

– For the lead layer:

$x=e_1$, $T=T_{i1}$

$x=e_1+e$, $T=T_{i2}$

b) Conditions on fluxes

– Convections (internal and external):

$\varphi=h_i\,(T_i-T_{e1})$ and $\varphi=h_e\,(T_{e2}-T_a)$

– Conduction in steel:

$$x=0:\ \varphi=h_i\left(T_i-T_{e1}\right)=-\lambda_{\text{Steel}}\frac{T_{i1}-T_{e1}}{e_1}$$

– Conduction in lead:

$$\varphi=-\lambda_{\text{Lead}}\frac{dT}{dx}$$

3) Temperature profiles

a) In steel

Integrating the equation of conduction gives: $T(x) = A\,x + B$.

Thus, using the boundary conditions on the temperatures generates the following profiles:

– For the first layer of steel:

$$x = e_l : T = T_{el} \Rightarrow B = T_{el}$$

$$x = e_l : T = T_{il} \Rightarrow A = \frac{T_{il} - T_{el}}{e_l}$$

The profile in the first layer of steel is linear and given by:

$$0 \le x \le e_l : \quad T(x) = \frac{T_{il} - T_{el}}{e_l} x + T_{el}$$

– For the second layer of steel:

$$x = e_l + e,\ T = T_{i2} \Rightarrow B = T_{i2} A\,(e_l + e)$$

$$x = e + 2e_l,\ T = T_{e2} \Rightarrow A = \frac{T_{e2} - T_{i2}}{2}$$

Hence: $B = T_{i2} - \frac{T_{e2} - T_{i2}}{2}(e_l + e)$

i.e.: $B = T_{i2}\left(1 + \frac{e_l + e}{2}\right) - \frac{T_{e2}}{2}(e_l + e)$.

The profile in the second layer of steel is linear and given by:

$$e + e_l \le x \le e + 2e_l : \quad T(x) = \frac{T_{el} - T_{i2}}{e_l} x + T_{i2}\left(1 + \frac{e_l + e}{2}\right) - \frac{T_{e2}}{2}(e_l + e)$$

b) In lead

$$\frac{d^2T}{dx^2} + \frac{E}{\lambda_{Lead}} = 0$$

with: $E = e^{\varepsilon_1 - \varepsilon_2 x} = E_0\, e^{-\alpha x}$,

where: $E_0 = e^{\varepsilon_1}$ and $\alpha = \varepsilon_2$.

By replacing E with its expression, the differential equation becomes:

$$\frac{d^2T}{dx^2} + \frac{E_0}{\lambda_{Lead}} e^{-\alpha x} = 0$$

This equation admits a general solution of the type:

$$T(x) = -\frac{E_0}{\lambda_{Lead}\alpha^2} e^{-\alpha x} + K_1 x + K_2$$

Using the boundary conditions enables the constants to be determined as follows:

$$x = e_1 : T = T_{i1} \Rightarrow T_{i_1} = -\frac{E_0}{\lambda_{Lead}\alpha^2} e^{-\alpha e_1} + K_1 e_1 + K_2$$

$$x = e_1 + e : T = T_{i2} \Rightarrow T_{i_2} = -\frac{E_0}{\lambda_{Lead}\alpha^2} e^{-\alpha(e_1 + e)} + K_1 (e_1 + e) + K_2$$

Hence:

$$K_1 = \frac{T_{i_2} - T_{i_1}}{e} + \frac{E_0}{e\,\lambda_{Lead}\alpha^2}\left(e^{-\alpha(e_1+e)} - e^{-\alpha e_1}\right)$$

$$K_2 = \left(1 + \frac{e_1}{e}\right) T_{i_1} - T_{i_2}\left(\frac{e_1}{e}\right) + \frac{E_0}{\lambda_{Lead}\alpha^2} e^{-\alpha e_1}\left(\frac{e_1}{e} e^{-\alpha e} - 1\right)$$

Hence the temperature distribution in lead:

$$e_1 \le x \le e + e_1 \ : \ T(x) = T_{i_1}\left(1 + \frac{e_1 - x}{e}\right) + T_{i_2}\left(\frac{x - e_1}{e}\right) - \frac{E_0}{\lambda_{Lead}\alpha^2}\left[e^{-\alpha x} - e^{-\alpha e_1}\left\{\left(\frac{e_1 + x}{e}\right) e^{-\alpha e} - \frac{x}{e} - 1\right\}\right]$$

Does the temperature in lead admit a maximum?

The temperature admits a maximum if its derivative is canceled out and if its second derivative is negative:

$$\frac{dT}{dx} = 0 \text{ and } \frac{d^2T}{dx^2} = 0$$

To simplify the presentation, let us consider this temperature distribution in lead in the following form:

$$T(x) = -\frac{E_0}{\lambda_{Lead}\alpha^2} e^{-\alpha x} + K_1 x + K_2$$

Hence:. $\frac{dT}{dx} = \frac{E_0}{\lambda_{Lead}\alpha} e^{-\alpha x} + K_1$

As K_1<0, the derivative is canceled out for $x^* = \frac{1}{\alpha} \mathrm{Ln}\left(-\frac{E_0}{\alpha K_1 \lambda_{Lead}}\right)$.

Moreover, $\frac{d^2T}{dx^2} = -\frac{E_0}{\lambda_{Lead}} e^{-\alpha x}$ is indeed negative as E_0 and λ_{lead} are positive.

Thus, the temperature distribution in lead *admits a maximum.*

4) Calculating the maximum power admissible in this castle

This power is obtained for $T_i = T_{max}$.

It corresponds to the flux ϕ_{Max}, transfered for $T_i = T_{max}$.

This flux can be expressed as a function of the different resistances.

a) As a function of the internal-convection/steel-conduction resistance

As there is no generation in steel, the flux transferred can therefore be expressed as a function of T_i and T_{max} as follows:

$$\phi_{Max} = \frac{T_{max} - T_{i1}}{\frac{e_1}{\lambda_{Steel}\Sigma} + \frac{1}{h_i\Sigma}}$$

Hence: $$T_{max}-T_{i1}=\phi_{Max}\left[\frac{e_1}{\lambda_{Steel}\Sigma}+\frac{1}{h_i\Sigma}\right].$$

b) As a function of the resistance to conduction in lead

We have generation in lead. The transferred flux is then given by:

$$\phi_{Max}=-\lambda_{Lead}\Sigma\left(\frac{dT}{dx}\right)_{x=e_1}$$

Yet: $$\left(\frac{dT}{dx}\right)_{x=e_1}=\frac{E_0}{\lambda_{Lead}\alpha}e^{-\alpha e_1}+K_1,$$

where: $$K_1=\frac{T_{i_2}-T_{i_1}}{e}+\frac{E_0}{e\,\lambda_{Lead}\alpha^2}\left(e^{-\alpha(e_1+e)}-e^{-\alpha e_1}\right)$$

i.e.: $$\phi_{Max}=-\lambda_{Lead}\Sigma\left[\frac{E_0}{\lambda_{Lead}\alpha}e^{-\alpha e_1}+\frac{T_{i_2}-T_{i_1}}{e}+\frac{E_0}{e\,\lambda_{Lead}\alpha^2}\left(e^{-\alpha(e_1+e)}-e^{-\alpha e_1}\right)\right].$$

Therefore: $$T_{i_1}-T_{i_2}=\frac{e\phi_{Max}}{\lambda_{Lead}\Sigma}+\frac{e\,E_0e^{-\alpha e_1}}{\lambda_{Lead}\alpha}+\frac{E_0e^{-\alpha e_1}}{\lambda_{Lead}\alpha^2}\left(e^{-\alpha e}-1\right).$$

c) As a function of the steel-conduction/external-convection resistance

As there is no generation in this thickness, the flux transferred can therefore be expressed as a function of T_{i2} and T_a as follows:

$$\phi_{Max}=\frac{T_{i2}-T_a}{\dfrac{e_1}{\lambda_{Steel}\Sigma}+\dfrac{1}{h_e\Sigma}}$$

Hence: $$T_{i2}-T_a=\phi_{Max}\left[\frac{e_1}{\lambda_{Steel}\Sigma}+\frac{1}{h_e\Sigma}\right].$$

Thus, we have:

$$T_{max}-T_{i_1}=\phi_{Max}\left[\frac{e_1}{\lambda_{Steel}\Sigma}+\frac{1}{h_i\Sigma}\right]$$

$$T_{i_1} - T_{i_2} = \frac{e\phi_{Max}}{\lambda_{Lead}\Sigma} + \frac{eE_0 e^{-\alpha e_1}}{\lambda_{Lead}\alpha} + \frac{E_0 e^{-\alpha e_1}}{\lambda_{Lead}\alpha^2}\left(e^{-\alpha e} - 1\right)$$

$$T_{i_2} - T_a = \phi_{Max}\left[\frac{e_1}{\lambda_{Steel}\Sigma} + \frac{1}{h_e\Sigma}\right]$$

By adding one member after another we obtain:

$$T_{max} - T_a = \phi_{Max}\left[\frac{e_1}{\lambda_{Steel}\Sigma} + \frac{1}{h_i\Sigma}\right] + \frac{e\phi_{Max}}{\lambda_{Lead}\Sigma} + \frac{eE_0 e^{-\alpha e_1}}{\lambda_{Lead}\alpha} + \frac{E_0 e^{-\alpha e_1}}{\lambda_{Lead}\alpha^2}\left(e^{-\alpha e} - 1\right) + \phi_{Max}\left[\frac{e_1}{\lambda_{Steel}\Sigma} + \frac{1}{h_e\Sigma}\right]$$

Hence:

$$\phi_{Max} = \frac{T_{max} - T_a - \dfrac{E_0 e^{-\alpha e_1}}{\lambda_{lead}\alpha^2}\left[\alpha e + \left(e^{-\alpha e} - 1\right)\right]}{\dfrac{2e_1}{\lambda_{steel}} + \dfrac{e}{\lambda_{lead}} + \dfrac{1}{h_i} + \dfrac{1}{h_e}}\Sigma$$

NUMERICAL APPLICATION.–

$T_{max} = 65$ °C	$T_a = 22$ °C	$\varepsilon_1 = 27.5$	$\varepsilon_2 = 5\ 10^{-8}$ m^{-1}
$L = 0.1$ m	$W = 0.08$ m	$H = 0.04$ m	$e = 0.06$ m
$\lambda_{lead} = 35$ W/m°C	$e_1 = 0.003$ m	$\lambda_{steel\ (25\%\ Cr,\ 20\%\ Ni)} = 13$ W/m°C	
$h_i = 10.25$ W/m^2°C	$h_e = 18.50$ W/m^2°C		
Σ	$\boldsymbol{\phi_{Max}}$ **= 51.9 MW**	**Thus: P_{max} = 1.3 MW/cm^3**	

5) Flux with no radioactive source

With no radioactive source, the flux is simply given by:

$$\phi = \frac{T_{max} - T_a}{\dfrac{2e_1}{\lambda_{Steel}} + \dfrac{e}{\lambda_{Lead}} + \dfrac{1}{h_i} + \dfrac{1}{h_e}}\Sigma$$

NUMERICAL APPLICATION.–			
$T_{max} = 65\ °C$	$T_a = 22\ °C$	$e_1 = 0.003$ m	$e = 0.06$ m
Σ	$\lambda_{Lead} = 35$ W/m°C	$\lambda_{Steel} = 13$ W/m°C	
$h_i = 10.25$ W/m^2 °C	$h_e = 18.50$ W/m^2 °C		
$\phi = 8.50$ W	P = $2.12\ 10^{-7}$ MW/cm^3.		

EXERCISE 18. Conductive/convective balance on a fin

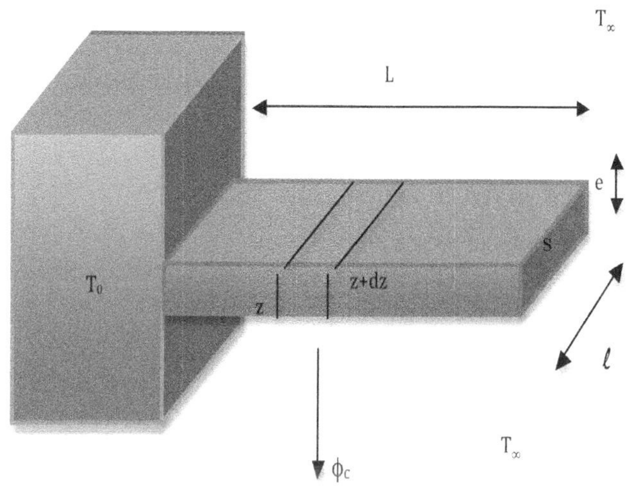

Let us consider the fin represented in the figure above, where we will assume that conduction occurs in the z direction only and a convection heat transfer takes place between the fin and the surrounding fluid. The fin cross-section is noted S and the perimeter of this cross section is noted p.

The convection heat transfer coefficient between the fin and the external environment is h.

We will also assume that: $T_0 > T_\infty$.

Questions

1) Starting from the general expression of the energy balance on the volume element, $S\,dz$, show that, in steady state, if ϕ_z designates the conductive flux passing through the cross-section, S at z, the balance is written:

$$\phi_z - \phi_{z+dz} - d\phi_{conv} = 0,$$

where $d\,\phi_{conv}$ is the flux element transferred by convection by the lateral surface, $L\,dz$.

2) Deduce therefrom that this balance can be written in the following form:

$$-\left(\lambda S\frac{dT}{dz}\right)_z+\left(\lambda S\frac{dT}{dz}\right)_{z+dz}-h\ L\,dz\ \left[T(z)-T_\infty\right]=0$$

3) Show that, for constant values of λ and S, the latter equation can be put in the form: $\dfrac{d^2\theta}{dz^2}-\omega^2\theta=0$.

4) Deduce therefrom the expression of ω.

5) Knowing that the following are reasonable boundary conditions:

– CL_1: $z=0$; $T=T0$;

– CL_2: $z=L$; $T=T_\infty$,

show that the temperature profile is given by:

$$\frac{T-T_\infty}{T_0-T_\infty}=\frac{e^{\omega z}-e^{\omega(2L-z)}}{1-e^{2\omega L}}$$

6) Deduce therefrom that the convective flux, ϕ_{conv}, is given by:

$$\phi_{conv}=\sqrt{\lambda h S p}\,(T_0-T_\infty)e^{2\omega L}$$

Solutions

1) Expressing the balance as a function of ϕ_z and $d\phi_{conv}$

Generally speaking, the balance on the volume element comprised between z and $z+dz$ is written as follows in stationary state:

Input – output + generation = 0

For this volume element, the input is ϕ_z.

Yet there are two outputs, which are ϕ_{z+dz} and $d\,\phi_{conv}$, where $d\phi_{conv}$ is the flux element transferred by convection via the lateral surface, $L\,dz$, with a transfer coefficient, h.

The balance equation is therefore: $\phi_z-\phi_{z+dz}-d\,\phi_{conv}=0$.

2) Differential form of the balance equation

We have: $\phi_z - \phi_{z+dz} - d\,\phi_{conv} = 0$

where:

$$\phi_{z+dz} = -\left(\lambda S \frac{dT}{dz}\right)_{z+dz}$$

$$\phi_{z+dz} = -\left(\lambda S \frac{dT}{dz}\right)_{z+dz}$$

$$d\phi_{conv} = h\ (2\ell + 2e)dz\ \left[T(z) - T_\infty\right]$$

The balance equation thus becomes:

$$-\left(\lambda S \frac{dT}{dz}\right)_z + \left(\lambda S \frac{dT}{dz}\right)_{z+dz} - 2h\ (\ell + e)\ dz\ \left[T(z) - T_\infty\right] = 0$$

3) Shaping of the balance equation

Since λ and S are constant, the balance equation can be written in the following form:

$$\frac{\left(\frac{dT}{dz}\right)_{z+dz} - \left(\frac{dT}{dz}\right)_z}{dz} = \frac{hp}{\lambda S}\left[T(z) - T_\infty\right]$$

i.e.: $\frac{d^2T}{dz^2} - \frac{hp}{\lambda S}(T - T_\infty)$ with the following boundary conditions:

- CL_1: $z=0$; $T=T0$;
- CL_2: $z=L$; $T=T_\infty$.

By changing the variable: $\theta = T - T_\infty$ and by writing: $\omega^2 = \frac{hp}{\lambda S}$, the differential equation becomes: $\frac{d^2\theta}{dz^2} - \omega^2\theta = 0$.

4) Expression of ω

$$\omega = \sqrt{\frac{hp}{\lambda s}}$$

5) Temperature profile

The general solution to this equation is given by: $\theta = Ae^{\omega z} + Be^{-\omega z}$, with the following boundary conditions:

– CL_1: $z=0$; $\theta = T0 - T\infty = \theta 0$;

– CL_2: $z=L$; $\theta = 0$.

Hence the integration-constant values: $A = \frac{\theta_0}{1-e^{2\omega L}}$ and $B = -\frac{\theta_0 e^{2\omega L}}{1-e^{2\omega L}}$.

Ultimately: $\theta = \frac{\theta_0}{1-e^{2\omega L}}\left[e^{\omega z} - e^{\omega(2L-z)}\right]$

or, returning to the initial variable: $\frac{T-T_\infty}{T_0-T_\infty} = \frac{e^{\omega z} - e^{\omega(2L-z)}}{1-e^{2\omega L}}$.

6) Expression of the convective flux

The flux transferred by convection, from the fin to the fluid, can therefore be calculated. We thus note that it is equal to the conductive flux at $z = 0$, i.e.:

$$\phi_{conv} = \phi_{z_{|z=0}} = -\lambda S\left(\frac{dT}{dz}\right)_{|z=0}$$

$$\left(\frac{dT}{dz}\right)_{|z=0} = -\omega e^{2\omega L}(T_0 - T_\infty)$$

Hence: $\phi_{conv} = \lambda S\sqrt{\frac{hp}{\lambda S}}e^{2\omega L}(T_0 - T_\infty)$

i.e.: $\phi_{conv} = \sqrt{\lambda hSp}(T_0 - T_\infty)e^{2\omega L}$.

EXERCISE 19. Efficiency of a fin

The fin represented in the following figure is encased in a plane wall of the same heat conductivity.

The plane-wall/fin assembly is at a uniform temperature, T_0, and is immersed in a fluid, the temperature of which is maintained at T_∞.

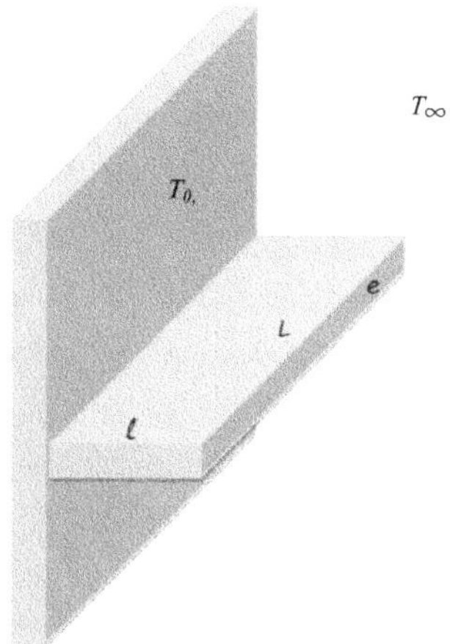

Questions

In stationary state, you are asked to:

1) Calculate the flux transferred by convection between the fin and the fluid;

2) Calculate the gain in transfer efficiency obtained by adding the fin.

Data:

$L = 12$ cm	$e = 1.5$ cm	$W = 8$ m	$T_0 = 78$ °C
$T_\infty = 27$ °C	$\lambda = 53$ W/m °C	$h = 40$ W/m² °C	

Solutions

1) Calculating the flux transferred by the fin

In this case, the flux is given by: $\phi_{conv} = \sqrt{\lambda h S p}(T_0 - T_\infty)e^{2\omega \ell}$, with:

$$\omega^2 = \frac{hp}{\lambda S} = \frac{2h\ (e + L\ )}{\lambda\ e\ L}$$

NUMERICAL APPLICATION.–

$L = 0.12$ m	$e = 0.015$ m	$W = 0.08$ m
$\lambda = 53$ W/m °C	$h = 40$ W/m²°C	
$\omega^2 = 113.21$	$\omega = 10.64$	
ϕ_c **= 284 W**		

2) Gain in transfer efficiency

The gain in efficiency corresponds to the relationship between the flux transferred by convection through the fin and the flux transmitted by convection, with no fin:

$$\eta = \frac{\varphi_c}{\varphi}$$

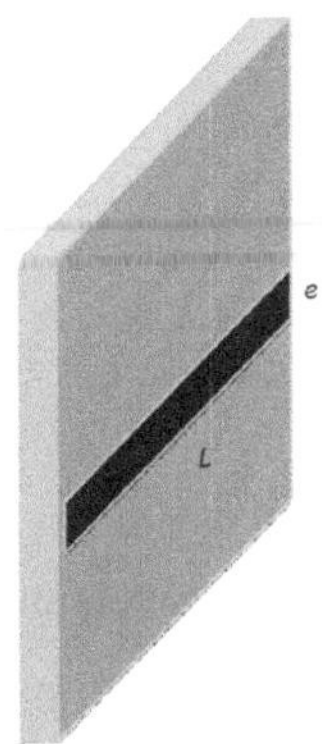

The flux transferred by convection, with no fin, would simply be that which would take place via the same contact surface area occupied by the fin, namely:

$$S = e \times L$$

This flux is therefore given by: $\phi = h\, S\, (T_0 - T_\infty)$

$$\phi = h\, (e \times L)\, (T_0 - T_\infty)$$

Hence: $\eta = \sqrt{\frac{\lambda p}{hS}}\; e^{2\omega l}$.

NUMERICAL APPLICATION.–	
$\phi_c = 284$ W	$\phi = 3.67$ W
ϕ_c is therefore more than seventy-seven times' larger than ϕ.	

EXERCISE 20. Heating a lemonade can

In a café, the waiter has just served you a can of your favorite lemonade. As he puts it on the table, he says:

> "I know you like it really cold! Here you are, it's cooled to 5°C!"

Surprised, you answer:

> "Given how warm the weather is, I'd better drink it up in less than 5 minutes, otherwise it will soon reach 10°C!"

But as he is used to doing, the waiter insists:

> "From experience, I can assure you that it will take well over a quarter of an hour for it to get to 10°C. You're an engineer, aren't you? So, do the maths!"

Question

Given that the ambient temperature on this summer's day is T_a and that under these conditions, the convection heat transfer coefficient between the can and the surrounding environment is equal to h_e, was the waiter right?

In order to provide the waiter with a quick answer, you assume that the temperature of the lemonade, whilst variable over time, is uniform throughout the can, and that the exchange with the surrounding environment occurs in the same way on the lateral surface of the can as on its two upper and lower surfaces. You do decide, however, to take into account the resistance to conduction of aluminum, of which the can is composed.

Data:

$T_a = 35$ °C	$h_e = 23$ W/m^2 °C	
Lemonade:	$C_p = 4.3$ kJ/kg °C	$\rho = 1{,}150$ kg/m^3
Aluminum can:	$R_1 = 3.5$ cm	$R_2 = 3.6$ cm
	$L = 9$ cm	$\lambda = 204$ W/m °C

Solution

Clearly, in this problem, the transfer of heat from the surrounding environment to the can occurs in non-stationary state. Indeed, the can's temperature is variable during this transfer. We will note it $T(t)$.

Thus, an overall energy balance on the lemonade gives:

Input flux = variation in internal energy

The input flux is given by:

$$\text{Input flux} = 2\pi R_2 (L + R_2) \frac{Ta - T(t)}{\frac{R_2 - R_1}{\lambda} Ln\left(\frac{R_2}{R_1}\right) + \frac{1}{h_e}}$$

The variation in the lemonade's internal energy is given by:

$$\text{Variation in internal energy} = \rho V C_p \frac{dT(t)}{dt},$$

where: $V = \pi R_1^2 L$.

Hence the balance equation is:

$$2 \frac{R_2 (L + R_2)}{L R_1^2} \frac{Ta - T(t)}{\frac{R_2 - R_1}{\lambda} Ln\left(\frac{R_2}{R_1}\right) + \frac{1}{h_e}} = \rho C_p \frac{dT(t)}{dt}.$$

We deduce therefrom the following expression of $\frac{dT(t)}{dt}$:

$$\frac{dT(t)}{dt} = \frac{2R_2 (L + R_2)}{\rho L R_1^2 C_p} \frac{Ta - T(t)}{\frac{R_2 - R_1}{\lambda} Ln\left(\frac{R_2}{R_1}\right) + \frac{1}{h_e}}$$

$$\text{Hence: } \frac{d\left[T(t) - Ta\right]}{\left[T(t) - Ta\right]} = -\frac{2R_2 (L + R_2)}{\rho L R_1^2 C_p} \frac{dt}{\frac{R_2 - R_1}{\lambda} Ln\left(\frac{R_2}{R_1}\right) + \frac{1}{h_e}}.$$

Through integration, we obtain:

$$Ln\left[T(t) - Ta\right] = -\frac{2R_2 (L + R_2)}{\rho L R_1^2 C_p \left[\frac{R_2 - R_1}{\lambda} Ln\left(\frac{R_2}{R_1}\right) + \frac{1}{h_e}\right]} t + C,$$

with the boundary condition: $t = 0$; $T = T_0$.

Hence: $C = \text{Ln}\left[T_0 - Ta\right]$.

The temperature-time profile is then given by:

$$\text{Ln}\left[\frac{T(t) - Ta}{T_0 - Ta}\right] = -\frac{2R_2(L + R_2)}{\rho L R_1^2 C_p \left[\frac{R_2 - R_1}{\lambda}\text{Ln}\left(\frac{R_2}{R_1}\right) + \frac{1}{h_e}\right]} t$$

i.e.: $\frac{T(t) - Ta}{T_0 - Ta} = e^{-\tau t}$.

or: $T(t) = T_a + (T_a - T_0)e^{-\tau t}$,

where: $\tau = \frac{2R_2(L + R_2)}{\rho L R_1^2 C_p \left[\frac{R_2 - R_1}{\lambda}\text{Ln}\left(\frac{R_2}{R_1}\right) + \frac{1}{h_e}\right]}$.

Hence the expression of the elapsed time, *t*, to arrive at temperature *T(t)*:

$$t = -\frac{1}{\tau}\text{Ln}\left(\frac{Ta - T(t)}{Ta - T_0}\right)$$

NUMERICAL APPLICATION.–

$T_0 = 5$ °C	$T_a = 35$ °C	$T_f = 10$ °C	$R_1 = 3.5$ cm
$R_2 = 3.6$ cm	$L = 9$ cm	$h_e = 23$ W/m^2 °C	$C_p = 4{,}300$ J/kg°C
$\rho = 1{,}150$ kg/m^3		$\lambda = 204$ W/m °C	

Hence: $\tau = 3.83\ 10^{-4}$ s.

The time needed to arrive at 10 °C is: $t_{10°C} = -\frac{1}{3.83\ 10^{-4}}\text{Ln}\left(\frac{35 - 10}{35 - 5}\right)$

i.e.: $t_{10°C} = 8$ min.

The temperature reached after five minutes:

$$T_{5mn} = 35 + (35 - 5)e^{-(3.83\ 10^{-4})\ 300}$$

i.e.: $T_{5min} = 8$ °C.

Therefore, the waiter was wrong.

EXERCISE 21. Estimating ice fusion time

A large amount of ice urgently needs to be transported to a fishing village located 110 km from Dakhla, in southern Morocco. For the time being, however, there is no refrigerated truck available to perform this transportation. We do nevertheless have a truck with a simple trailer and a type of polyethylene cabinet of thickness, e_l, which is usually used to store ice at the production site. We therefore decide to place the ice in this cabinet, which is then loaded into the truck. Yet without heat lagging, we will have nothing left but water upon arrival! The problem is that there is no heat lagging ready to use on-site; a few planks have simply been assembled, just enough to cover the truck floor in order to minimize heat losses through the bottom of the cabinet.

After some searching, a roll was found, on which there is still 32.4 meters left of cardboard bitumen, of thickness e_2, and of width W_b, left behind by the ice-plant construction company that carried out the roof's waterproofing. Knowing that cardboard bitumen has a relatively low heat conductivity, we decide to apply it to the side walls and to the lid of the plastic cabinet.

Once loaded onto the truck, the cabinet is completely filled with stacked-up bars of ice, leaving no free space. It is placed onto the truck floor, which we take care to cover with planks of wood. Under these conditions, the heat losses from the cabinet to the floor are negligible; in other words, the majority of the losses occur through the lateral surfaces and the ceiling.

Questions

1) How many layers of cardboard bitumen can be applied to the cabinet?

2) Calculate how long it will take for the entirety of the ice to melt if the bitumen is not fixed to the free surfaces of the cabinet.

3) Assuming that the maximum number of layers of cardboard bitumen is applied, calculate how long it will take for the entirety of the ice to melt.

4) Imagining that the maximum number of layers of cardboard bitumen is used, what percentage of unmelted ice would we obtain upon arriving at the fishing village?

<u>*Data:*</u>

– Polyethylene cabinet:

Dimensions: $W = 1.60$ m $L = 3.20$ m $H = 0.80$ m

Wall thickness: $e_l = 3$ mm

Heat conductivity: $\lambda_p = 0.56$ W/m °C

– Cardboard bitumen:

Thickness: $e_2 = 1$ mm $W_b = 80$ cm $L_{available} = 32.4$ m

Heat conductivity: $\lambda_B = 0.23$ W/m °C

– Water:

Latent heat of fusion: $\Lambda = 330$ kJ/kg

$\rho = 1{,}000$ kg/m^3

– $T_a = 45$ °C

– Average truck speed for the journey being considered: $v = 65$ kph

– Convection heat transfer coefficient at this speed: $h_e = 345$ w/m^2 °C

Solutions

1) Number of layers of cardboard bitumen to be applied to the cabinet

The total transfer area is composed of the lateral surface area (S_L) and the lid (ceiling) surface area (S_C). Hence:

$$S = S_L + S_C = 2(W+L)\,H + W\,L,$$

where, for the present case: $W = 2H$.

Hence: $S = 4H(H+L)$

The necessary length for a layer of cardboard bitumen is:

$$L_b = \frac{S}{W_b} = \frac{S}{H} = \frac{4H(H+L)}{H} = 4(H+L)$$

The number of layers is then given by:

$$N = \frac{L_{available}}{L_b} = \frac{L_{available}}{4(H+L)}$$

NUMERICAL APPLICATION.–

$L = 3.20$ m $H = 0.80$ m $L_{available} = 32.4$ m

Hence: $N = \dfrac{32.4}{4(0.8+3.2)}$ $\qquad N = 2.03$

We will only be able to apply *two layers of cardboard bitumen.*

2) *Time taken for the entirety of the ice to melt: scenario with no bitumen*

The ice is going to melt owing to the flux of the frigorie losses during transportation. This flux is given by:

$$\phi_1 = S\frac{T_a - T_G}{\frac{e_1}{\lambda_p} + \frac{1}{h_e}},$$

where:

S is the sum of the lateral and lid surface areas: $S = 2(W+L)H + W\,L$

T_G is the ice fusion temperature = 0 °C

The flux is then constant and equal to: $\phi_1 = S\dfrac{T_a}{\frac{e_1}{\lambda_p} + \frac{1}{h_e}}$.

Q_M is the amount of heat needed to melt a mass, M, of ice. Q_M is given by: $Q_M = M\,\Lambda$.

The time needed to melt a mass, M, of ice is then given by:

$$\tau_1 = \frac{Q_M}{\phi_1} = \frac{\frac{e_1}{\lambda_p} + \frac{1}{h_e}}{ST_a} M\Lambda$$

For the entirety of the ice to melt (i.e. a mass, M = ρWLH), it will thus take a time, τ_1, given by:

$$\tau_1 = \frac{\frac{e_1}{\lambda_p} + \frac{1}{h_e}}{ST_a} \rho WLH\Lambda$$

NUMERICAL APPLICATION.–

$W = 1.6$ m | $L = 3.2$ m | $H = 0.8$ m

$e_1 = 3$ mm | $\lambda_P = 0.56$ W/m °C | $\Lambda = 330$ kJ/kg °C

$\rho = 1{,}000$ kg/m^3 | $T_a = 45$ °C | $h_e = 345$ w/m^2 °C

Hence:

$V = 4.096$ m^3 | $M = 4{,}096$ kg

$Q_M = 1.35\ 10^6$ kJ | $\phi_1 = 69.8$ kW

And: $\tau_1 = 5.38$ hrs

3) Time taken for the entirety of the ice to melt: scenario with bitumen

The flux in this case is given by: $\phi_2 = S\dfrac{T_a}{\dfrac{e_1}{\lambda_p} + \dfrac{Ne_2}{\lambda_B} + \dfrac{1}{h_e}}$ and $\tau_2 = \dfrac{M\ \Lambda}{\phi_2}$.

NUMERICAL APPLICATION.–

$T_a = 45$ °C | $S = 12.8$ m^2 | $e_1 = 3$ mm | $e_2 = 1$ mm

$\lambda_P = 0.56$ W/m °C | $\lambda_B = 0.23$ W/m °C | $h_e = 845$ w/m^2 °C

$N = 2$ | $\Lambda = 330$ kJ/kg | $M = 4{,}096$ kg

Hence: $\boldsymbol{\phi_2}$ **= 34.0 kW** and $\boldsymbol{\tau_2}$ **= 11 hrs**

4) What amount of unmelted ice would we obtain upon arriving at the fishing village?

In this case, the loss flux is the same as that calculated for the previous question: ϕ_2.

The journey time is: $\tau_3 = \dfrac{D}{v}$, where D is the distance between Dakhla and the fishing village and v is the average speed.

During this time, the mass of ice that will have melted will be M_3, such that:

$$\tau_3 = \frac{M_3 \Lambda}{\phi_2}$$

Hence, by equating the two expressions of τ_3: $M_3 = \frac{D\ \phi_2}{v\ \Lambda}$.

The percentage of unmelted ice is then given by:

$$\pi = 100\left(1 - \frac{M_3}{M}\right) = 100\left(1 - \frac{D\ \phi_2}{v\ \Lambda\ M}\right)$$

NUMERICAL APPLICATION.–

$D = 110$ km	$v = 1.81\ 10^{-2}$ km/sec	$\phi_2 = 34$ kW
$\Lambda = 330$ kJ/kg	$M = 4{,}096$ kg	

Hence: $\boldsymbol{\pi} =$ **85 %**

EXERCISE 22. No, no, it's too imprecise, my dear Watson!

Despite having lived in Marrakesh for more than ten years, Sir Peters had fully retained his English habits. In particular, he accorded special importance to his five o'clock tea, which he was accustomed to taking in the riad of his traditional house, located in the very heart of the Medina.

Very meticulous in nature, he always took his tea at the same time of day. "Seventeen zero zero", he would often repeat to his cook, Lalla Fatéma; so much so that the latter came to be a model for punctuality, in what was a relatively relaxed town, thus earning her the nickname, *"Ftitima Lmagana"* ("Ftitima the clock").

Ftitima Lmagana always arrived at work at eight o'clock in the morning and opened the door with the duplicate keys that Sir Peters had entrusted to her so as not to wake him in the morning. Every day, like clockwork, she would leave at 17:05 after serving Sir Peters his tea.

Whatever the weather, be it heatwave or wind, every day at the same time Sir Peters was at his riad: at 17:00 on the dot, he would be sat before the tea tray that had just been brought to him by Ftitima Lmagana, who, every time, her mission accomplished, would proudly announce: "Here you go, sir: seventeen zero zero!" The ritual was implacable, uncompromising, unshakable, no matter what! "Seventeen zero zero": tea served; "seventeen zero five": Ftitima leaves the house.

Sir Peters was just as meticulous in the manner in which he took his tea; he wanted it to always be served in his teapot, at 95 °C. To achieve this, even his teapot was special: it was equipped with a dial thermometer that indicated the average temperature of the tea inside the teapot.

Before pouring his tea, he patiently waited for its temperature to reach *65 °C*. "Above that, just a hot drink, and below that, better to the sink!", he would explain to those unfamiliar with the custom.

That evening, Inspector Holmes and his assistant Watson, having been dispatched from London in a great hurry during the day, were in the riad in the presence of Ftitima Lmagana. And for good reason: Sir Peters had been found dead next to his tea tray, the teapot overturned and the thermometer broken. Early that morning, when Ftitima arrived to conduct her daily tasks, she had discovered him lying on the ground.

For Inspector Holmes and the insurance company that sent him, it was crucial to determine the time at which the crime took place.

The forensic services had already passed on to Inspector Holmes the specialists' findings: "Death must have occurred immediately after the blow received to the head with the teapot. Examination of the remains showed that this no doubt occurred between 18:00 and 01:30".

Annoyed by this conclusion, which seemed too imprecise to him, Inspector Holmes, pacing around the riad, yelled at his assistant: "It's just not good enough, my dear Watson, between 18:00 and 01:30, that's *seven* hours and *thirty* minutes, in that time we would be able to travel from Marrakech to London and back! No, no, it's too imprecise, my dear Watson".

The assistant, who had spent some time examining the broken teapot, looked to his boss, a smile on his face and his eyes shining: "Look, Mr Holmes, the broken thermometer indicates 63 °C. Given that the death occurred at the exact moment the teapot broke, we can therefore estimate the time of the crime more precisely".

Astounded, Inspector Holmes answered: "Wonderful, my dear Watson, but how will you go about determining the time of the crime from a broken thermometer?"

Watson then explained: "Listen up, Inspector, we know that the tea was at 95 °C at 17:00, and as the teapot has small legs that raise it above the tray, the heat exchange with the surrounding environment could therefore only have occurred by convection between the walls of the teapot and the air".

Inspector Holmes then replies: "But of course, but of course, my dear Watson!"

Questions

1) Give the expression of the thermal resistance between the teapot and the surounding environment, overlooking the conduction in the metal constituting the teapot.

2) Give the expression of the thermal resistance between the teapot and the surrounding environment, taking into consideration the conduction in the metal constituting the teapot.

3) Can we overlook conduction in the energy exchanges between the teapot and the external environment?

4) Calculate the time at which the crime took place.

Data:

The temperature of the tea varies as a function of time, but it is uniform in the teapot.

The teapot can be likened to a cylinder of diameter d and height H

The teapot is made of bronze and is covered with a minute layer of silver

The heat conductivity of the teapot wall, of thickness e, is λ

The convection heat transfer coefficient between the teapot and the room is equal to h_e

The surrounding temperature is uniform and equal to θ_a

$\theta_0 = 95$ °C $\quad\quad \theta_a = 28$ °C $\quad\quad \theta_m = 63$ °C

$d = 12$ cm $\quad\quad e = 2$ mm

$\lambda = 55$ W/m °C $\quad\quad h_e = 23$ W/m^2 °C

For the tea: $C_p = 4.8$ kJ/kg °C $\quad\quad \rho = 1{,}110$ kg/m^3

By calculating $\dfrac{\Delta\phi}{\phi}$, we will verify, before any calculation of time, that it is possible to overlook the conductive resistance.

Solutions

1) Thermal resistance, taking conduction into consideration

Let R_1 be the thermal resistance, taking into consideration the conduction in the metal constituting the teapot. R_1 is given by:

$$R_1 = \frac{e}{\lambda A} Ln\left(\frac{r+e}{r}\right) + \frac{1}{h_e A}$$, where A is the transfer area.

2) Thermal resistance, overlooking conduction

Let R_2 be used to designate the thermal resistance, without taking into consideration the conduction in the metal constituting the teapot. R_2 is given by:

$$R_2 = \frac{1}{h_e A},$$

where A is the transfer area.

3) Are we able to overlook conduction?

The fluxes are then given by:

$$\phi_1(t) = \frac{\theta(t) - \theta_a}{\frac{e}{\lambda A} Ln\left(\frac{r+e}{r}\right) + \frac{1}{h_e A}}$$

$$\phi_2(t) = \frac{\theta(t) - \theta_a}{\frac{1}{h_e A}}$$

Hence: . $\frac{\Delta\phi}{\phi} = 1 - \frac{1}{1 + \frac{eh_e}{\lambda} Ln\left(\frac{r+e}{r}\right)}$

NUMERICAL APPLICATION.–			
$r = 0.06$ m	$e = 0.002$ m	$\lambda = 55$ W/m °C	$h_e = 23 W/m^2 °C$
i.e.: $\frac{\Delta\phi}{\phi} = 2.7\ 10^{-5}$.			

Conduction can therefore be overlooked.

4) Calculating the time of the crime

Likening the teapot to a cylinder, we will analyze the heat transfer to the surrounding area.

This transfer occurs in transient state. We will use $\theta(t)$ to refer to *the temperature of the tea at instant* t.

An overall energy balance on the tea gives:

Variation in internal energy = output flux

The variation in the tea's internal energy is given by:

$$\text{Variation in internal energy} = -\rho V C_p \frac{d\theta(t)}{dt},$$

where: $V = \pi r^2 H$.

The output flux is given by: $\phi_O = 2\,\pi\; H\, h_e\,[\theta(t) - \theta_a]$

Hence the balance equation yields: $-\rho\pi r^2 H C_p \dfrac{d\theta(t)}{dt} = 2\pi r H\; h_e\left[\theta(t)-\theta_a\right]$.

We thus deduce the following expression of $\dfrac{d\theta(t)}{dt}$:

$$\frac{d\theta(t)}{dt} = -\frac{2h_e}{\rho C_p r}\left[\theta(t)-\theta_a\right]$$

Hence: $\dfrac{d\left[\theta(t)-\theta_a\right]}{\left[\theta(t)-\theta_a\right]} = -\dfrac{2h_e}{\rho C_p r}dt$.

By writing $\tau = \dfrac{\rho C_p r}{2h_e}$ and integrating, we obtain: $\text{Ln}\left(\dfrac{\theta(t)-\theta_a}{\theta_0-\theta_a}\right) = -\dfrac{t}{\tau}$

i.e.: $\dfrac{\theta(t)-\theta_a}{\theta_0-\theta_a} = e^{-\frac{t}{\tau}}$ or: $\theta(t) = \theta_a - (\theta_a - \theta_0)e^{-\frac{t}{\tau}}$.

Hence the time, t_m, that elapsed between temperature θ_0 and temperature θ_m, where the death occurred:

$$t_m = \tau \; Ln\left(\frac{\theta_0 - \theta_a}{\theta_m - \theta_a}\right)$$

NUMERICAL APPLICATION.–

$\theta_0 = 95$ °C $\theta_a = 28$ °C $\theta_m = 63$ °C $r = 0.06$ m

$h_e = 23$ W/m^2 °C $C_p = 4.8$ J/kg °C $\rho = 1{,}110$ kg/m^3

Hence: $\tau = 6{,}949.57$ secs.

The elapsed time is therefore: $t_m = 4{,}513$ seconds, that is: $\boldsymbol{t_m}$ **= 1.25 hrs**.

The death therefore occurred at 18:15.

EXERCISE 23. Heating a copper plate

A copper plate of thickness e is initially at a homogeneous temperature, θ_i.

At instant $t = 0$, we immerse this plate into a liquid, the temperature of which we maintain at the constant value, θ_b.

Questions

1) After how long will the temperature on the central plane of the plate reach 60 °C?

2) What would the temperature of the outer surfaces of the plate be at this instant?

Data:

$e = 12$ cm $\rho = 8{,}954$ kg/m^3

$\theta_i = 20$ °C $\theta_b = 100$ °C

$\lambda = 386$ W/m °C $C_p = 383$ J/kg°C $h = 1{,}285$ W/m^2 °C

Solutions

1) When will the central temperature reach 80 °C?

To determine the temperature at the center of the plate, we firstly calculate the Biot number and the Fourier number, then we use the Heisler chart, which gives the central temperatures for a plane conductor (see Chapter 6, Figure 6.5).

We have: $\text{Bi} = \dfrac{hs}{\lambda}$, where $s = \dfrac{e}{2}$:

$$s = 0.06 \text{ m} \Rightarrow \quad \text{Bi} = \frac{1285(0.06)}{386}$$

$$Bi = 0.1997 \Rightarrow \quad \frac{1}{\text{Bi}} = 5.0$$

We then use the curve relating to $\dfrac{1}{\text{Bi}} = 5$ in the Heisler diagram (see red curve in the figure below).

The temperature on the central plane (θ_0) is known, therefore we can calculate the ratio $\dfrac{T_0}{T_i}$.

We have: $T_0 = \theta_0 - \theta_b$ and $T_i = \theta_i - \theta_b$.

Hence: $\dfrac{T_0}{T_i} = \dfrac{\theta_0 - \theta_b}{\theta_i - \theta_b}$.

$$\theta_0 = 60 \text{ °C}; \ \theta_i = 20\text{°C and } \theta_b = 100 \text{ °C} \Rightarrow \ \frac{T_0}{T_i} = 0.5$$

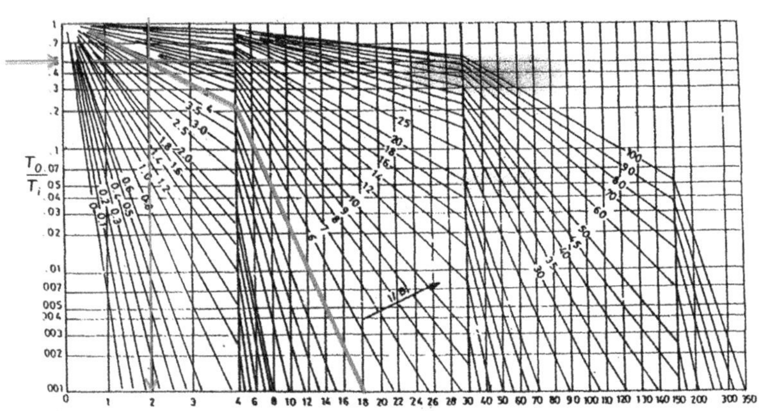

Central temperatures as a function of Fo *and of* $\frac{1}{Bi}$ *(plane conductor).*

For a color version of this figure, see www.iste.co.uk/benallou/energy2.zip

We then place 0.5 on the y-axis, then we determine the corresponding *Fo* number.

We obtain: $Fo = 2$.

The expression of *Fo* ($\mathrm{Fo} = \frac{\alpha t}{s^2}$) allows us to retrieve *t*.

Hence the time: $t = \frac{s^2 \mathrm{Fo}}{\alpha}$ with: $\alpha = \frac{\lambda}{\rho C_p}$.

$Fo = 2 \Rightarrow t = 64$ secs

2) Temperature of the outer surfaces of the plate

To determine this temperature, we use the Heisler chart, which reveals the temperature distribution in a plane conductor as a function of the central temperature (Chapter 6, Figure 6.8).

Indeed, this diagram, which is scaled according to $\frac{x}{s}$, gives the ratio of the reduced temperature, $\frac{T}{T_0}$, as a function of $\frac{1}{Bi}$.

For the outer surfaces, we have: $x = s \Rightarrow \frac{x}{s} = 1$.

We will therefore use the curve corresponding to $\frac{x}{s} = 1$: the red curve.

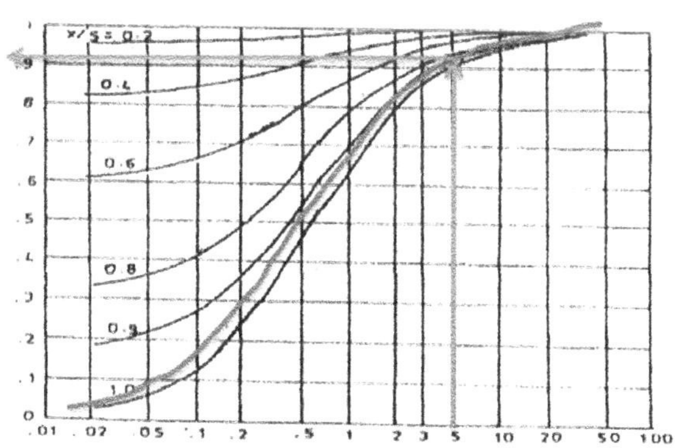

Temperature distribution in a plane conductor. For a color version of this figure, see www.iste.co.uk/benallou/energy2.zip

By plotting the vertical line passing through $\frac{1}{Bi} = 5$, we obtain: $\frac{T(s)}{T_0} = 0.9$.

Hence: $T(s) = 0.9\ T_0$.

or: $\theta_s - \theta_b = 0.9(\theta_0 - \theta_b)$.

i.e.: $\theta_s = \theta_b + 0.9(\theta_0 - \theta_b)$.

NUMERICAL APPLICATION.–

$\theta_0 = 60$ °C $\theta_b = 100$ °C

$\theta_s = 64$ °C

EXERCISE 24. Cooling a steel cable

During its manufacture, a steel cable of radius s leaves the production apparatus at speed v. Before being rolled up, it needs to be cooled in order to pass from its production temperature, θ_i, to an acceptable temperature for the rewinder: θ_e.

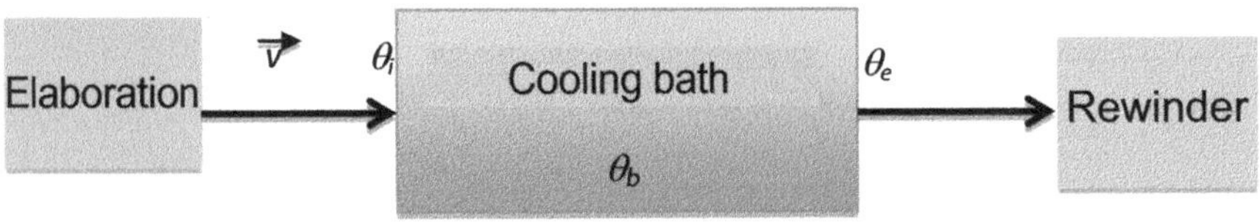

To do this, as it travels towards the rewinder, the cable is immersed in a bath at constant, uniform temperature: θ_b.

Question

How long does this bath need to take, if we consider that the cable production process, cooling included, needs to be continuous?

Data:

$v = 1.46\ 10^{-3}$ m/sec

$\theta_i = 250$ °C

$\theta_e = 40$ °C

$\theta_b = 35$ °C

For the steel considered:

Heat conductivity: $\lambda = 60$ kcal/h.m.°C

Heat capacity: $C_p = 0.5$ kJ/kg°C

Density: $\rho = 7{,}800$ kg/m^3

Convection heat transfer coefficient between the cable and the bath: $h = 250$ W/m² °C

Cylinder radius: $s = 0.12$ m

Solution

We calculate the Biot number, $\mathrm{Bi} = \dfrac{hs}{\lambda}$.

NUMERICAL APPLICATION.–		
$h = 250$ W/m² °C	$s = 0.12$ m	$\lambda = 60$ kcal/h.m °C
$\lambda = 60 \text{ x } 1.1611 = 69.67$ W/m °C		
i.e.: $Bi = 1.998$ and $\dfrac{1}{\mathrm{Bi}} = 0.5$		

We then use the curve relating to $\dfrac{1}{\mathrm{Bi}} = 0.5$ in the Heisler diagram (see red curve in the figure below).

Furthermore, for the cable cooling, we will consider the temperature of its center as the objective, given that when the temperature at the center is at θ_e, the entire cable will be at a temperature that is $\leq \theta_e$, therefore acceptable for the rewinder. We then calculate the ratio of the reduced temperatures as follows:

$$\frac{T_{e0}}{T_i} = \frac{\theta_e - \theta_b}{\theta_i - \theta_b} = \frac{40-35}{250-35} = 0.023$$

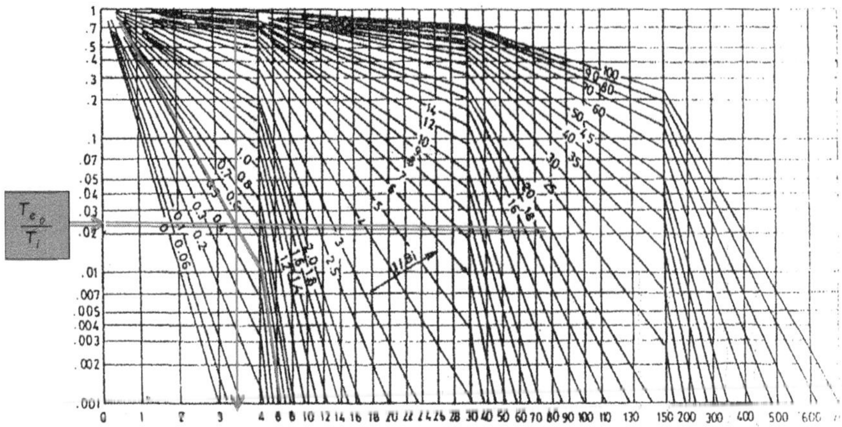

Central temperatures as a function of Fo *and of* $\frac{1}{Bi}$ *(cylindrical conductor).*

For a color version of this figure, see www.iste.co.uk/benallou/energy2.zip

For $\frac{T_{f0}}{T_i} = 0.023$ (see intersection with green horizontal line), we find: $Fo = 3.4$.

The expression of Fo ($Fo = \frac{\alpha t}{s^2}$) allows us to retrieve t.

Hence the time: $t = \frac{s^2 Fo}{\alpha}$,

where: $\alpha = \frac{\lambda}{\rho C_p}$.

$\alpha = 1.79\ 10^{-5}$ m²/sec and $Fo = 3.4$ ⇨ $t = 2{,}740$ secs

$v = 1.46\ 10^{-3}$ m/sec and $w = v\,t$ ⇨ $w = 4$ m

EXERCISE 25. Cooling a sphere in free fall

A tin sphere is dropped in free fall from a pelletizing furnace at a uniform temperature, θ_s. It must undergo tempering in a bath located at a distance, H, below the furnace (see diagram below).

Questions

1) During free fall, which modes of heat transfer occur? (tick all applicable boxes):

natural convection forced convection

conduction radiation

2) Using a straightforward calculation, show which mode is predominant.

3) Taking into consideration the predominant heat transfer mode, establish an energy balance that will govern the temperature variation of the sphere between the instant when it leaves the furnace and its entry into the bath. We will assume that the predominant flux will remain constant during the fall and that the temperature of the sphere will be uniform.

4) Deduce therefrom the temperature, θ_i, of the sphere as it enters the cooling bath. Let us suppose that the free fall lasts a time interval, Δt, which we will calculate.

5) As the bath is considered to be at uniform, constant temperature, calculate the temperature at the center of the sphere, five minutes, ten minutes, fifteen minutes and thirty minutes after entering tempering.

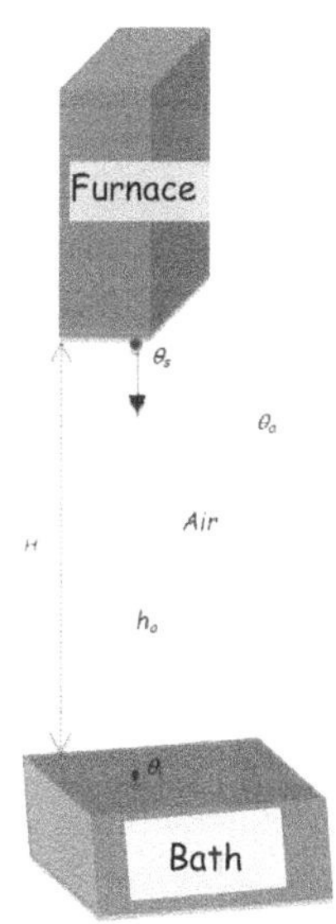

6) Calculate, at the same instants, the temperature at the half-radius of the sphere.

Data:

$H = 15$ m

Sphere temperature: $\theta_s = 750$ °C

Air temperature: $\theta_a = 30$ °C

Bath temperature: $\theta_b = 120$ °C

Tin sphere:

Gray body: $\varepsilon = 0.75$

Density: $\rho = 7{,}304$ kg/m^3

Heat capacity: $C_p = 226$ J/kg °C

Radius: $R = 10$ cm

Heat conductivity: $\lambda = 64$ W/m °C

Initial uniform temperature: $\theta_{i0} = 500$ °C

Convection heat transfer coefficient between the air and the sphere:

$h_a = 12$ W/m² °C

Convection heat transfer coefficient between the bath and the sphere:

$h_b = 250$ W/m² °C

Stefan-Boltzmann constant: $\sigma = 5.67\ 10^{-8}$ W/m² °K^{4}g = 9.81 m/s^2

Solutions

1) Modes of heat transfer

☐ natural convection ☒ forced convection

☐ conduction ☒ radiation

2) Predominant mode of transfer

Compare the flux densities at furnace output:

$$\varphi_{conv} = h_a\,(\theta_s - \theta_a)$$

$$\varphi_{rad} = \sigma(T_s^{\,4} - T_a^{\,4})$$

NUMERICAL APPLICATION.–

$\theta_i = 750$ °C	$\theta_a = 30$ °C
$h_a = 12$ W/m² °C	$\sigma = 5.67\ 10^{-8}$ W/m² °K^4
$\varphi_{conv} = 8{,}640$ W/m²	$\varphi_{rad} = 61{,}621$ W/m²
$\boldsymbol{\varphi_{rad} \sim 7\ \varphi_{conv}}$	Radiation is then predominant.

3) Energy balance on the sphere between t = 0 and entry into the bath

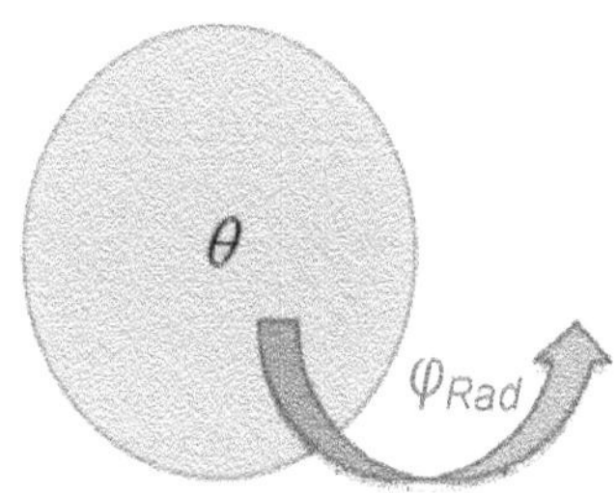

The temperature of the sphere being assumed to be uniform and the flux constant, the balance on the sphere is written:

$$\phi_{rad} S \Delta t = \rho V C_p \Delta\theta,$$

where Δt is the time taken by the sphere to reach the furnace.

Hence: $\phi_{rad}\, 4\pi R^2 \Delta t = \frac{4}{3}\pi R^3 \rho C_p \left(\theta_s - \theta_i\right)$.

i.e.: $3\phi_{rad} \Delta t = R\rho C_p \left(\theta_s - \theta_i\right)$.

4) Temperature of the sphere as it enters the cooling bath

$$\theta_i = \theta_s - \frac{3\phi_{rad}}{R\rho C_p}\Delta t,$$

where: $H = \frac{1}{2} g (\Delta t)^2$ or: $\Delta t = \sqrt{\frac{2H}{g}}$.

i.e.: $\theta_i = \theta_s - \frac{3\varphi_{Ray}}{R\rho\ C_p}\sqrt{\frac{2H}{g}}$.

NUMERICAL APPLICATION.–	
$R = 0.1$ m	$\rho = 7{,}304$ kg/m^3
$C_p = 226$ J/kg °C	$g = 9.81$ m/s^2
$\theta_s = 750$ °C	$\theta_a = 30$ °C
Φ_{rad} 61,621 W/m^2	$H = 15$ m
θ_i = 748 °C	

5) Temperature at the center of the sphere at different instants

The Biot number is given by: $Bi = \frac{h_b R}{\lambda}$.

In the present case: $h_b = 250$ W/m² °C, $R = 0.1$ m, $\lambda = 64$ W/m °C.

Hence: $Bi = 0.39 \Rightarrow \frac{1}{Bi} = 2.56$.

We will therefore use the Heisler diagram for the spherical conductors (see Chapter 6, Figure 6.7). Note that the curves for $\frac{1}{Bi} = 2.4$ and $\frac{1}{Bi} = 2.6$ are so close (see figure below) that interpolation will not be necessary: we will read directly between the two curves.

Moreover, given that we are interested in the central temperature at different instants, t, we will express the Fourier number as a function of t:

$$Fo = \frac{\alpha}{R^2} t, \text{ with } \alpha = \frac{\lambda}{\rho C_p}$$

$\alpha = 3.88\ 10^{-5}$ m^2/sec and $R = 0.1$ m

Hence: $Fo = 3.88\ 10^{-3}$t.

Thus, for each value of *t*, we calculate the Fourier number, which is added to the diagram to deduce therefrom the corresponding value of $\frac{T_0}{T_i}$ (see table and figure below).

t (seconds)	60	600	900	1,800
Fo	0.23	2.33	3.49	6.98
$\frac{T_0}{T_i}$	0.85	0.1	0.04	0.025

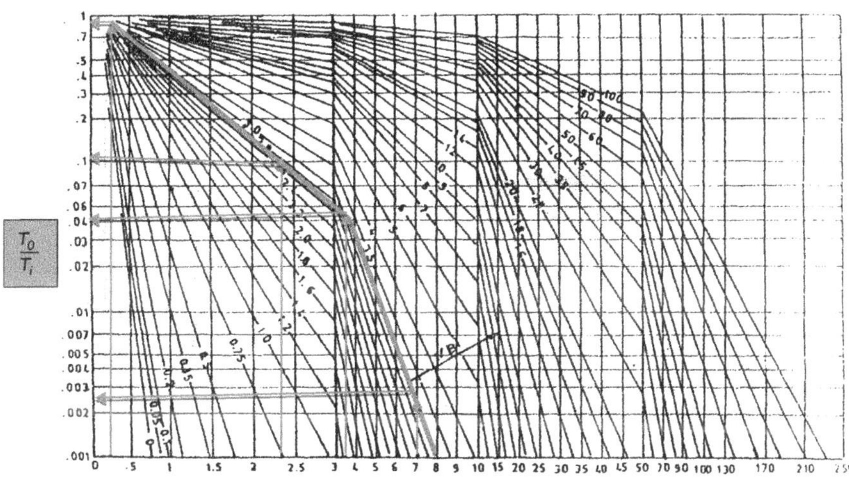

The central temperatures as a function of Fo and of $\frac{1}{Bi}$ (spherical conductor).

For a color version of this figure, see www.iste.co.uk/benallou/energy2.zip

Hence the temperatures at the center at different instants:

$$\theta_0(t) = \theta_b + (\theta_i - \theta_b)\left(\frac{T_0}{T_i}\right)(t)$$

t (minutes)	1	10	15	30
$\frac{T_0}{T_i}$	0.85	0.1	0.04	0.025
$\theta_0(t)$	654 °C	183 °C	145 °C	136 °C

6) Temperatures at half-radius

These temperatures are deduced from the values of $T_0\,(t)$ using the Heisler chart, which gives the temperature distribution in a spherical conductor as a function of the central temperature (see Chapter 6, Figure 6.10).

Given that there is no curve corresponding to $\frac{r}{R} = 0.5$, we will conduct the calculations for the two curves, $\frac{r}{R} = 0.4$ and $\frac{r}{R} = 0.6$, then we will interpolate.

We then plot the vertical line passing through $\frac{1}{Bi} = 2.56$, and we obtain:

$$\frac{T_{0.5R}}{T_0} = 0.91 \text{ and } \frac{T_{0.5R}}{T_0} = 0.96$$

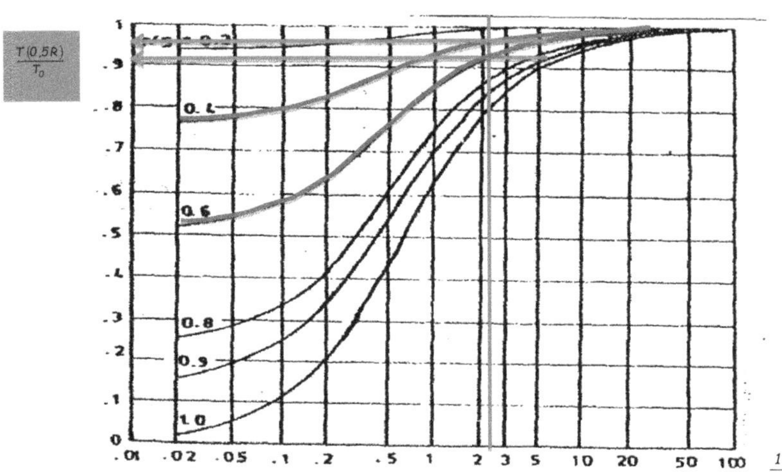

Temperature distribution in a spherical conductor. For a color version of this figure, see www.iste.co.uk/benallou/energy2.zip

Hence: $\frac{T_{0.5R}}{T_0} = 0.91 + \frac{0.96-0.91}{0.2} * 0.1 = 0.935$.

i.e.: $T_{0.5R}(t) = 0.935 T_0(t)$ or: $\theta_{0.5R}(t) = \theta_b + 0.935\left[\theta_0(t) - \theta_b\right]$

t (minutes)	1	10	15	30
$\theta_0(t)$	654 °C	183 °C	145 °C	136 °C
$\theta_{0,5R}(t)$	619 °C	179 °C	143 °C	135 °C

EXERCISE 26. Tempering a cupronickel sphere

As it exits the production furnace, a cupronickel sphere (70%Cu, 30%Ni) is at uniform initial temperature θ_i. Its temperature must be brought back to θ_0 precisely. Yet we only have one bath at uniform, constant temperature, θ_b. Nevertheless, we decide to use this bath, even if θ_b is too close to θ_0.

The sphere is then tempered suddenly inside the bath.

Question

Calculate the time it takes for the temperature at the center of the sphere to reach θ_0.

<u>*Data:*</u>

Sphere radius: $s = 0.1$ m

$\theta_i = 160$ °C $\theta_b = 80$ °C $\theta_0 = 80.15$ °C

Chrome: $\lambda = 29.3$ W/m °C $\rho = 8{,}900$ kg/m^3 $C_p = 377$ J/kg °C

Sphere/bath convection heat transfer coefficient: $h = 500$ W/m^2 °C

Solution

We calculate the Biot number: $\text{Bi} = \frac{hs}{\lambda}$.

NUMERICAL APPLICATION.–

$h = 500$ W/m² °C $s = 0.1$ m $\lambda = 29.3$ W/m °C

i.e.: $Bi = 1.71$ and $\frac{1}{\text{Bi}} = 0.59$.

In the Heisler diagram for spherical conductors (Chapter 6, Figure 6.7), there is no curve relative to $\frac{1}{\mathrm{Bi}}=0.59$ or to a near enough value such as, for example, $\frac{1}{\mathrm{Bi}}=0.6$. The curves closest to the desired value are those corresponding to $\frac{1}{\mathrm{Bi}}=0.5$ and to $\frac{1}{\mathrm{Bi}}=0.75$, respectively (see curves in red in the figure that follows). As these two curves are fairly distant from one another, we will perform the calculation for the two situations and we will interpolate the results.

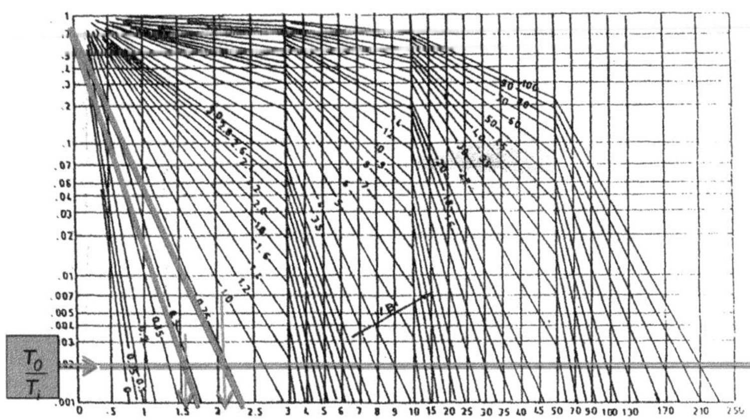

The central temperatures as a function of Fo and of $\frac{1}{Bi}$ (spherical conductor).

For a color version of this figure, see www.iste.co.uk/benallou/energy2.zip

We have: $\frac{T_0}{T_i}=\frac{\theta_0-\theta_b}{\theta_i-\theta_b}=\frac{80.15-80}{160-80}=0.002$.

For $\frac{T_0}{T_i}=0.002$ (see intersections with the green horizontal line), we find the following two values of Fo:

$1/Bi$	Fo
0.5	1.6
0.75	2.1

The expression of *Fo* ($Fo = \frac{\alpha t}{s^2}$) allows us to retrieve *t*:

$$t = \frac{s^2 Fo}{\alpha}, \text{ where } \alpha = \frac{\lambda}{\rho C_p}$$

$$\alpha = 8.73\ 10^{-6}\ \text{m}^2/\text{sec}$$

Hence *t* for each value of *Fo*:

$1/Bi$	Fo	t (seconds)
0.5	1.6	1,832
0.75	2.1	2,405

Hence, after interpolation: $t = 1832 + (0.59 - 0.5)\frac{2405 - 1832}{0.75 - 0.5}$.

i.e.: $t = 2{,}038$ seconds.

EXERCISE 27. Tempering a jewel

Upon exiting a treatment furnace, a cylindrical metal part, of radius R, must be tempered in a bath, the temperature of which, θ_b, is uniform and constant. This part is composed entirely of gold. It is initially at a uniform temperature, θ_i.

Questions

A) Part one: estimations using the graphical method

1) After how long will the temperature at the center of the cylinder reach θ^*?

2) What is the temperature on the outer surface of the cylinder at this instant?

3) Calculate, in kJ/cm, the linear amount of energy exchanged between the cylinder and the bath during the first thirty seconds of tempering.

B) Part two: calculations using the numerical method

In order to estimate the temperature profile more precisely, we now wish to develop the equations that will enable implementation of the calculations in a spreadsheet.

1) Write, in simplified cylindrical coordinates, the equation translating the energy balance in transient state.

2) What are the boundary conditions necessary for its resolution?

3) Develop the discrete iterative form of this equation.

4) Develop the discrete form of the boundary conditions and deduce therefrom the equation enabling the temperature at $r = R$ to be calculated, at each instant, as a function of θ_b.

5) Develop a spreadsheet implementation algorithm.

6) We decide to take an increment, $\Delta r = 2$ mm. Develop the condition on Δt in order to assure stability in the numerical integration.

7) Taking a value of Δt equal to one hundredth of a second, recalculate the time it will take for the temperature at the center of the cylinder to reach θ^*.

Data:

$R = 2$ cm $\quad \theta_i = 280$ °C $\quad \theta_b = 30$ °C $\quad \theta^* = 31.75$ °C

For gold: $\lambda_{Au} = 312$ W/m °C $\quad \rho_{Au} = 19{,}300$ kg/ m^3 $\quad C_{pAu} = 128$ J/kg °C

Cylinder/bath convection heat transfer coefficients: $h = 975$ W/m^2 °C

Solutions

A) Part one: estimations using the graphical method

1) Time for θ_0 to reach θ^*

This time is estimated from the Heisler chart, which gives the central temperatures for a cylindrical conductor as a function of the Fourier number, with a set according to $\frac{1}{\text{Bi}}$ (see Chapter 6, Figure 6.9).

To determine which curve to choose, we firstly calculate the Biot number:

$$\text{Bi} = \frac{\text{hR}}{\lambda}$$

$$R = 0.02 \text{ m} \Rightarrow \text{Bi} = \frac{975(0.02)}{312}$$

$$Bi = 0.0625 \Rightarrow \frac{1}{\text{Bi}} = 16$$

We then use the red curve in the figure below.

We calculate the reduced temperatures:

$$T_i = \theta_i - \theta_b = 280 - 30 = 250\ °C$$

$$T_o = \theta^* - \theta_b = 31.75 - 30 = 1.75\ °C$$

Hence: $\frac{T_0}{T_i} = 0.007$.

For this value of $\frac{T_0}{T_i}$ (see intersection with the green horizontal line), we find: $Fo = 80$.

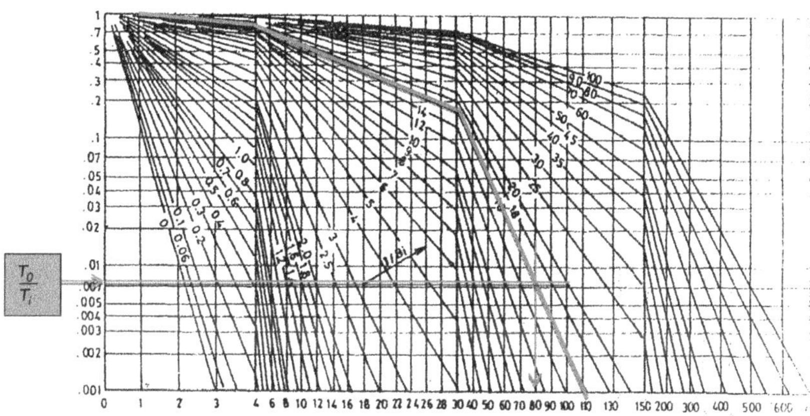

The central temperatures as a function of Fo *and of* $\frac{1}{Bi}$ *(cylindrical conductor).*

For a color version of this figure, see www.iste.co.uk/benallou/energy2.zip

The expression of Fo ($Fo = \frac{\alpha t}{s^2}$) allows us to retrieve t:

$$t = \frac{R^2 Fo}{\alpha},\ \text{where}\ \alpha = \frac{\lambda}{\rho C_p}$$

$\alpha = 1.26\ 10^{-4}$ m²/sec

$Fo = 80 \ \Rightarrow\ t = 253$ secs

2) Temperature of the outer surface of the cylinder

To determine this temperature, we use the Heisler chart, which gives the temperature distribution in a cylindrical conductor as a function of the central temperature, T_0, and of $\frac{1}{\mathrm{Bi}}$ (Chapter 6, Figure 6.9).

For the outer surface, we have: $r = R \Rightarrow \frac{\mathrm{r}}{\mathrm{R}} = 1$.

We will therefore use the curve corresponding to: $\frac{\mathrm{r}}{\mathrm{R}} = 1$.

By plotting the vertical line passing through $\frac{1}{\mathrm{Bi}} = 16$, we obtain: $\frac{\mathrm{T(R)}}{\mathrm{T_0}} = 0.96$.

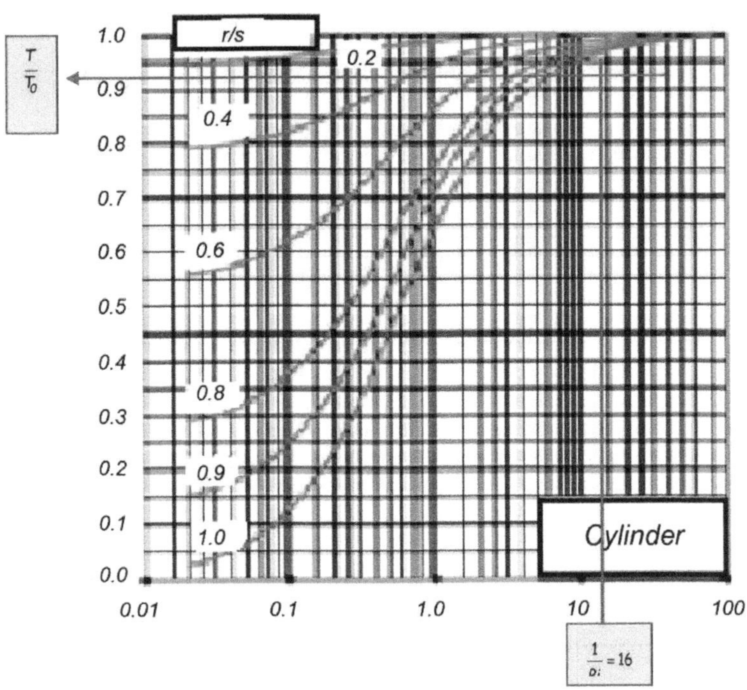

For a color version of this figure, see www.iste.co.uk/benallou/energy2.zip

Hence: $T(R) = 0.96\, T_0$, or: $\theta_R = \theta_b + 0.96\,(\theta^* - \theta_b)$.

NUMERICAL APPLICATION.–

$\theta^* = 31.75\ °C$ $\theta_b = 30\ °C$

$\boldsymbol{\theta_R = 31.68\ °C}$

3) Linear amount of energy exchanged during the first thirty seconds of tempering

To calculate the fluxes exchanged with the bath, we use the Heisler diagram presenting the ratio of the energy, Q, exchanged between the conductor and the bath (see Chapter 6, Figure 6.12), in relation to the initial internal energy, Q_0, defined by:

$$Q_0 = \rho V C_p (\theta_i - \theta_b),$$

where V is the volume of the conductor, θ_i the initial temperature and θ_b the bath temperature.

This diagram gives $\frac{Q}{Q_0}$ as a function of $Bi^2 Fo$.

We therefore have: $Bi = 0.0625$ and $Fo = \frac{\alpha t}{s^2}$.

For $t = 30$ seconds: $Fo = 9.47$.

Hence: $Bi^2 Fo = 3.7\ 10^{-2}$.

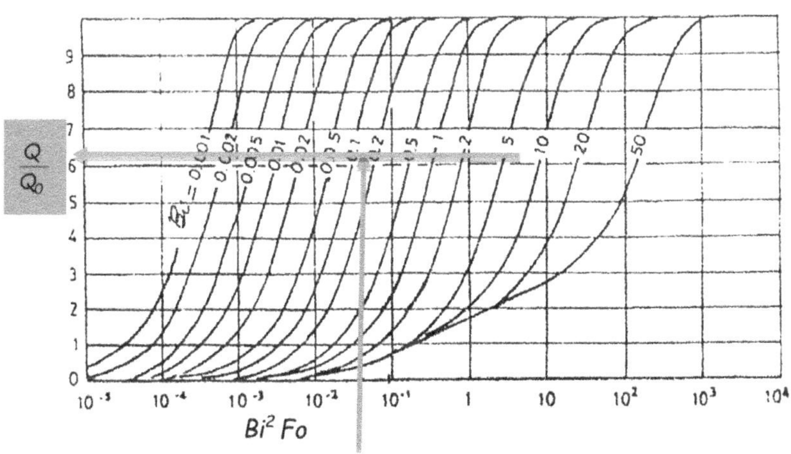

Flux exchanged with the outside (cylindrical conductor)

We can read on the curve corresponding to $Bi = 0.0625$: $\frac{Q}{Q_0} = 0.7$.

Hence the linear amount, Q_H: $Q_H = \frac{Q}{H} = 0.7\frac{Q_0}{H} = 0.7\rho\pi R^2 C_p(\theta_i - \theta_b)$.

NUMERICAL APPLICATION.–		
$R = 2$ cm	$\theta_i = 280$ °C	$\theta_b = 30$ °C
$\rho_{Au} = 19{,}300$ kg/ m^3		$C_{pAu} = 128$ J/kg °C
$Q_H = 5.43$ kJ/cm		

B) Part two: calculations using the numerical method

1) Equation giving the energy balance in transient state

In simplified cylindrical coordinates, this equation is:

$$\frac{1}{\alpha}\frac{\partial\theta}{\partial t} = \frac{\partial^2\theta}{\partial r^2} + \frac{1}{r}\frac{\partial\theta}{\partial r}$$

2) Boundary conditions

Two boundary conditions with r and one condition with t are needed:

$$CL1:\ t = 0:\ \forall r\ :\ \theta(r, 0) = \theta_i$$

$$CL2:\ r = 0: \forall t\ :\ \left(\frac{\partial\theta}{\partial r}\right)_{r=0} = 0$$

$$CL3:\ r = R: \forall t\ :\ -\lambda\left(\frac{\partial\theta}{\partial r}\right)_{r=R} = h\ (\theta - \theta_b)$$

3) Discrete form of the balance equation and the boundary conditions

We will use θ_j^n to note the temperature, at instant $t = n\ \Delta t$, at point M of the space defined by $r = j\ \Delta r$.

Consequently, we will develop the finite-difference approximations of the partial derivatives appearing in the conduction equation:

$$\left[\frac{\partial\theta}{\partial r}\right]_j^n = \frac{\theta_{j+1}^n - \theta_j^n}{\Delta r}$$

$$\left[\frac{\partial^2\theta}{\partial r^2}\right]_j^n = \frac{\theta_{j+1}^n - 2\theta_j^n + \theta_{j-1}^n}{(\Delta r)^2}$$

$$\left[\frac{\partial\theta}{\partial t}\right]_j^n = \frac{\theta_j^{n+1} - \theta_j^n}{\Delta t}$$

By replacing in the balance equation, we find:

$$\frac{1}{\alpha}\frac{\theta_j^{n+1} - \theta_i^n}{\Delta t} = \frac{\theta_{j+1}^n - 2\theta_j^n + \theta_{j-1}^n}{(\Delta r)^2} + \frac{1}{r}\frac{\theta_{j+1}^n - \theta_j^n}{\Delta r}$$

Hence, by multiplying by $(\Delta r)^2$ and by replacing r with $j\,\Delta r$:

$$\frac{(\Delta r)^2}{\alpha}\frac{\theta_j^{n+1} - \theta_j^n}{(\Delta t)} = \theta_{j+1}^n - 2\theta_j^n + \theta_{j-1}^n + \frac{\Delta r}{j(\Delta r)}\left(\theta_{j+1}^n - \theta_j^n\right)$$

Hence: $\theta_j^{n+1} - \theta_j^n = \dfrac{\alpha(\Delta t)}{(\Delta r)^2}\left[\theta_{j+1}^n\left(1+\dfrac{1}{j}\right) - \left(2+\dfrac{1}{j}\right)\theta_j^n + \theta_{j-1}^n\right]$.

Yet: $\mathrm{Fo}_\Delta = \alpha\dfrac{(\Delta t)}{(\Delta r)^2}$.

We then obtain the discrete iterative form of the balance equation:

$$j > 0:\ \theta_j^{n+1} = \theta_j^n + \mathrm{Fo}_\Delta\left[\frac{1+j}{j}\theta_{j+1}^n - \frac{1+2j}{j}\theta_j^n + \theta_{j-1}^n\right]$$

For $r = 0$: we need to come back to the continuous equation and consider the fact that:

$$\forall t\ :\ \left(\frac{\partial\theta}{\partial r}\right)_{r=0} = 0$$

The equation to be discretized is thus: $\frac{1}{\alpha}\left(\frac{\partial\theta}{\partial t}\right)_{r=0} = \left(\frac{\partial^2\theta}{\partial r^2}\right)_{r=0}$.

i.e., after discretization: $\frac{1}{\alpha}\frac{\theta_j^{n+1}-\theta_i^n}{\Delta t} = \frac{\theta_{j+1}^n - 2\theta_j^n + \theta_{j-1}^n}{(\Delta r)^2}$.

Hence: $\theta_0^{n+1} = \theta_0^n\left(1-2Fo_\Delta\right) + Fo_\Delta\left(\theta_1^n + \theta_{-1}^n\right)$.

4) Discretization of the boundary conditions

If we recall that θ_j^n designates the temperature at instant $t = n\ \Delta t$ and at point M of the space defined by $r = j\ \Delta r$, the three boundary conditions can be discretized as follows:

CL1: $\forall j\ :\ \theta_j^0 = \theta_i$

CL2: $r=0$: $\forall n\ :\ \frac{\theta_1^n - \theta_{-1}^n}{2\Delta r} = 0$

Or: $\forall n\ :\ \theta_1^n = \theta_{-1}^n$

CL3: $J = \frac{R}{\Delta r}$: $\forall n\ :\ \theta_J^n - \theta_{J-1}^n = -\frac{h\ \Delta r}{\lambda}\left(\theta_J^n - \theta_b\right)$

Hence, by taking into consideration the definition of the discrete Biot number ($Bi_\Delta = \frac{h\ (\Delta r)}{\lambda}$):

CL1: $\forall j\ \ 0 \le j \le J:\ \theta_j^0 = \theta_i$

CL2: $\forall n\ :\ \theta_1^n = \theta_{-1}^n \Rightarrow \forall n\ :\ \theta_0^{n+1} = \left(1-2Fo_\Delta\right)\theta_0^n + 2Fo_\Delta\theta_1^n$

CL3: $\forall n \rangle 0\ :\ \theta_J^n = \frac{1}{(1+Bi_\Delta)}\theta_{J-1}^n + \frac{Bi_\Delta}{(1+Bi_\Delta)}\theta_b$

5) Implementation algorithm

Step 1: define the parameters and the meshing: θ_i, θ_b, θ_0, R, λ, ρ, C_p, h, Δr, Δt.

Step 2: calculate $Fo_\Delta = \alpha \frac{(\Delta t)}{(\Delta r)^2}$; $Bi_\Delta = \frac{h\ (\Delta r)}{\lambda}$ and $J = \frac{R}{\Delta r}$.

Step 3: initialize the time and space counters: $n = 0$, $t = 0$ and $j = 0$, $r = 0$.

Step 4: initialize the temperatures: $\theta_j^0 = \theta_i$ for $0 \le j \le J$.

Step 5: increment the time: $n = n + 1$ and $t = t + \Delta t$.

Step 6:

– as long as $\theta_0^n \ge \theta_0$:

$$\theta_0 = (1 - 2Fo_\Delta)\theta_0 + 2Fo_\Delta\theta_1$$

– for $1 \le j \le J - 1$:

$$\theta_j = \theta_j + Fo_\Delta\left[\frac{1+j}{j}\theta_{j+1} - \frac{1+2j}{j}\theta_j + \theta_{j-1}\right]$$

$$\theta_J = \frac{1}{(1+Bi_\Delta)}\theta_{J-1} + \frac{Bi_\Delta}{(1+Bi_\Delta)}\theta_b$$

6) Choosing the integration increments

We are in a one-dimensional scenario, therefore Δt and Δr so as to keep $Fo_\Delta \le \frac{1}{2}$.

Hence: $\alpha \frac{(\Delta t)}{(\Delta r)^2} \le \frac{1}{2}$.

or:

$$\Delta t \le \frac{(\Delta r)^2}{2\alpha}$$

$$(\Delta r) \ge \sqrt{2\alpha(\Delta t)}$$

NUMERICAL APPLICATION.–

$\Delta r = 2\ 10^{-3}$ m ⇨ $\Delta t \leq 1.58\ 10^{-2}$ s

7) Calculating the time it takes for the temperature at the center of the cylinder to reach θ^*

The algorithm developed in Question 5) is implemented in a spreadsheet with a time increment $\Delta t = 10^{-2}$ sec and $\Delta r = 2$ mm. The results obtained are presented in the table below:

t (seconds)	θ_0
0.00	280.00
0.10	280.00
0.20	279.93
0.30	279.60
0.40	279.03
0.50	278.29
1.00	273.74
2.00	264.23
3.00	255.06
5.00	237.79
10.00	200.20
20.00	144.19
30.00	106.61
40.00	81.40
50.00	64.48
60.00	53.14
70.00	45.52
80.00	40.41
90.00	36.99
100.00	34.69
110.00	33.14
120.00	32.11
121.00	32.03
122.00	31.95
123.00	31.87
124.00	31.80
125.00	31.73
126.00	31.66
127.00	31.60
128.00	31.53
129.00	31.47
130.00	31.42

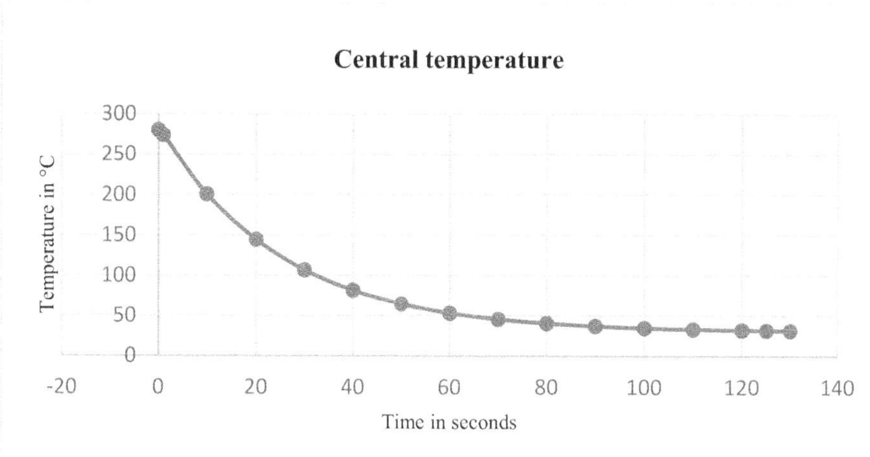

We find $t = 125$ sec.

Note that the graphical method gave $t = 253$ sec.

This enormous difference allows us to conclude that the numerical method is more precise.

A comparison between the accuracy of the graphical method and that of the numerical method is discussed in the case study presented in section 6.8 of Volume 1.

EXERCISE 28. Tempering a plane solid

Upon exiting a heat-treatment furnace, a planar metal part, of thickness $2e$, must be tempered in a bath, the temperature of which, θ_b, is uniform and constant. This part is made entirely of titanium, and is initially at a uniform temperature, θ_i.

Questions

A) Part one: estimations using the graphical method.

1) After how long will the temperature at the center of the solid reach θ_0^*?

2) What is the temperature of the outer surfaces of the solid at this instant?

3) Calculate the energy flux exchanged between the cylinder and the bath starting from the initial instant.

B) Part two: calculations using the numerical method.

We now wish to develop more accurate estimations of the temperature profile.

1) Knowing that a single propagation direction is considered, develop the discrete form of the equation that reflects the energy balance in transient state, as well as the discrete boundary conditions required for its resolution.

2) Develop a spreadsheet implementation algorithm.

3) In this question, we wish to implement the calculation algorithm developed above, with a spatial integration step, $\Delta x = 1$ cm. What constraint does this impose on Δt in order for the numerical-integration calculations to be stable?

4) We decide to take $\Delta t = 5$ seconds, with $\Delta x = 1$ cm. In a spreadsheet, recalculate the time needed for the temperature at the center of the solid to reach θ_0^*.

5) Recalculate the temperature of the outer surfaces of the solid at this instant.

C) Part three: calculations from the analytical solution.

In this part, we wish to conduct the same calculations from the analytical expression of the temperature field. We therefore carry out the following variable and function changes:

$$T = \frac{\theta - \theta_e}{\theta_i - \theta_e} \text{ and } \overline{x} = \frac{x}{e}$$

1) Give the general form of the temperature field $T(\overline{x}, Fo)$ and the boundary conditions.

2) According to the values of the Fourier and Biot numbers, what approximations can you make to simplify the analytical solution?

3) Calculate the first two solutions to the equation, $\omega tg(\omega) = Bi$.

4) Deduce therefrom the expression enabling us to obtain a good approximation of $T(\overline{x}, Fo)$.

5) From this expression, recalculate the time needed for the temperature at the center of the solid to reach θ_0^*, as well as the temperature of the outer surfaces at this instant?

6) Compare the results obtained from this expression with those calculated in parts 1 and 2 above.

Data:

Plane thickness: $2e = 14$ cm

$\theta_i = 175$ °C $\theta_b = 60$ °C $\theta_0^* = 62.3$ °C

Titanium: $\lambda_t = 20.9$ W/m °C $\rho_t = 4{,}500$ kg/ m^3 $C_{pt} = 523$ J/kg °C

Cylinder/bath convection heat transfer coefficients: $h = 250$ W/m^2 °C

Solutions

A) Part one: estimations using the graphical method

1) When will the central temperature reach θ_0^*?

To determine the central temperature, we firstly calculate the Biot number and the Fourier number, then we use the Heisler chart, which gives the central temperatures for a plane conductor (Chapter 6, Figure 6.5).

We have:

$$\text{Bi} = \frac{\text{he}}{\lambda}$$

$$Bi = 0.84 \quad \Rightarrow \quad \frac{1}{\text{Bi}} = 1.19$$

We then use the curve corresponding to $\frac{1}{\text{Bi}} = 1.2$.

Moreover, as the central temperature is known (θ_0), we can calculate the ratio $\frac{T_0}{T_i}$.

We have: $T_0 = \theta_0 - \theta_b$ and $T_i = \theta_i - \theta_b$.

Hence:

$$\frac{T_0}{T_i} = \frac{\theta_0 - \theta_b}{\theta_i - \theta_b}$$

$$\theta_0^* = 62.3\ ^\circ\text{C},\ \theta_I = 175\ ^\circ\text{C and } \theta_b = 60\ ^\circ\text{C} \Rightarrow \frac{T_0^*}{T_i} = 0.02$$

We then place 0.02 on the y-axis, then we determine the corresponding *Fo* number:

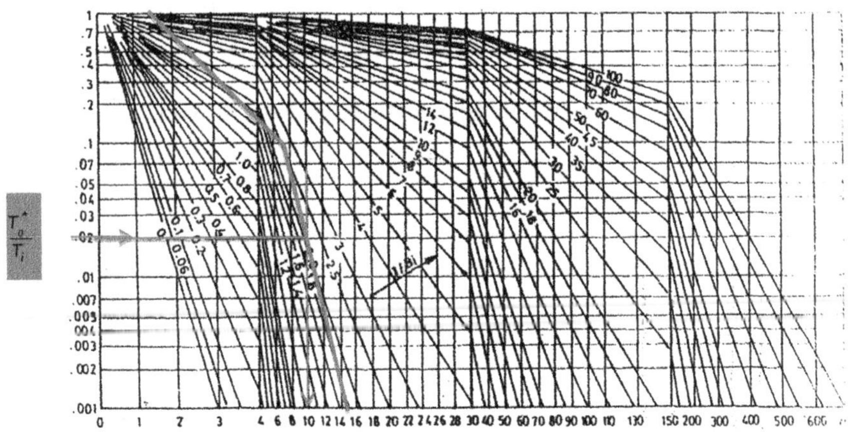

Central temperatures as a function of Fo *and of* $\frac{1}{Bi}$ *(plane conductor).*

For a color version of this figure, see www.iste.co.uk/benallou/energy2.zip

We obtain: $Fo = 6$.

The expression of *Fo* ($Fo = \frac{\alpha t}{e^2}$) allows us to retrieve *t*:

$$t = \frac{e^2 Fo}{\alpha} \text{ with } \alpha = \frac{\lambda}{\rho C_p}$$

NUMERICAL APPLICATION.–

$e = 0.07$ m $\qquad$ $\lambda_t = 20.9$ W/m °C

$\rho_t = 4{,}500$ kg/ m^3 $\qquad$ $C_{pt} = 523$ J/kg °C $\qquad$ $Fo = 3$

$\alpha = 8.88\ 10^{-6}$ m^2/sec ⇨ $\boldsymbol{t = 3{,}311}$ **seconds**

$\boldsymbol{t = 55.18}$ **mins**

2) Temperature of outer surfaces

To determine this temperature, we use the Heisler chart, which gives the temperature distribution in a plane conductor as a function of the central temperature (see Chapter 6, Figure 6.8). Indeed, this diagram, which is parameterized according to $\frac{x}{e}$, gives $\frac{T}{T_0}$ as a function of $\frac{1}{Bi}$.

For the outer surfaces, we have: $x = e \Rightarrow \frac{x}{e} = 1$.

We will therefore use the curve corresponding to $\frac{x}{s} = 1$: the red curve.

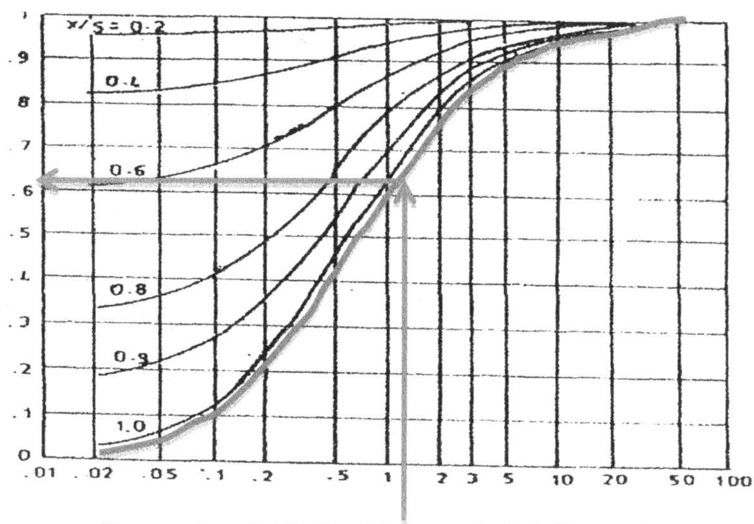

Temperature distribution (plane conductor). For a color version of this figure, see www.iste.co.uk/benallou/energy2.zip

By plotting the vertical line passing through $\frac{1}{Bi} = 1.2$, we obtain: $\frac{T(e)}{T_0} = 0.63$.

Hence: $T(e) = 0.63\ T_0$, or: $\theta_e - \theta_b = 0.63(\theta_0^* - \theta_b)$.

I.e.: $\theta_e = \theta_b + 0.63(\theta_0^* - \theta_b)$.

NUMERICAL APPLICATION.–

$\theta_0^* = 62.3$ °C $\qquad \theta_b = 60$ °C

$\mathbf{\theta_e = 61.45}$ **°C**

3) Energy flux exchanged with the bath

To calculate the energy exchanged between the conductor and the bath, we use Figure 6.11 (see Chapter 6), which gives, as a function of $Bi^2\,Fo$, the $\frac{Q}{Q_0}$ ratios of the energy exchanged, Q, to the initial internal energy, Q_0, defined by: $Q_0 = \rho V C_p(\theta_i - \theta_b)$, where V is the volume of the conductor, θ_i is the initial temperature and θ_e is the bath temperature.

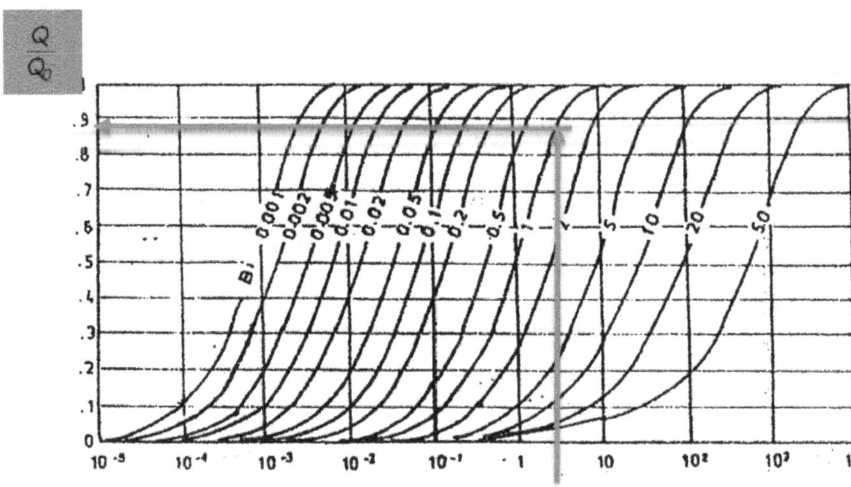

Flux exchanged with the outside (plane conductor). For a color version of this figure, see www.iste.co.uk/benallou/energy2.zip

$$Bi = 0.84 \text{ and } Bi^2\,Fo = 4.21 \;\Rightarrow\; \frac{Q}{Q_0} = 0.88$$

Using φ to describe the flux density, we have: $\varphi = \frac{0.88 Q_0}{S\Delta t}$, where S is the transfer area and Δt is the time interval considered.

Hence: $\varphi = 0.88 \frac{(2e)S}{(\Delta t)S} \rho C_p(\theta_i - \theta_b)$

i.e.: $\varphi = 1.76 \frac{e}{\Delta t} \rho C_p(\theta_i - \theta_b)$.

NUMERICAL APPLICATION.–

$\theta_I = 175$ °C $\theta_b = 60$ °C $\Delta t = 3{,}311$ sec $e = 0.07$ m

$\rho_t = 4{,}500$ kg/m^3 $C_{pt} = 523$ J/kg °C

$\varphi = 10$ **kW**

B) Part two: calculations using the numerical method

1) Equation giving the energy balance in transient state

In rectangular coordinates, assuming a single heat propagation direction, the conduction equation is given by:

$$\frac{1}{\alpha}\frac{\partial \theta}{\partial t} = \frac{\partial^2 \theta}{\partial x^2},$$

with the boundary conditions:

$$CL_1 \; t = 0: \; \forall x \; : \; \theta(x, 0) = \theta_i$$

$$CL_2 \; x = \pm e: \; \forall t \; : \; -\lambda\left(\frac{\partial \theta}{\partial x}\right)_{x=\pm e} = h(\theta - \theta_b)$$

For the discretization of this equation and of its boundary conditions, we will use θ_j^n to note the temperature, at instant $t = n\,\Delta t$, at point M of the space defined by $x = j\,\Delta x$.

The finite-difference approximations of the partial derivatives appearing in the conduction equation can then be written as follows:

$$\left[\frac{\partial \theta}{\partial t}\right]_j^n = \frac{\theta_j^{n+1} - \theta_j^n}{\Delta t}$$

$$\left[\frac{\partial^2 \theta}{\partial x^2}\right]_j^n = \frac{\theta_{j+1}^n - 2\theta_j^n + \theta_{j-1}^n}{(\Delta x)^2}$$

By replacing the partial derivatives with their expressions in the conduction equation, we find:

$$\frac{1}{\alpha}\frac{\theta_j^{n+1}-\theta_i^n}{\Delta t}=\frac{\theta_{j+1}^n-2\theta_j^n+\theta_{j-1}^n}{(\Delta x)^2}$$

If we multiply the two members of the equation by $(\Delta x)^2$, we obtain:

$$\frac{(\Delta x)^2}{\alpha}\frac{\theta_j^{n+1}-\theta_i^n}{\Delta t}=\theta_{j+1}^n-2\theta_j^n+\theta_{j-1}^n$$

Yet: $Fo_\Delta=\alpha\frac{(\Delta t)}{(\Delta x)^2}$.

Hence: $\theta_j^{n+1}=\theta_j^n+Fo_\Delta\left(\theta_{j+1}^n-2\theta_j^n+\theta_{j-1}^n\right)$

or: $\theta_j^{n+1}=(1-2Fo_\Delta)\theta_j^n+Fo_\Delta\left(\theta_{j+1}^n+\theta_{j-1}^n\right)$.

Discretization of the boundary conditions:

CL1: $\forall j\ :\ \theta_j^0=\theta_i$

CL2: @ $x=+e$: $\forall n\ :-\lambda\left(\frac{\theta_J^n-\theta_{J-1}^n}{\Delta x}\right)=h(\theta_J-\theta_b)$ *o*

or: $\theta_J^n=\frac{\theta_{J-1}^n+Bi_\Delta\theta_b}{1+Bi_\Delta}$ with: $Bi_\Delta=\frac{h(\Delta x)}{\lambda}$

CL2: @ x = - e: Symmetry ⇨ $\forall n\ :\ \theta_{-1}^n=\theta_1^n$

Hence: $\theta_0^{n+1}=(1-2Fo_\Delta)\theta_0^n+2Fo_\Delta\theta_1^n$.

2) Spreadsheet implementation algorithm

Step 1: define the parameters and the meshing: θ_i, θ_b, θ_0, e, λ, ρ, C_p, h, Δx, Δt. *Δx must be a divisor of* e.

Step 2: calculate $Fo_\Delta = \alpha \frac{(\Delta t)}{(\Delta x)^2}$; $Bi_\Delta = \frac{h\ (\Delta x)}{\lambda}$ and $J = \frac{R}{\Delta r}$.

Step 3: initialize the time and space counters: $n = 0$, $t = 0$ and $j = 0$, $x = 0$.

Step 4: initialize the temperatures: $\theta_j^0 = \theta_i$ for $0 \leq j \leq J$;

Step 5: increment the time: $n = n + 1$ and $t = t + \Delta t$;

Step 6:

– as long as $\theta_0^n \geq \theta_0 : \theta_0^{n+1} = (1 - 2Fo_\Delta)\theta_0^n + 2Fo_\Delta \theta_1^n$;

– for $1 \leq j \leq J - 1$: $\theta_j^{n+1} = (1 - 2Fo_\Delta)\theta_j^n + Fo_\Delta\left(\theta_{j+1}^n + \theta_{j-1}^n\right)$;

– for $j = J$: $\theta_J^n = \frac{\theta_{J-1}^n + Bi_\Delta \theta_b}{1 + Bi_\Delta}$.

3) Constraint on Δt for numerical integration to be stable

The stability of the numeral-integration calculations requires that $Fo_\Delta \leq \frac{1}{2}$ (see section 6.4.2):

$$\alpha \frac{(\Delta t)}{(\Delta x)^2} \leq \frac{1}{2} \Rightarrow \Delta t \leq \frac{(\Delta x)^2}{2\alpha}$$

NUMERICAL APPLICATION.–
$\Delta x = 0.01$ m ⇨ $\Delta t \leq 5.63$ sec

4) Calculate in a spreadsheet the time it takes for the temperature at the center to be equal to θ_0^*

$\Delta x = 0.01$ m $\Delta t = 5$ sec

We then have: $Bi_\Delta = 0.1196$ and $Fo_\Delta = 0.4440$.

Implementing on a spreadsheet enables us to calculate the entire temperature field at each instant. The central temperature profile is presented more specifically in the figure below.

The central temperature will reach $\theta_0^* = 62.3$ °C at: $\boldsymbol{t}$ **= 3,260 seconds.**

I.e. **54.33 min.**

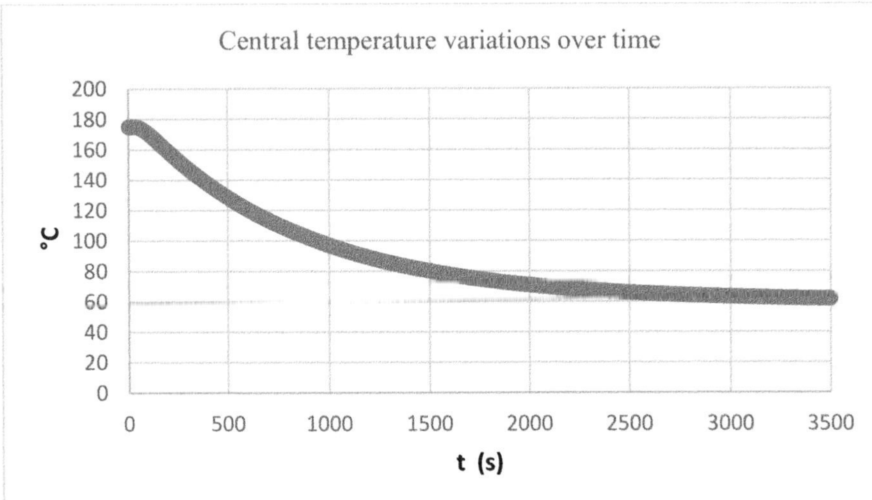

5) Temperature of outer surfaces

Numerical integration gives us the entire temperature field (see table below); in particular $\theta_e = 61.56$ °C.

x (cm)	0	1	2	3	4	5	6	7
t (seconds)	Temperatures (*x*,*t*) in °C							
0	175.00	175.00	175.00	175.00	175.00	175.00	175.00	175.00
300	147.57	146.96	145.16	142.18	138.07	132.88	126.68	119.56
600	120.55	120.13	118.88	116.82	113.97	110.38	106.09	101.17
900	101.86	101.57	100.71	99.28	97.32	94.83	91.87	88.46
1,200	88.94	88.74	88.14	87.16	85.80	84.08	82.03	79.68
1,500	80.01	79.87	79.46	78.78	77.84	76.65	75.23	73.60
1,800	73.83	73.74	73.45	72.98	72.33	71.51	70.53	69.41
2,100	69.56	69.50	69.30	68.98	68.53	67.96	67.28	66.50
2,400	66.61	66.57	66.43	66.21	65.89	65.50	65.03	64.50
2,700	64.57	64.54	64.45	64.29	64.08	63.80	63.48	63.11
3,000	63.16	63.14	63.07	62.97	62.82	62.63	62.41	62.15
3,260	62.30	62.28	62.23	62.15	62.05	61.91	61.75	61.56
3,300	62.19	62.17	62.13	62.05	61.95	61.82	61.66	61.49

C) Part three: calculations from the analytical solution

1) Expression of the analytical solution and the boundary conditions

In this case, *Bi* = 0.8373, the temperature is a function of *x* and *t*: θ(*x*, *t*).

We write: $T = \dfrac{\theta - \theta_b}{\theta_i - \theta_b}$ and $\overline{x} = \dfrac{x}{e}$.

The general conduction solution is given by:

$$T(\overline{x}, Fo) = \sum_{j>0} A_{\omega j}\, e^{-\omega_j^2 Fo} \cos(\omega_j \overline{x}) \ ,$$

where ω_j and $A_{\omega j}$ are determined from the boundary conditions:

$$CL1\ \forall\ Fo > 0 \quad : \omega_j\, tg(\omega_j) = Bi$$

$$CL2\text{: } Fo = 0\text{: } T(\overline{x}, 0) = 1 \quad \Rightarrow A_{\omega j} = \frac{2\sin(\omega_j)}{\omega_j + \sin(\omega_j)\cos(\omega_j)}$$

2) Simplification for an approximate analytical solution

As we are interested in *values of Fo above 0.2*, it is important to know that a good approximation of the temperature field is obtained by keeping only the first two solutions to the equation $\omega\, tg\, (\omega) = Bi$: Consequently, a good approximation of $T(\overline{x}, Fo)$ is given by:

$$T(\overline{x}, Fo) \approx A_{\omega 1}\, e^{-\omega_1^2 Fo} \cos(\omega_1 \overline{x}) + A_{\omega 2}\, e^{-\omega_2^2 Fo} \cos(\omega_2 \overline{x})$$

3) The first two solutions to the equation $\omega\, tg\, (\omega) = Bi$

For *Bi* = 0.8373, the first two solutions to the equation $\omega\, tg\, (\omega) = Bi$ are:

$\omega_1 = 0.8050466$

$\omega_2 = 3.3841458$

4) Approximate expression of $T(\bar{x}, Fo)$

Thus, a good approximation of the temperature field is obtained by:

$$T(\bar{x}, Fo) \approx A_{\omega 1}\, e^{-\omega_1^2 Fo} \cos(\omega_1 \bar{x}) + A_{\omega 2}\, e^{-\omega_2^2 Fo} \cos(\omega_2 \bar{x}) ,$$

where $A_{\omega j} = \dfrac{2 \sin(\omega_j)}{\omega_j + \sin(\omega_j)\cos(\omega_j)}$:

Coefficients $A_{\omega 1}$ and $A_{\omega 2}$ are given in the table below:

j	ω_j	$A_{\omega j}$	ω_j^2
1	0.8050466	1.10505820	0.6481000
2	3.3841458	-0.1327963	11.4524428

5) Calculations from the analytical expression

a) Time it takes for the temperature at the center to be equal to θ_0^*

Central temperature: $\bar{x} = 0 \Rightarrow T(0, Fo) \approx A_{\omega 1}\, e^{-\omega_1^2 Fo} + A_{\omega 2}\, e^{-\omega_2^2 Fo}$.

We find: $Fo = 6.19 \Rightarrow t = \dfrac{e^2 Fo}{\alpha} \Rightarrow t_{\theta_0^*} \cong 3415.6$ s.

NOTE.–

We could have calculated $t_{\theta_0^*}$ by further simplifying the expression of $T(\bar{x}, Fo)$..

Indeed, by assessing the contributions of ω_1 and ω_2, we would easily be able to see that $\dfrac{A_{\omega 2}\, e^{-\omega_2^2 Fo}}{A_{\omega 1}\, e^{-\omega_1^2 Fo}} \langle 1.5\%$.

$Fo = 0.2$

j	$A_{\omega j}\, e^{-\omega_j^2 Fo}$
1	0.970715353
2	-0.013441242

Consequently, without much error, we could limit the development of $T(\overline{x}, Fo)$ to the first term, namely:

$$T_0^* \cong A_{\omega 1} \, e^{-\omega_1^2 Fo}$$

Hence: $e^{\omega_1^2 Fo} \cong \dfrac{A_{\omega 1}}{T_0^*}$

or: $Fo \cong \dfrac{1}{\omega_1^2} Ln\left[\dfrac{A_{\omega 1}}{T_0^*}\right]$.

Hence: $t = \dfrac{e^2 Fo}{\alpha}$

i.e.: $t_{\theta_0^*} \cong 3416s$.

b) Temperature of the outer surfaces at the same instant

To calculate this temperature, we use $(\overline{x} = 1 \text{ and } Fo = 6.19)$:

$$T(1, Fo) \approx A_{\omega 1} \, e^{-\omega_1^2 Fo} \cos(\omega_1) + A_{\omega 2} \, e^{-\omega_2^2 Fo} \cos(\omega_2)$$

$$T(1,6,19) = 0.01386 \Rightarrow \theta = \theta_b + T(\theta_i - \theta_b)$$

i.e.: θ_e= 61.59 °C.

6) Comparison of results

The following table assembles the results of the analytical, numerical and graphical methods.

Method	$t_{\theta_{0*}}$ (seconds)	θ_e (°C)
Analytical	3,416	61.59
Numerical	3,260	61.56
Graphical	3,311	61.45

EXERCISE 29. Numerical-integration stability in the heat equation

In this exercise, we wish to examine the influence that the choice of increments Δx and Δt has on stability in the numerical-integration calculations for the heat equation.

In this perspective, let us consider the cooling of an aluminum plate of thickness $2L$ in a bath of uniform, constant temperature, θ_b. This solid, initially at a uniform temperature, θ_i, is suddenly tempered in the bath.

We will consider a single propagation direction.

Questions

1) Develop the discrete form of the equation that reflects the energy balance in transient state, as well as the discrete boundary conditions required for its resolution.

2) Develop a spreadsheet implementation algorithm.

3) For the implementation of the calculations to be effective, the choice of Δx and Δt must adhere to the following stability criterion (see Chapter 6, section 6.4.2): $Fo_{\Delta} \leq \frac{1}{2}$. We decide to take $\Delta t_0 = 1$ minute and $\Delta x_0 = 10$ cm.

a) Do these increments satisfy the stability condition?

b) Calculate in a spreadsheet the temperature field between $t_i = 0$ and $t_f = 30$ minutes.

4) We now decide to keep $\Delta x = \Delta x_0$, but to push Δt to the limit, by each time increasing it by $n\,\tau$ seconds, with $\tau = 1$ and $n = 0, 1, 2$, etc.

By repeating the temperature-field calculations for $\Delta t_i = \Delta t_0 + n\,\tau$, determine the smallest natural integer, which leads to instable calculations. This instability will firstly appear in the form of oscillations in the central temperature, $\theta_0(t)$.

5) In this question, we wish to compare the solutions obtained for different values of Δt, whilst keeping $\Delta x = \Delta x_0$.

a) Repeat the calculations of the temperature field for $\Delta t_1 = \dfrac{\Delta t_0}{2}$ and $\Delta t_2 = \dfrac{\Delta t_0}{4}$.

b) For $\theta\,(0,t)$, compare the results obtained to those given by the approximate analytical solution. We will compare the approximate analytical solution to the numerical results obtained with Δt_0, Δt_1, then with Δt_2. *In this comparison, we are only interested in the values obtained for* t = *30 minutes, 40 minutes, 50 minutes and 60 minutes.*

Data:

$L = 80$ cm $\quad \theta_i = 180$ °C $\quad \theta_b = 80$ °C

Aluminum: $\lambda_A = 204$ W/m °C $\quad \rho_A = 2{,}755$ kg/ m^3 $\quad C_{pA} = 920$ J/kg °C

Cylinder/bath convection heat transfer coefficients: $h = 1{,}275$ W/m^2 °C

Solutions

1) Equation of the energy balance in transient state and boundary conditions

In rectangular coordinates, assuming a single heat propagation direction, the conduction equation is given by:

$$\frac{1}{\alpha}\frac{\partial\theta}{\partial t} = \frac{\partial^2\theta}{\partial x^2},$$

with the boundary conditions:

$$CL1\text{: } t = 0\text{: } \forall x \;:\; \theta\left(x,\ 0\right) = \theta_i$$

$$CL2\text{: } x = \pm L \;:\; \forall t \;:\; -\lambda\left(\frac{\partial\theta}{\partial x}\right)_{x=\pm L} = h\left(\theta - \theta_b\right)$$

If θ_j^n designates the temperature, at point M of the space defined by $x = j\,\Delta x$, at instant $t = n\,\Delta t$, then the finite-difference approximations of the partial derivatives appearing in the conduction equation can be written as follows:

$$\left[\frac{\partial\theta}{\partial t}\right]_j^n = \frac{\theta_j^{n+1} - \theta_j^n}{\Delta t}$$

$$\left[\frac{\partial^2\theta}{\partial x^2}\right]_j^n = \frac{\theta_{j+1}^n - 2\theta_j^n + \theta_{j-1}^n}{\left(\Delta x\right)^2}$$

By replacing the partial derivatives with their expressions in the conduction equation, we find:

$$\frac{1}{\alpha}\frac{\theta_j^{n+1}-\theta_i^n}{\Delta t}=\frac{\theta_{j+1}^n-2\theta_j^n+\theta_{j-1}^n}{(\Delta x)^2}$$

By multiplying by $(\Delta x)^2$, we therefore obtain:

$$\frac{(\Delta x)^2}{\alpha}\frac{\theta_j^{n+1}-\theta_i^n}{\Delta t}=\theta_{j+1}^n-2\theta_j^n+\theta_{j-1}^n$$

By writing $Fo_\Delta=\alpha\frac{(\Delta t)}{(\Delta x)^2}$, we therefore obtain:

$$\theta_j^{n+1}=\theta_j^n+Fo_\Delta\left(\theta_{j+1}^n-2\theta_j^n+\theta_{j-1}^n\right)$$

or: $\theta_j^{n+1}=(1-2Fo_\Delta)\theta_j^n+Fo_\Delta\left(\theta_{j+1}^n+\theta_{j-1}^n\right)$.

Discretization of the boundary conditions:

CL1: $\forall j\ :\ \theta_j^0=\theta_i$

CL2: @ $x=+L$: $\forall n\ :-\lambda\left(\frac{\theta_J^n-\theta_{J-1}^n}{\Delta x}\right)=h(\theta_J-\theta_b)$

or: $\theta_J^n=\frac{\theta_{J-1}^n+Bi_\Delta\theta_b}{1+Bi_\Delta}$,

where: $Bi_\Delta=\frac{h(\Delta x)}{\lambda}$

CL3: @ $x=-L$: Symmetry ⇨ $\forall n\ :\ \theta_{-1}^n=\theta_1^n$

Hence: $\theta_0^{n+1}=(1-2Fo_\Delta)\theta_0^n+2Fo_\Delta\theta_1^n$.

2) Spreadsheet implementation algorithm

Step 1: define the parameters and the meshing: θ_i, θ_b, L, λ, ρ, C_p, h, Δx, Δt, t_f. Δx *must be a divisor of* L.

Step 2: calculate $Fo_\Delta = \alpha \frac{(\Delta t)}{(\Delta x)^2}$; $Bi_\Delta = \frac{h\ (\Delta x)}{\lambda}$ and $J = \frac{e}{\Delta x}$.

Step 3: initialize the time and space counters: $n = 0$, $t = 0$ and $j = 0$, $x = 0$

Step 4: initialize the temperatures: $\theta_j^0 = \theta_i$ for $0 \le j \le J$.

Step 5: increment the time: $n = n + 1$ and $t = t + \Delta t$.

Step 6:

– as long as $t \ge t_f : \theta_0^{n+1} = (1 - 2Fo_\Delta)\theta_0^n + 2Fo_\Delta \theta_1^n$;

– for $1 \le j \le J - 1$: $\theta_j^{n+1} = (1 - 2Fo_\Delta)\theta_j^n + Fo_\Delta(\theta_{j+1}^n + \theta_{j-1}^n)$;

– for $j = J$: $\theta_J^n = \frac{\theta_{J-1}^n + Bi_\Delta \theta_b}{1 + Bi_\Delta}$.

3) Numerical-integration stability constraints

We will take $\Delta t_0 = 1$ minute and $\Delta x_0 = 10$ cm.

a) Stability condition for Δt_0 and Δx_0

Calculating Fo_Δ leads to: $\alpha \frac{(\Delta t)}{(\Delta x)^2} \le 0.483$.

Hence: $Fo_\Delta \le \frac{1}{2}$.

The numerical calculations will therefore be stable with $\Delta t_0 = 1$ minute and $\Delta x_0 = 10$ cm.

b) Calculating the temperature field between 0 and 30 minutes

We will calculate the discrete Biot number: $Bi_\Delta = \frac{h(\Delta x)}{\lambda}$ ⇨ $Bi_\Delta = 0.625$.

$Fo_\Delta = \alpha \frac{(\Delta t)}{(\Delta x)^2}$ previously calculated at 0.483.

$\Delta x_0 = 10$ cm			$\Delta t_0 = 1$ min			$Bi_{=0.625}$		$Fo_{=0.483}$	
x (cm)	0	10	20	30	40	50	60	70	80
t (minutes)	Temperatures (x,t) in °C								
0.00	180.00	180.00	180.00	180.00	180.00	180.00	180.00	180.00	180.00
1.00	180.00	180.00	180.00	180.00	180.00	180.00	180.00	180.00	141.54
2.00	180.00	180.00	180.00	180.00	180.00	180.00	180.00	161.43	130.11
3.00	180.00	180.00	180.00	180.00	180.00	180.00	171.03	155.27	126.32
4.00	180.00	180.00	180.00	180.00	180.00	175.67	167.75	148.90	122.40
5.00	180.00	180.00	180.00	180.00	177.91	173.94	162.47	145.21	120.13
6.00	180.00	180.00	180.00	178.99	177.00	170.32	159.67	141.43	117.80
7.00	180.00	180.00	179.51	178.52	174.73	168.40	156.01	138.83	116.20
8.00	180.00	179.76	179.27	177.17	173.50	165.47	153.70	136.20	114.58
9.00	179.77	179.64	178.49	176.41	171.40	163.67	150.93	134.21	113.36
10.00	179.64	179.15	178.04	175.00	170.09	161.25	149.01	132.22	112.13
11.00	179.17	178.85	177.11	174.10	168.19	159.60	146.81	130.63	111.15
12.00	178.86	178.16	176.50	172.70	166.90	157.57	145.17	129.04	110.18
13.00	178.19	177.70	175.47	171.73	165.20	156.09	143.37	127.72	109.37
14.00	177.71	176.86	174.74	170.38	163.95	154.34	141.95	126.41	108.56
15.00	176.88	176.25	173.66	169.38	162.42	153.00	140.43	125.30	107.88
16.00	176.27	175.30	172.84	168.08	161.23	151.48	139.19	124.19	107.20
17.00	175.34	174.58	171.73	167.07	159.83	150.26	137.88	123.23	106.60
18.00	174.61	173.57	170.86	165.82	158.70	148.90	136.78	122.28	106.02
19.00	173.61	172.76	169.74	164.82	157.41	147.78	135.63	121.43	105.49
20.00	172.79	171.71	168.82	163.62	156.34	146.56	134.64	120.59	104.98
21.00	171.75	170.84	167.70	162.62	155.13	145.53	133.61	119.84	104.52
22.00	170.87	169.76	166.76	161.46	154.11	144.41	132.71	119.09	104.06
23.00	169.80	168.85	165.65	160.47	152.98	143.44	131.78	118.41	103.64
24.00	168.88	167.76	164.69	159.35	151.99	142.42	130.96	117.73	103.22
25.00	167.80	166.82	163.60	158.38	150.92	141.51	130.11	117.11	102.84
26.00	166.85	165.74	162.63	157.30	149.97	140.55	129.34	116.49	102.46
27.00	165.77	164.78	161.55	156.34	148.96	139.69	128.55	115.92	102.10
28.00	164.81	163.70	160.59	155.29	148.04	138.79	127.83	115.35	101.75
29.00	163.74	162.73	159.54	154.35	147.07	137.96	127.09	114.81	101.42
30.00	162.77	161.68	158.58	153.34	146.19	137.11	126.41	114.27	101.09

4) Destabilizing the calculations

$\Delta x = \Delta x_0 = 10$ cm $\Rightarrow$ Bi_{Δ} will not change

$\Delta t_i = \Delta t_0 + n\,\tau$

a) For $n = 1$

$\Delta x_0 = 10$ cm			$\Delta t_1 = 61$ seconds		$Bi_{=0.625}$		$Fo_{=0.49096}$		
x (cm)	0	10	20	30	40	50	60	70	80
t (minutes)	Temperatures $\theta(x,t)$ in °C								
0.00	180.00	180.00	180.00	180.00	180.00	180.00	180.00	180.00	180.00
1.02	180.00	180.00	180.00	180.00	180.00	180.00	180.00	180.00	141.54
2.03	180.00	180.00	180.00	180.00	180.00	180.00	180.00	161.12	129.92
3.05	180.00	180.00	180.00	180.00	180.00	180.00	170.73	155.07	126.20
4.07	180.00	180.00	180.00	180.00	180.00	175.45	167.59	148.58	122.20
5.08	180.00	180.00	180.00	180.00	177.77	173.83	162.12	144.97	119.98
6.10	180.00	180.00	180.00	178.90	176.93	170.01	159.44	141.12	117.61
7.12	180.00	180.00	179.46	178.47	174.50	168.22	155.63	138.57	116.05
8.13	180.00	179.74	179.24	177.01	173.37	165.13	153.44	135.89	114.39
9.15	179.74	179.62	178.39	176.32	171.11	163.43	150.56	133.95	113.20
10.17	179.62	179.07	177.98	174.78	169.90	160.88	148.73	131.92	111.95
11.18	179.08	178.81	176.94	173.95	167.87	159.34	146.44	130.37	110.99
12.20	178.81	178.03	176.39	172.43	166.67	157.19	144.88	128.75	110.00
13.22	178.04	177.61	175.25	171.54	164.85	155.80	143.01	127.46	109.21
14.23	177.62	176.66	174.59	170.08	163.69	153.96	141.66	126.13	108.39
15.25	176.68	176.11	173.39	169.16	162.05	152.70	140.07	125.04	107.72
16.27	176.12	175.06	172.65	167.75	160.95	151.09	138.89	123.92	107.03
17.28	175.08	174.40	171.42	166.82	159.45	149.94	137.53	122.98	106.45
18.30	174.41	173.27	170.62	165.46	158.40	148.51	136.48	122.01	105.85
19.32	173.29	172.53	169.39	164.53	157.01	147.46	135.28	121.18	105.34
20.33	172.54	171.36	168.54	163.22	156.01	146.17	134.34	120.33	104.82
21.35	171.38	170.56	167.32	162.29	154.72	145.19	133.27	119.59	104.36
22.37	170.57	169.37	166.44	161.04	153.76	144.02	132.41	118.83	103.90
23.38	169.39	168.52	165.23	160.12	152.55	143.10	131.44	118.16	103.49
24.40	168.54	167.33	164.34	158.91	151.63	142.02	130.65	117.48	103.06
25.42	167.35	166.45	163.14	158.00	150.49	141.15	129.76	116.87	102.69
26.43	166.47	165.27	162.24	156.84	149.59	140.14	129.02	116.24	102.30
27.45	165.29	164.37	161.07	155.93	148.51	139.32	128.20	115.67	101.95
28.47	164.39	163.21	160.17	154.81	147.64	138.37	127.51	115.09	101.59
29.48	163.23	162.30	159.03	153.92	146.61	137.59	126.75	114.56	101.27
30.50	162.31	161.15	158.13	152.84	145.77	136.70	126.09	114.02	100.93

The calculations are therefore stable for $n = 1$.

b) For $n = 2$

$\Delta x_0 = 10$ cm			$\Delta t_2 = 62$ seconds		$Bi_\Delta = 0.625$		$Fo_\Delta = 0.49090$		
x (cm)	0	10	20	30	40	50	60	70	80
t (minutes)	Temperatures $\theta(x,t)$ in °C								
0.00	180.00	180.00	180.00	180.00	180.00	180.00	180.00	180.00	180.00
1.03	180.00	180.00	180.00	180.00	180.00	180.00	180.00	180.00	141.54
2.07	180.00	180.00	180.00	180.00	180.00	180.00	180.00	160.81	129.73
3.10	180.00	180.00	180.00	180.00	180.00	180.00	170.42	154.88	126.08
4.13	180.00	180.00	180.00	180.00	180.00	175.22	167.44	148.26	122.01
5.17	180.00	180.00	180.00	180.00	177.62	173.72	161.75	144.73	119.84
6.20	180.00	180.00	180.00	178.81	176.86	169.69	159.23	140.80	117.42
7.23	180.00	180.00	179.41	178.43	174.26	168.05	155.25	138.33	115.90
8.27	180.00	179.70	179.22	176.83	173.24	164.76	153.20	135.58	114.20
9.30	179.70	179.61	178.27	176.23	170.80	163.22	150.18	133.70	113.05
10.33	179.61	178.99	177.92	174.54	169.73	160.50	148.47	131.62	111.76
11.37	178.99	178.77	176.77	173.83	167.52	159.10	146.06	130.12	110.84
12.40	178.77	177.88	176.30	172.15	166.47	156.80	144.61	128.45	109.82
13.43	177.88	177.53	175.02	171.38	164.48	155.54	142.63	127.22	109.06
14.47	177.53	176.45	174.46	169.75	163.46	153.56	141.38	125.85	108.21
15.50	176.45	176.00	173.10	168.96	161.66	152.43	139.70	124.80	107.57
16.53	176.00	174.78	172.48	167.38	160.70	150.68	138.61	123.64	106.85
17.57	174.78	174.24	171.08	166.59	159.04	149.66	137.16	122.74	106.30
18.60	174.24	172.94	170.42	165.06	158.13	148.10	136.20	121.73	105.68
19.63	172.94	172.33	169.00	164.27	156.59	147.16	134.92	120.94	105.20
20.67	172.33	170.97	168.30	162.80	155.72	145.76	134.05	120.06	104.65
21.70	170.98	170.32	166.89	162.01	154.28	144.89	132.91	119.35	104.22
22.73	170.32	168.94	166.17	160.59	153.45	143.60	132.12	118.57	103.73
23.77	168.94	168.24	164.76	159.81	152.10	142.79	131.08	117.93	103.34
24.80	168.25	166.85	164.03	158.43	151.30	141.59	130.36	117.21	102.90
25.83	166.86	166.14	162.65	157.67	150.01	140.83	129.40	116.63	102.54
26.87	166.14	164.75	161.90	156.33	149.25	139.71	128.73	115.98	102.14
27.90	164.76	164.02	160.55	155.58	148.02	138.99	127.85	115.44	101.81
28.93	164.03	162.65	159.80	154.29	147.29	137.94	127.22	114.83	101.43
29.97	162.66	161.92	158.47	153.55	146.11	137.25	126.38	114.33	101.12
31.00	161.92	160.57	157.73	152.30	145.40	136.25	125.79	113.75	100.77

The calculations always remain stable for $n = 2$.

c) For $n = 3$

$\Delta x_0 = 10$ cm			$\Delta t_3 = 63$ seconds		$Bi_\Delta = 0.625$			$Fo_\Delta = 0.5071$	
x (cm)	0	10	20	30	40	50	60	70	80
t (minutes)	Temperatures $\theta(x,t)$ in °C								
0.00	180.00	180.00	180.00	180.00	180.00	180.00	180.00	180.00	180.00
1.05	180.00	180.00	180.00	180.00	180.00	180.00	180.00	180.00	141.54
2.10	180.00	180.00	180.00	180.00	180.00	180.00	180.00	160.50	129.54
3.15	180.00	180.00	180.00	180.00	180.00	180.00	170.11	154.69	125.96
4.20	180.00	180.00	180.00	180.00	180.00	174.99	167.30	147.94	121.81
5.25	180.00	180.00	180.00	180.00	177.46	173.63	161.38	144.51	119.70
6.30	180.00	180.00	180.00	178.71	176.81	169.36	159.04	140.48	117.22
7.35	180.00	180.00	179.35	178.40	174.00	167.90	154.86	138.10	115.75
8.40	180.00	179.67	179.20	176.65	173.14	164.38	152.97	135.27	114.01
9.45	179.66	179.60	178.14	176.16	170.48	163.04	149.78	133.47	112.90
10.50	179.60	178.89	177.88	174.28	169.59	160.09	148.23	131.31	111.58
11.55	178.88	178.73	176.57	173.72	167.15	158.89	145.66	129.88	110.70
12.60	178.73	177.71	176.22	171.83	166.30	156.37	144.37	128.16	109.63
13.65	177.70	177.47	174.75	171.25	164.07	155.32	142.23	126.99	108.91
14.70	177.47	176.21	174.36	169.39	163.27	153.12	141.14	125.55	108.03
15.75	176.19	175.91	172.78	168.81	161.23	152.19	139.31	124.57	107.43
16.80	175.91	174.46	172.35	166.98	160.49	150.24	138.37	123.35	106.68
17.85	174.44	174.13	170.70	166.41	158.58	149.42	136.78	122.51	106.16
18.90	174.12	172.55	170.26	164.61	157.91	147.65	135.95	121.45	105.51
19.95	172.53	172.19	168.56	164.08	156.11	146.92	134.53	120.72	105.06
21.00	172.18	170.52	168.13	162.31	155.49	145.30	133.81	119.78	104.48
22.05	170.49	170.15	166.39	161.80	153.78	144.64	132.52	119.14	104.08
23.10	170.14	168.42	165.97	160.06	153.21	143.13	131.88	118.29	103.56
24.15	168.39	168.05	164.21	159.58	151.57	142.54	130.70	117.71	103.21
25.20	168.05	166.28	163.81	157.87	151.05	141.11	130.12	116.94	102.73
26.25	166.25	165.93	162.05	157.43	149.47	140.58	129.01	116.42	102.41
27.30	165.92	164.13	161.67	155.74	149.00	139.22	128.49	115.70	101.97
28.35	164.10	163.79	159.91	155.33	147.46	138.74	127.45	115.22	101.68
29.40	163.79	161.98	159.55	153.66	147.03	137.43	126.97	114.55	101.26
30.45	161.95	161.67	157.79	153.29	145.53	136.99	125.98	114.11	100.99
31.50	161.66	159.85	157.47	151.64	145.13	135.74	125.55	113.48	100.60

The calculations are always stable for $n = 3$.

d) For $n = 4$

$\Delta x_0 = 10$ cm			$\Delta t_4 = 64$ seconds		$Bi_\Delta = 0.625$		$Fo_\Delta = 0.5151$	
x (cm)	0	10	20	30	40	50	60	70	80
t (minutes)	Temperatures $\theta(x,t)$ in °C								
0.00	180.00	180.00	180.00	180.00	180.00	180.00	180.00	180.00	180.00
1.07	180.00	180.00	180.00	180.00	180.00	180.00	180.00	180.00	141.54
2.13	180.00	180.00	180.00	180.00	180.00	180.00	180.00	160.19	129.35
3.20	180.00	180.00	180.00	180.00	180.00	180.00	169.79	154.51	125.85
4.27	180.00	180.00	180.00	180.00	180.00	174.74	167.18	147.62	121.61
5.33	180.00	180.00	180.00	180.00	177.29	173.55	161.00	144.30	119.57
6.40	180.00	180.00	180.00	178.61	176.76	169.01	158.86	140.16	117.02
7.47	180.00	180.00	179.28	178.37	173.72	167.78	154.46	137.88	115.62
8.53	180.00	179.63	179.18	176.44	173.06	163.98	152.78	134.95	113.82
9.60	179.62	179.59	178.00	176.11	170.13	162.88	149.36	133.25	112.77
10.67	179.59	178.79	177.85	174.00	169.48	159.65	148.03	131.00	111.38
11.73	178.76	178.72	176.35	173.65	166.75	158.72	145.24	129.67	110.56
12.80	178.71	177.52	176.18	171.48	166.17	155.91	144.16	127.85	109.45
13.87	177.49	177.44	174.45	171.17	163.62	155.14	141.81	126.77	108.78
14.93	177.44	175.92	174.30	168.97	163.14	152.64	140.93	125.25	107.85
16.00	175.88	175.87	172.39	168.71	160.74	152.02	138.89	124.36	107.30
17.07	175.87	174.08	172.29	166.50	160.35	149.75	138.17	123.06	106.50
18.13	174.03	174.08	170.23	166.32	158.06	149.25	136.35	122.31	106.04
19.20	174.08	172.07	170.20	164.08	157.77	147.14	135.76	121.16	105.33
20.27	172.01	172.14	168.01	163.98	155.54	146.76	134.10	120.53	104.94
21.33	172.14	169.95	168.06	161.71	155.36	144.76	133.63	119.49	104.30
22.40	169.88	170.11	165.76	161.71	153.17	144.49	132.08	118.95	103.97
23.47	170.12	167.75	165.92	159.40	153.10	142.57	131.71	118.00	103.38
24.53	167.68	168.02	163.50	159.51	150.92	142.40	130.24	117.53	103.10
25.60	168.03	165.52	163.77	157.14	150.96	140.53	129.96	116.64	102.55
26.67	165.44	165.92	161.26	157.37	148.77	140.45	128.54	116.24	102.30
27.73	165.93	163.27	161.66	154.94	148.92	138.60	128.34	115.40	101.78
28.80	163.19	163.81	159.03	155.30	146.71	138.63	126.96	115.05	101.57
29.87	163.83	161.03	159.57	152.80	146.97	136.78	126.84	114.24	101.07
30.93	160.95	161.72	156.83	153.28	144.72	136.91	125.47	113.95	100.89
32.00	161.74	158.80	157.52	150.70	145.11	135.04	125.43	113.16	100.40

Instability appears in the last values of θ_0. n is therefore equal to 4.

The following figure illustrates the emergence of the instability with an integration step $\Delta t = 64$ seconds.

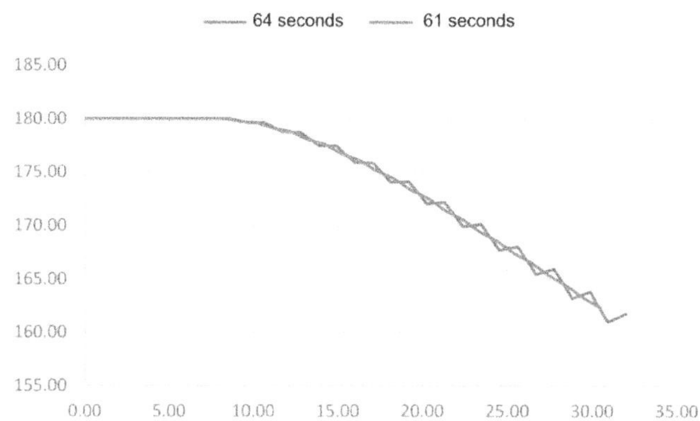

For better illustration, the calculations were continued for $n = 10$. The following figure shows the amplification of the stability as a function of the integration increment, Δt.

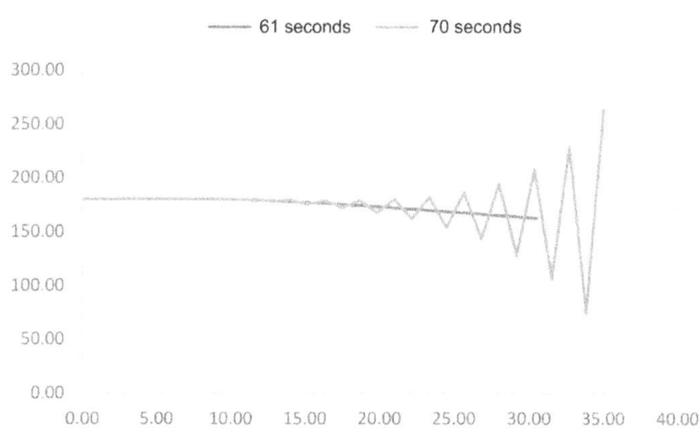

5) Comparison of results

In this part, we keep $\Delta x = \Delta x_0$.

a) Calculations of the temperature field for $\Delta t_1 = \dfrac{\Delta t_0}{2}$ *and* $\Delta t_2 = \dfrac{\Delta t_0}{4}$

i) For $\Delta t = 30$ seconds

$\Delta x_0 = 10$ cm			$\Delta t = 30$ seconds			$Bi_{= 0.625}$		$Fo_{= 0.2415}$	
x (cm)	0	10	20	30	40	50	60	70	80
t (minutes)	Temperatures (x,t) in °C								
0.00	180.00	180.00	180.00	180.00	180.00	180.00	180.00	180.00	180.00
0.50	180.00	180.00	180.00	180.00	180.00	180.00	180.00	180.00	141.54
1.00	180.00	180.00	180.00	180.00	180.00	180.00	180.00	170.71	135.82
2.00	180.00	180.00	180.00	180.00	180.00	179.46	175.11	159.87	129.15
3.00	180.00	180.00	180.00	179.97	179.58	177.40	170.00	153.01	124.93
4.00	180.00	180.00	179.97	179.73	178.60	174.82	165.53	147.99	121.84
5.00	180.00	179.98	179.84	179.24	177.26	172.17	161.69	144.05	119.41
6.00	179.97	179.90	179.58	178.52	175.72	169.59	158.34	140.82	117.43
7.00	179.89	179.76	179.19	177.65	174.10	167.16	155.40	138.09	115.75
8.00	179.74	179.53	178.69	176.66	172.46	164.87	152.79	135.74	114.30
9.00	179.51	179.21	178.09	175.60	170.83	162.74	150.46	133.69	113.04
10.00	179.19	178.79	177.41	174.49	169.24	160.75	148.34	131.87	111.92
10.50	179.00	178.56	177.04	173.93	168.46	159.80	147.36	131.03	111.40
11.00	178.79	178.30	176.65	173.36	167.69	158.89	146.42	130.23	110.91
12.00	178.29	177.72	175.84	172.21	166.19	157.14	144.66	128.76	110.00
13.00	177.72	177.07	174.98	171.06	164.74	155.51	143.04	127.41	109.18
14.00	177.08	176.36	174.09	169.90	163.34	153.97	141.54	126.18	108.42
15.00	176.37	175.60	173.16	168.76	162.00	152.52	140.15	125.04	107.72
16.00	175.61	174.78	172.21	167.62	160.69	151.14	138.85	123.99	107.07
17.00	174.80	173.93	171.23	166.49	159.44	149.83	137.63	123.01	106.47
18.00	173.95	173.04	170.24	165.38	158.22	148.59	136.48	122.10	105.91
19.00	173.06	172.12	169.24	164.27	157.03	147.41	135.40	121.24	105.38
20.00	172.15	171.18	168.23	163.18	155.89	146.27	134.37	120.43	104.88
21.00	171.21	170.22	167.21	162.10	154.77	145.18	133.40	119.66	104.41
22.00	170.25	169.24	166.19	161.03	153.69	144.13	132.47	118.94	103.96
23.00	169.27	168.25	165.17	159.98	152.63	143.12	131.58	118.25	103.54
23.50	168.78	167.76	164.66	159.46	152.11	142.63	131.14	117.91	103.33
24.00	168.29	167.26	164.15	158.94	151.59	142.14	130.72	117.59	103.13
24.50	167.79	166.76	163.64	158.43	151.09	141.67	130.31	117.27	102.93
25.00	167.29	166.25	163.14	157.91	150.58	141.20	129.90	116.96	102.74
25.50	166.79	165.75	162.63	157.40	150.09	140.74	129.50	116.65	102.55
26.00	166.29	165.25	162.12	156.90	149.60	140.28	129.11	116.35	102.37
26.50	165.79	164.74	161.61	156.40	149.11	139.83	128.73	116.06	102.19
27.00	165.28	164.24	161.11	155.90	148.63	139.39	128.35	115.77	102.01
28.00	164.27	163.23	160.11	154.91	147.68	138.53	127.61	115.21	101.66
29.00	163.27	162.23	159.11	153.93	146.76	137.68	126.90	114.66	101.33
30.00	162.26	161.22	158.12	152.97	145.84	136.86	126.20	114.14	101.01

ii) For Δt = 15 seconds

Δx_0 = 10 cm			Δt = 15 seconds		$Bi_{= 0.625}$		$Fo_{= 0.1207}$	
x (cm)	0	10	20	30	40	50	60	70	80
t (minutes)	Temperatures (x,t) in °C								
0.00	180.00	180.00	180.00	180.00	180.00	180.00	180.00	180.00	180.00
0.25	180.00	180.00	180.00	180.00	180.00	180.00	180.00	180.00	141.54
0.50	180.00	180.00	180.00	180.00	180.00	180.00	180.00	175.36	138.68
1.00	180.00	180.00	180.00	180.00	180.00	179.93	178.55	168.20	134.28
2.00	180.00	180.00	180.00	179.99	179.84	178.77	173.81	158.62	128.38
3.00	180.00	180.00	179.98	179.86	179.21	176.64	169.03	152.20	124.43
4.00	180.00	179.98	179.90	179.52	178.14	174.13	164.79	147.39	121.47
5.00	179.97	179.93	179.71	178.96	176.78	171.56	161.09	143.59	119.13
6.00	179.91	179.82	179.41	178.21	175.27	169.07	157.85	140.44	117.20
7.00	179.80	179.64	178.99	177.32	173.68	166.70	154.99	137.78	115.56
8.00	179.61	179.37	178.46	176.33	172.06	164.47	152.44	135.48	114.14
9.00	179.35	179.02	177.84	175.28	170.47	162.38	150.15	133.46	112.90
10.00	178.99	178.58	177.14	174.18	168.90	160.43	148.07	131.67	111.80
11.00	178.56	178.06	176.38	173.05	167.37	158.60	146.18	130.06	110.80
12.00	178.04	177.46	175.56	171.91	165.90	156.88	144.45	128.60	109.91
13.00	177.46	176.80	174.70	170.77	164.47	155.27	142.85	127.27	109.09
14.00	176.80	176.08	173.80	169.63	163.08	153.74	141.36	126.05	108.34
15.00	176.09	175.31	172.88	168.49	161.75	152.31	139.98	124.93	107.65
16.00	175.32	174.49	171.92	167.36	160.46	150.94	138.69	123.88	107.00
17.00	174.51	173.64	170.95	166.24	159.21	149.65	137.48	122.91	106.41
18.00	173.65	172.75	169.97	165.13	158.00	148.41	136.35	122.00	105.85
19.00	172.77	171.83	168.97	164.03	156.82	147.23	135.27	121.15	105.32
20.00	171.86	170.89	167.96	162.94	155.68	146.10	134.25	120.34	104.83
21.00	170.92	169.94	166.95	161.87	154.57	145.02	133.28	119.58	104.36
22.00	169.96	168.96	165.94	160.80	153.49	143.98	132.35	118.86	103.91
23.00	168.99	167.98	164.92	159.76	152.44	142.97	131.46	118.17	103.49
24.00	168.01	166.99	163.90	158.72	151.41	142.00	130.61	117.51	103.08
25.00	167.02	165.99	162.89	157.70	150.40	141.06	129.79	116.88	102.70
26.00	166.02	164.99	161.88	156.69	149.42	140.14	129.01	116.28	102.32

27.00	165.03	163.99	160.88	155.69	148.46	139.25	128.25	115.69	101.97
28.00	164.02	162.99	159.88	154.71	147.51	138.39	127.51	115.13	101.62
29.00	163.02	161.99	158.88	153.73	146.59	137.55	126.80	114.59	101.29
30.00	162.02	160.99	157.90	152.78	145.68	136.73	126.10	114.07	100.96

b) Assessing the approximate analytical solution

For Fourier numbers > 0.2, the approximate analytical solutions to the heat equation are given below as a function of the Fourier and Biot numbers (see Chapter 6, section 6.6.2.2):

Bi	$T(\overline{x},Fo)$
0	The temperature remains constant
$0 < Bi < 0.1$	$\theta = \theta_e + (\theta_0 - \theta_e)\ e^{-Bi\ Fo}$
0.1	$T(\overline{x},Fo) \approx 1.0161\ e^{-0.09678\ Fo} \cos(0.3111\ \overline{x})$
1	$T(\overline{x},Fo) \approx 1.1191\ e^{-0.7401\ Fo} \cos(0.8603\overline{x}) - 0.1517\ e^{-11.7347\ Fo} \cos(3.4256\overline{x})$
5	$T(\overline{x},Fo) \approx 1.2402\ e^{-1.7261\ Fo} \cos(1.3138\overline{x}) - 0.3442\ e^{-16.2697\ Fo} \cos(4.0336\overline{x})$
10	$T(\overline{x},Fo) \approx 1.2621\ e^{-2.0449\ Fo} \cos(1.43\overline{x}) - 0.3935\ e^{-18.5416\ Fo} \cos(4.306\overline{x})$
100	$T(\overline{x},Fo) \approx 1.2731\ e^{-2.4180\ Fo} \cos(1.555\overline{x}) - 0.4240\ e^{-21.7716\ Fo} \cos(4.666\overline{x})$
∞	$T(\overline{x},Fo) \approx \frac{4}{\pi} e^{-\frac{\pi^2}{4}Fo} \cos\left(\frac{\pi}{2}\overline{x}\right) - \frac{4}{3\pi} e^{-\frac{9\pi^2}{4}Fo} \cos\left(\frac{3\pi}{2}\overline{x}\right)$

where: $T = \frac{\theta - \theta_b}{\theta_i - \theta_b}$; $\overline{x} = \frac{x}{L}$; $Bi = \frac{h\ L}{\lambda}$ and $Fo = \frac{\alpha t}{L^2}$.

In the scenario of interest to us:

$$Bi = 5 \text{ and } Fo > 0.2 \Rightarrow t > 0.2\frac{L^2}{\alpha}$$

$$t > 26.6 \text{ mins}$$

The expression of the approximate analytical solution to be used is then:

$$T(\overline{x},Fo) \approx 1.2402\ e^{-1.7261\ Fo} \cos(1.3138\overline{x}) - 0.3442\ e^{-16.2697\ Fo} \cos(4.0336\overline{x})$$

i.e. for $x = 0$: $\bar{x} = 0 \Rightarrow T(0, Fo) \approx 1.2402\ e^{-1.7261\ Fo} - 0.3442\ e^{-16.2697\ Fo}$.

Calculating $T(0, Fo)$ enables us to determine $\theta(0, t)$ by:

$$\theta(0,t) = \theta_b + (\theta_i - \theta_b)\ T(0, Fo)$$

The following table presents the values of θ_0 obtained for t = 30 minutes, 40 minutes, 50 minutes and 60 minutes, for different integration increments, τ.

		(0,t)				
t (min)	*Fo*	*t* = 60 seconds	*t* = 30 seconds	*t* = 15 seconds		(0,*Fo*)
30.00	2.26E-01	162.77	162.26	162.02	8.30E-01	163.04
40.00	3.02E-01	152.91	152.53	152.34	7.34E-01	153.41
50.00	3.77E-01	144.00	143.71	143.56	6.46E-01	144.59
60.00	4.53E-01	136.12	135.89	135.78	5.67E-01	136.75

Note that the approximate analytical method gives results close to those obtained for Δt = 60 seconds.

Appendix

Database

A.1. Introduction

This appendix presents the physical data encountered in Volumes 1 and 2 of the *Energy Engineering* series. The following parameters provide the basis upon which the global database, containing the most-used data in energy and mass transfer calculations, relies upon: viscosity, thermal conductivity, diffusion coefficients, heat transfer coefficients, properties of molten salts, etc.

Let us recall that this database has been constructed progressively. Indeed, each volume in the series contributes to its development by "feeding" it with the parameters encountered along the way.

This appendix also includes unit conversion tables, which are exceptionally useful when handling different unit systems, as well as a table of fundamental constants.

If the exactitude of the data discussed here does not seem to be sufficient, please refer to the physical data sources in the Bibliography.

However, on the other hand, if the desired data discusses uncommon materials or components that are not referred to in the sources, then the values of the physical properties can be estimated using the methods and correlations presented in Volume 5.

A.2. Collision diameters and interaction energies

The collision diameters and interaction energy values to be used in Chapman Enskog equations and in Bird equations (see Volume 5) are presented in Table A.1:

Formula	Name	σ_A (Angström)	$\frac{\varepsilon}{k}$ (°K)
A	Argon	3.542	93.3
He	Helium	2.551	10.22
Kr	Krypton	3.655	178.9
Ne	Neon	2.820	32.8
Xe	Xenon	4.047	1.0
Air	Air	3.711	78.6
BCl_3	Boron Chloride	5.127	337.7
BF	Boron Monofluoride	4.198	186.3
$B(OCH3)_3$	Trimethyl Borate	5.503	396.7
Br_2	Bromine	4.296	507.9
CCl_4	Carbon Tetrachloride	5.947	322.7
CF_4	Carbon Tetrafluoride	4.662	134.0
$CHCL_3$	Chloroform	5.389	340.2
CH_2Cl_2	Methylene Chloride	4.898	356.3
CH_3Br	Methyl Boride	4.118	449.2
CH_3Cl	Methyl Chloride	4.182	350
CH_3OH	Methanol	3.626	481.8
CH_4	Methane	3.758	148.6
CO	Carbon Monoxide	3.690	91.7
CO_2	Carbon Dioxide	3.941	195.2
CS_2	Carbon Disulfide	4.483	467
C_2H_2	Acetylene	4.033	231.8
C_2H_4	Ethylene	4.163	224.7
C_2H_6	Ethane	4.443	215.7
C_2H_5Cl	Ethyl Chloride	4.898	300
C_2H_5CH	Ethanol	4.530	362.6
CH_3OCH_3	Dimethyl Ether	4.307	395.0
CH_2CHCH_3	Propylene	4.678	298.9

Formula	Name	σ_A (Angström)	$\frac{\varepsilon}{k}$ (°K)
CH_3CCH	Methylacetylene	4.761	251.8
C_3H_6	Cyclopropane	4.807	248.9
C_3H_8	Propane	5.118	237.1
n-C_3H_7OH	n-Propyl Alcohol	4.549	576.7
CH_3COCH_3	Acetone	4.600	560.2
CH_3COOCH_3	Methyl Acetate	4.936	469.8
n-C_4H_{10}	n-Butane	4.687	531.4
i-C_4H_{10}	Isobutane	5.278	330.1
$C_2H_5OC_2H_5$	Ethel Ether	5.678	313.8
$CH_3COOC_2H_5$	Ethel Acetate	5.205	521.3
n-C_5H_{12}	n-Pentane	5.784	341.1
$C(CH_3)_4$	2,2-Dimethylpropane	6.464	193.4
C_6H_6	Benzene	5.349	412.3
C_6H_{12}	Cyclohexane	6.182	297.1
n-C_6H_{14}	n-Hexane	5.949	399.3
Cl_2	Chlorine	4.217	316.0
F_2	Fluorine	3.357	112.6
HBr	Hydrogen Bromide	3.353	449
HCl	Hydrogen Chloride	3.339	344.7
HF	Hydrogen Fluoride	3.148	330
HI	Hydrogen Iodide	4.211	288.7
H_2	Hydrogen	2.827	59.7
H_2O	Water	2.641	809.1
H_2O_2	Hydrogen Peroxide	4.196	289.3
H_2S	Hydrogen Sulfide	3.623	301.1
Hg	Mercury	2.969	750
$HgBr_2$	Mercuric Bromide	5.080	686.2
$HgCl_2$	Mercuric Chloride	4.550	750

Formula	Name	σ_A (Angström)	$\frac{\varepsilon}{k}$ (°K)
HgI_2	Mercuric Iodide	5.625	695.6
I_2	Iodine	5.160	474.2
NH_3	Ammonia	2.900	558.3
NO	Nitric Oxide	3.492	116.7
N_2	Nitrogen	3.798	71.4
N_2O	Nitrogen Oxide	3.828	232.4
O_2	Oxygen	3.467	106.7
PH_3	Phosphine	3.981	251.5
SF_6	Sulfur Hexafluoride	5.128	222.1
SO_2	Sulfur Dioxide	4.112	335.4
SiF_4	Silicon Tetrafluoride	4.880	171.9
UF_6	Uranium Hexafluoride	5.967	236.8

Table A.1. *Collision diameters and interaction energies*

A.3. Ω_{AB} values based on Lennard-Jones potential

To be used in Chapman–Enskog equations and in Bird equations (see Volume 5).

$\frac{kT}{\varepsilon_{AB}}$	Ω_{AB}
0.30	2.662
0.35	2.476
0.40	2.218
0.45	2.184
0.50	2.066
0.55	1.966
0.60	1.877
0.65	1.798
0.70	1.729
0.75	1.667

$\frac{kT}{\varepsilon_{AB}}$	Ω_{AB}
0.80	1.612
0.85	1.562
0.90	1.517
0.95	1.476
1.00	1.439
1.05	1.406
1.10	1.375
1.15	1.346
1.20	1.320
1.25	1.296
1.30	1.273
1.35	1.253
1.40	1.233
1.45	1.215
1.50	1.198
1.55	1.182
1.60	1.167
1.65	1.153
1.70	1.140
1.75	1.128
1.80	1.116
1.85	1.105
1.90	1.094
1.95	1.084
2.00	1.075
2.1	1.057
2.2	1.041
2.3	1.026
2.4	1.012
2.5	0.9996
2.6	0.9878
2.7	0.9770

$\frac{kT}{\varepsilon_{AB}}$	Ω_{AB}
2.8	0.9672
2.9	0.9576
3.0	0.9490
3.1	0.9406
3.2	0.9328
3.3	0.9256
3.4	0.9186
3.5	0.9120
3.6	0.9058
3.7	0.8998
3.8	0.8942
3.9	0.8888
4.0	0.8836
4.1	0.8788
4.2	0.8740
4.3	0.8694
4.4	0.8652
4.5	0.8610
4.6	0.8568
4.7	0.8530
4.8	0.8492
4.9	0.8456
5.0	0.8422
6	0.8124
7	0.7896
8	0.7712
9	0.7556
10	0.7424
20	0.6640
30	0.6232
40	0.5960
50	0.5756

$\frac{kT}{\varepsilon_{AB}}$	Ω_{AB}
60	0.5596
70	0.5464
80	0.5352
90	0.5256
100	0.5130
200	0.4644
400	0.4170

Table A.2. *Ω_{AB} as a function of $\frac{kT}{\varepsilon_{AB}}$*

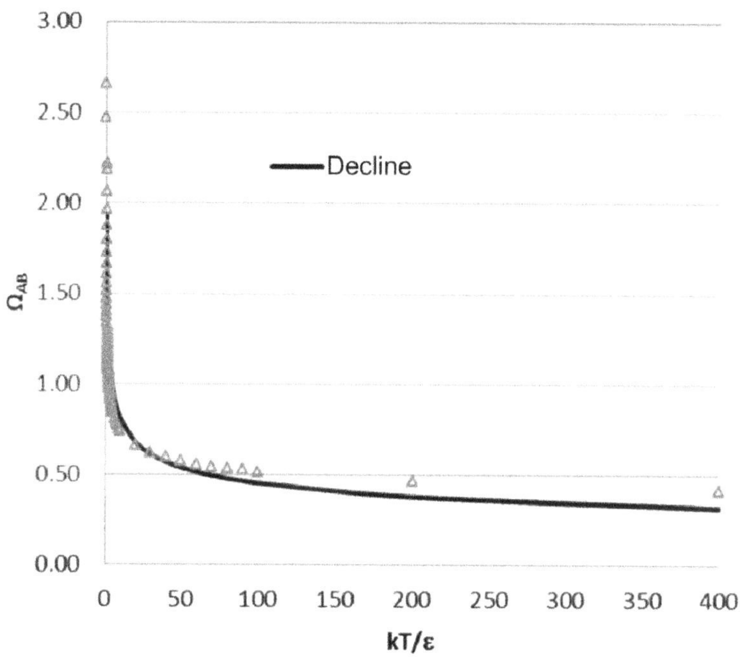

Figure A.1. $\Omega_{AB} = 1.435\left(\frac{kT}{\varepsilon_{AB}}\right)^{-0,25}$

A.4. Diffusion volumes

Values of diffusion volumes to be used with Fuller, Schettler and Giddings relations (see Volume 5).

Constituants	V (cm^3/mol)
C	16.5
$C\ Cl_2\ F_2$	114.8
Cl	19.5
Cl_2	37.7
N	5.69
S	17.0
SF_6	69.7
SO_2	41.1
Xe	37.9
Air	20.1
Br_2	67.2
CO	18.9
CO_2	26.9
Aromatic compounds	-20.2
D_2	6.70
H	1.98
H_2	7.07
H_2O	12.7
He	2.88
Heterocyclic compounds	-20.2
Kr	22.8
N_2	17.9
N_2O	35.9
NH_3	14.9
O	5.48
O_2	16.6

Table A.3. *Diffusion volumes (cm^3/mol)*

A.5. Densities

A.5.1. *Densities for certain metals and alloys*

Metals and alloys	ρ (kg/m^{-3})
Carbon steel	7,833
Stainless steel 15% Cr, 10% Ni	7,864
Stainless steel 18% Cr, 8% Ni	7,816
Stainless steel 25% Cr, 20% Ni	7,864
Aluminium	2,707
Silver	10,525
Bronze 75% Cu, 25% Sn	8,800
Bronze 92% Cu, 8% Al	7,900
Carbon graphite	2,250
Chrome	2,118
Constantan 60% Cu, 40% Ni	8,922
Copper	8,954
Cupronickel 70% Cu, 30% Ni	8,900
Duralumin	2,787
Tin	7,304
Iron	7,870
Cast iron	7,849
Brass 70% Cu, 30% Zn	8,522
Magnesium	1,740
Gold	19,300
Platinum	21,400
Lead	11,373
Liquid sodium	930
Titanium	4,500
Tungsten	19,350
Zinc	7,144

Table A.4. *Densities for metals and alloys (kg/m^3)*

A.5.2. *Densities for certain construction materials*

Construction materials	ρ (kg m^{-3})
Asbestos[1]	575
Slate	2,400
Asphalt	2,115
Basalt	2,850
Carnevous concrete	1,900
Solid concrete	2,300
Bitumen (cardboard)	1,050
Low-density hardwood	525
Medium-density hardwood	675
Very-low-density hardwood	375
Low-density softwood	375
Medium-density softwood	500
Very-low-density softwood	375
Clay brick	1,800
Hard limestone	2,450
Soft limestone	1,650
Rubber (natural)	1,150
Rubber (vulcanized)	1,100
Tile	2,400
Cardboard	650

1 Warning: inhaling the dust of this product can have carcinogenic consequences.

Construction materials	ρ (kg m^{-3})
Okoumé plywood	400
Pine plywood	500
Leather	998
Ice	920
Granite	2,600
Gravel (loose)	1,800
Sandstone	2,500
Lava	2,350
Marble	2,700
Plaster	1,440
Plexiglass	1,190
Porcelain	2,400
Polyethylene	929
PVC	1,459
Sand	1,515
Shale	2,400
Teflon	2,170
Wet earth	1,900
Dry earth	1,500
Glass	2,300
Pyrex glass	2,220

Table A.5. *Densities for construction materials*

A.5.3. *Densities for certain thermal insulation materials*

Thermal insulation materials	Density	ρ (kg m^{-3})
Balsa	High	140
Cotton	High	80
Kapok	Average	50
Rock wool	Low	20
	Average	55
	High	135
Glass wool	Low	8
	Average	10
	Average	15
	High	40
Expanded cork	High	120
Carpet	High	200
Polyurethane (foam)	Low	32
	Average	50
	High	85
PVC (rigid foam)	Average	30
	High	40
Expanded polystyrene	Low	12
	Low	14
	High	18
Styrofoam	Average	30

Table A.6. *Densities according to manufacture (kg/m^3)*

A.6. Heat capacities

A.6.1. *Sensible heats for certain metals and alloys*

Material	C_p (J kg^{-1} °C^{-1})
Carbon steel	465
Stainless steel 15% Cr, 10% Ni	460
Stainless steel 18% Cr, 8% Ni	460
Stainless steel 25% Cr, 20% Ni	460
Aluminium	896
Silver	234
Bronze 75% Cu, 25% Sn	377
Bronze 92% Cu, 8% Al	377
Carbon graphite	707
Chrome	7,160
Constantan 60% Cu, 40% Ni	410
Copper	383
Cupronickel 70% Cu, 30% Ni	377
Duralumin	883
Tin	226
Iron	452
Cast iron	460
Brass 70% Cu, 30% Zn	385
Magnesium	1,004
Gold	128
Platinum	140
Lead	130
Liquid sodium	1,381
Titanium	523
Tungsten	134
Zinc	384

Table A.7. *Sensible heats for metals and alloys (J kg^{-1}°C^{-1})*

A.6.2. *Sensible heats of certain construction materials*

Construction materials	C_p (J kg^{-1} °C^{-1})
Asbestos[2]	1,046
Slate	879
Asphalt	920
Basalt	881
Cavernous concrete	879
Solid concrete	878
Bitumen (cardboard)	1,305
Low-density hardwood	3,143
Medium-density hardwood	3,156
Very-low-density hardwood	3,147
Low-density softwood	3,147
Medium-density softwood	3,160
Very-low-density softwood	3,147
Clay brick	878
Hard limestone	882
Soft limestone	879
Rubber (natural)	65
Rubber (vulcanized)	2,010
Tile	875
Cardboard	86

2 Warning: inhaling the dust of this product can have carcinogenic consequences.

Construction materials	C_p (J kg^{-1} °C^{-1})
Okoumé plywood	3,000
Pine plywood	3,000
Leather	120
Ice	2,040
Granite	881
Gravel (loose)	889
Sandstone	880
Lava	881
Marble	881
Plaster	840
Plexiglass	1,465
Porcelain	1,088
Polyethylene	1,830
PVC	930
Sand	800
Shale	879
Teflon	1,004
Wet earth	2,000
Dry earth	1,900
Glass	837
Pyrex glass	728

Table A.8. *Sensible heats for certain construction materials (J kg^{1}°C^{-1})*

A.6.3. *Sensible heats of certain thermal insulation materials*

Thermal insulation materials	C_p as a function of ρ	
	ρ (kg m^{-3})	C_p (J kg^{-1} °C^{-1})
Balsa	140	1,420
Cotton	80	1,300
Kapok	50	950
Rock wool	20	880
	55	880
	135	880
Glass wool	8	875
	10	880
	15	880
	40	880
Expanded cork	120	2,100
Carpet	200	1,300
Polyurethane (foam)	32	1,300
	50	1,360
	85	1,300
PVC (rigid foam)	30	1,300
	40	1,300
Expanded polystyrene	12	1,300
	14	1,300
	18	1,300
Styrofoam	30	1,250

Table A.9. *Sensible heats of certain thermal insulation materials*

A.7. Heat conductivities

A.7.1. *Heat conductivities of certain metals and alloys*

Metals and alloys	λ (W m^{-1} °C^{-1})
Carbon steel	54
Stainless steel 15% Cr, 10% Ni	20
Stainless steel 18% Cr, 8% Ni	16.3
Stainless steel 25% Cr, 20% Ni	13
Aluminium	204
Silver	407
Bronze 75% Cu, 25% Sn	188
Bronze 92% Cu, 8% Al	71
Carbon graphite	147
Chrome	449
Constantan 60% Cu, 40% Ni	22.7
Copper	386
Cupronickel 70% Cu, 30% Ni	29.3
Duralumin	164
Tin	64
Iron	73
Cast iron	59
Brass 70% Cu, 30% Zn	111
Magnesium	151
Gold	312
Platinum	69
Lead	35
Liquid sodium	84.5
Titanium	20.9
Tungsten	163
Zinc	112

Table A.10. *Heat conductivities (metals and alloys)*

A.7.2. *Heat conductivities of certain construction materials*

Construction materials	λ (W m^{-1} °C^{-1})
Asbestos[3]	0.15
Slate	2.2
Asphalt	0.062
Basalt	1.6
Cavernous concrete	1.4
Solid concrete	1.75
Bitumen (cardboard)	0.23
Low-density hardwood	0.15
Medium-density hardwood	0.23
Very-low-density hardwood	0.12
Low-density softwood	0.12
Medium-density softwood	0.15
Very-low-density softwood	0.12
Clay brick	1.15
Hard limestone	2.4
Soft limestone	1
Rubber (natural)	0.28
Rubber (vulcanized)	0.13

3 Warning: inhaling the dust of this product can have carcinogenic consequences.

Construction materials	λ (W m^{-1} °C^{-1})
Tile	2.4
Cardboard	0.09
Okoumé plywood	0.12
Pine plywood	0.15
Leather	0.159
Ice	1.88
Granite	3
Gravel (loose)	0.7
Sandstone	2.6
Lava	1.1
Marble	2.5
Plaster	0.48
Plexiglass	0.19
Porcelain	1.035
Polyethylene	0.46
PVC	0.21
Sand	0.6
Shale	2.2
Teflon	0.25
Wet earth	2
Dry earth	1
Glass	1.05
Pyrex glass	1.13

Table A.11. *Heat conductivities (construction materials)*

A.7.3. *Conductivities of certain thermal insulation materials*

Thermal insulation materials	λ as a function of ρ	
	ρ (kg m^{-3})	λ (W m^{-1} °C^{-1})
Balsa	140	0.054
Cotton	80	0.06
Kapok	50	0.035
Rock wool	20	0.047
	55	0.038
	135	0.041
Glass wool	8	0.051
	10	0.045
	15	0.041
	40	0.035
Expanded cork	120	0.044
Carpet	200	0.06
Polyurethane (foam)	32	0.03
	50	0.035
	85	0.045
PVC (rigid foam)	30	0.031
	40	0.041
Expanded polystyrene	12	0.047
	14	0.043
	18	0.041
Styrofoam	30	0.032

Table A.12. *Conductivities of certain thermal insulation materials*

A.8. Data on heat insulation materials

Thermal insulation materials	P (kg m^{-3})	C_p (Jkg^{-1} °C^{-1})	λ (Wm^{-1}°C^{-1})	T_{Max} (°C)	Durability[4]				Observations
					D	F	R	C	
Balsa	140	1,350	0.054	< 100	μ	μ	μ	A	– Comes in rolls – Renewable resource – High prices in areas of production
Cellulose (flakes)	200	890	0.045	< 100	μ	μ	μ	A	– Comes in flakes or rolls – Renewable resource – High prices in areas of production
Hemp	180	900	0.055	< 100	A	μ	G	A	– Comes in flakes or rolls – Renewable resource – High prices in areas of production

4 Durability: D: against dampness, F: against fire, R: against rodents, C: against chemical products. Durability scale: G: good durability, A: average durability, μ: poor durability.

<table>
<tr><td>Cotton</td><td>80</td><td>1,300</td><td>0.060</td><td>< 100</td><td>μ</td><td>μ</td><td>μ</td><td>A</td><td>– Comes in flakes or rolls
– Renewable resource
– High prices in areas of production</td></tr>
<tr><td rowspan="3">Rock wool</td><td>20</td><td>880</td><td>0.047</td><td>> 150</td><td>A[5]</td><td>G</td><td>G</td><td>G</td><td rowspan="3">– Excellent heat insulation at high temperatures
– Better crush resistance than glass wool
– Sold in/as:
- rolls
- soft felt
- felt mattresses, glued or bitumized on kraft or aluminium foil
- cylindrical shells of different diameters for piping
- flat rigid panels
– Non-renewable resource
– Good price–quality ratio</td></tr>
<tr><td>55</td><td>880</td><td>0.038</td><td>> 150</td><td>A[5]</td><td>G</td><td>G</td><td>G</td></tr>
<tr><td>135</td><td>880</td><td>0.041</td><td>> 150</td><td>A[5]</td><td>G</td><td>G</td><td>G</td></tr>
</table>

5 Requires a paper barrier; insulation panels or felts are glued onto bituminous kraft grids or aluminium foil.

Glass wool	8	875	0.051	> 150	A[5]	G	G	G	– Excellent heat insulation at high temperatures – Sold in/as: - rolls - soft felt - felt mattresses, glued or bitumized on kraft or aluminium paper - sleeves of different diameters for piping - flat rigid panels – Non-renewable resource – Good price/quality ratio
	10	880	0.045	> 150	A[5]	G	G	G	
	15	880	0.041	> 150	A[5]	G	G	G	
	40	880	0.035	> 150	A[5]	G	G	G	
Expanded cork	120	2,100	0.044	> 100	A[5]	μ	G	A	– Sold as chips or rigid panels – Does not stick to silicone, teflon and polythylene – Renewable resource – Good price–quality ratio

Carpet	200	1,300	0.06	< 60	μ	μ	A	μ	– Sold as rugs – Non-renewable resource – Good price–quality ratio
Polyvinyl chloride (PVC rigid foam)	30	1,300	0.031	≈ 100	G	μ	G	A	– Good insulation for medium to high temperatures – Sold as rigid panels – Non-renewable resource
	40	1,300	0.041	≈ 100	G	μ	G	A	
Polyurethane (foam)	32	1,300	0.03	≈ 100	G	μ[6]	A	A	– Excellent insulation for average–high temperatures – Sold in the following forms: - pressurized liquid spray: a 0.5 kg spray can fill, after expansion, a volume of about 20 liters - rigid panels - flexible plates - sleeves of different diameters – Non-renewable resource – Good quality–price ratio

6 Does not spread flames.

	50	1,360	0.035	≈ 100	G	μ[6]	A	A	
	85	1,300	0.045	≈ 100	G	μ[7]	A	A	
Expanded polystyrene obtained by modeling	15	1,300	0.040	≈ 85	A[5]	μ	μ	A[7]	– Can be moulded into different shapes: crates, boxes, rigid panels for walls, moulds for pipes, parts for tanks – Non-renewable resource – Good price–quality ratio
	20	1,300	0.041	≈ 85	A[5]	μ	μ	A[7]	
	25	1,300	0.041	≈ 85	A[5]	μ	μ	A[7]	
Compression moulded and expanded polystyrene	18	1,300	0.041	≈ 85	A[5]	μ	μ	A[7]	– Comes in the form of chips, panels or segments – Non-renewable resource – Good price–quality ratio
	26	1,300	0.041	≈ 85	A[5]	μ	μ	A[7]	
	30	1,300	0.041	≈ 85	A[5]	μ	μ	A[7]	
Sawdust or wood wool	160	190	0.059	< 100	μ	μ	G	A	– Advantageous if available – Renewable resource – Good price–quality ratio

Table A.13. *Physical data and mechanical properties of heat insulators*

7 Except in the presence of aggressive solvents.

A.9. Physical properties of water

θ	ρ	C_p	λ	$10^5\mu$	$10^7\alpha$	Pr
°C	$kg.m^{-3}$	$J.kg^{-1}.°C^{-1}$	$W.m^{-1}.°C^{-1}$	$Pa.s^{-1}$	$m^2.s^{-1}$	Adimensional
0	1,002	4,218	0.552	17.90	1.31	13.06
20	1,001	4,182	0.597	10.10	1.43	7.02
40	995	4,178	0.628	6.55	1.51	4.34
60	985	4,184	0.651	4.71	1.55	3.02
80	974	4,196	0.668	3.55	1.64	2.22
100	960	4,216	0.680	2.82	1.68	1.74
120	945	4,250	0.685	2.33	1.71	1.45
140	928	4,283	0.684	1.99	1.72	1.24
160	910	4,342	0.680	1.73	1.73	1.10
180	889	4,417	0.675	1.54	1.72	1.00
200	867	4,505	0.665	1.39	1.71	0.94
220	842	4,610	0.652	1.26	1.68	0.89
240	816	4,756	0.635	1.17	1.64	0.88
260	786	4,949	0.611	1.08	1.58	0.87
280	753	5,208	0.580	1.02	1.48	0.91
300	714	5,728	0.540	0.96	1.32	1.02

Table A.14. *Density (ρ), sensible heat (C_p), thermal conductivity (λ), thermal diffusivity (α) and Prandtl number (Pr) according to the temperature*

A.10. Physical properties of air

θ	ρ	C_p	λ	$10\ \mu$	$10\ \alpha$	Pr
°C	$kg.m^{-3}$	$J.kg^{-1}.°C^{-1}$	$W.m^{-1}.°C^{-1}$	$Pa.s^{-1}$	$m^2.s^{-1}$	Adimensional
0	1.292	1,006	0.0242	1.72	1.86	0.72
20	1.204	1,006	0.0257	1.81	2.1	0.71
40	1.127	1,007	0.0272	1.90	2.40	0.70
60	1.059	1,008	0.0287	1.99	2.69	0.70
80	0.999	1,010	0.0302	2.09	3.00	0.70
100	0.946	1,012	0.0318	2.18	3.32	0.69
120	0.898	1,014	0.0333	2.27	3.66	0.69
140	0.854	1,016	0.0345	2.34	3.98	0.69
160	0.815	1,019	0.0359	2.42	4.32	0.69
180	0.779	1,022	0.0372	2.50	4.67	0.69
200	0.746	1,025	0.0386	2.57	5.05	0.68
220	0.700	1,028	0.0399	2.64	5.43	0.68
240	0.688	1,032	0.0412	2.72	5.80	0.68
260	0.662	1,036	0.0425	2.79	6.20	0.68
280	0.638	1,040	0.0437	2.86	6.59	0.68
300	0.616	1,045	0.0450	2.93	6.99	0.68

Table A.15. *Density (ρ), sensible heat (C_p), thermal conductivity (λ), thermal diffusivity (α) and Prandtl number (Pr) according to the temperature*

A.11. ω_1 values for the analytical solution of heat equations

When exploring analytical solutions of heat equations, for one-dimensional planar, cylindrical or spherical configurations (see Chapter 6, section 7.2), the first solutions of the following equations are of major importance:

– for a solid plane: $\omega \, \mathrm{tg}(\omega) = \mathrm{Bi}$;

– for a cylindrical solid: $\omega \dfrac{J_1(\omega)}{J_0(\omega)} = \mathrm{Bi}$;

– for a spherical solid: $1 - \omega \cot(\omega) = \mathrm{Bi}$.

The first solutions, ω_1, of these equations are given in Table A.16, for different geometries and Biot numbers:

Bi	ω_1		
	Planar[8]	Cylindrical[9]	Spherical[10]
0.01	0.0998	0.1412	0.1731
0.02	0.1410	0.1995	0.2445
0.04	0.1987	0.2814	0.3450
0.06	0.2425	0.3438	0.4217
0.08	0.2791	0.3960	0.4860
0.1	0.3111	0.4417	0.5423
0.2	0.4328	0.6170	0.7593
0.3	0.5218	0.7465	0.9208
0.4	0.5932	0.8516	1.0528
0.5	0.6533	0.9408	1.1656
0.6	0.7051	1.0184	1.2644
0.7	0.7506	1.0873	1.3525
0.8	0.7910	1.1490	1.4320
0.9	0.8274	1.2048	1.5044
1	0.8603	1.2558	1.5708

8 ω_1 is the first solution to the equation: $\omega \, \mathrm{tg}(\omega) = \mathrm{Bi}$.

9 ω_1 is the first solution to the equation: $\omega \dfrac{J_1(\omega)}{J_0(\omega)} = \mathrm{Bi}$.

10 ω_1 is the first solution to the equation: $1 - \omega \cot(\omega) = \mathrm{Bi}$.

Bi	ω_1		
	Planar[8]	Cylindrical[9]	Spherical[10]
2	1.0769	1.5995	2.0288
3	1.1925	1.7887	2.2889
4	1.2646	1.9081	2.4556
5	1.3138	1.9898	2.5704
6	1.3496	2.049	2.6537
7	1.3766	2.0937	2.7165
8	1.3978	2.1286	2.7654
9	1.4149	2.1566	2.8044
10	1.4289	2.1795	2.8363
20	1.4961	2.288	2.9857
30	1.5202	2.3261	3.0372
40	1.5325	2.3455	3.0632
50	1.54	2.3272	3.0788
60	1.5552	2.3809	3.1102
∞	1.5708	2.4048	3.1416

Table A.16. *ω_1 solutions according to the Biot number, for planar, cylindrical or spherical conductors*

A.12. $A_{\omega 1}$ values

This section presents $A_{\omega 1}$ values which occur in the first term of the heat equation solutions (see Chapter 6, section 6.7). These values have been calculated from their expressions, namely:

– for a solid plane: $A_{\omega_1} = \dfrac{2\sin(\omega_1)}{\omega_1 + \sin(\omega_1)\cos(\omega_1)}$;

– for a cylindrical solid: $A_{\omega_1} = \dfrac{2}{\omega_1}\dfrac{J_1(\omega_1)}{J_0^2(\omega_1) + J_1^2(\omega_1)}$;

– for a spherical solid: $A_{\omega_1} = \dfrac{2\left[\sin(\omega_1) - \omega_1\cos(\omega_1)\right]}{\omega_1 - \sin(\omega_1)\cos(\omega_1)}$.

Bi	$A_{\omega 1}$		
	Planar	Cylindrical	Spherical
0.01	1.0017	1.0025	1.0030
0.02	1.0033	1.0050	1.0060
0.04	1.0066	1.0099	1.0120
0.06	1.0098	1.0148	1.0179
0.08	1.0130	1.0197	1.0239
0.1	1.0161	1.0246	1.0298
0.2	1.0311	1.0483	1.0592
0.3	1.0450	1.0712	1.0880
0.4	1.0580	1.0931	1.1164
0.5	1.0701	1.1143	1.1441
0.6	1.0814	1.1345	1.1713
0.7	1.0919	1.1539	1.1978
0.8	1.1016	1.1724	1.2236
0.9	1.1107	1.1902	1.2488
1	1.1191	1.2071	1.2732
2	1.1785	1.3384	1.4793
3	1.2102	1.4191	1.6227
4	1.2287	1.4698	1.7201
5	1.2402	1.5029	1.7870
6	1.2479	1.5253	1.8338
7	1.2532	1.5411	1.8674
8	1.2570	1.5526	1.8921
9	1.2598	1.5611	1.9106
10	1.2620	1.5677	1.9249
20	1.2699	1.5919	1.9781
30	1.2717	1.5973	1.9898
40	1.2723	1.5993	1.9942
50	1.2727	1.6002	1.9962
100	1.2731	1.6015	1.9990
∞	1.2732	1.6021	2.0000

Table A.17. *$A_{\omega l}$ coefficients according to the Biot number for planar, cylindrical or spherical conductors*

A.13. Γ function

The *Gamma* function is generally defined using the Euler integral, itself determined by the integral, from 0 to infinity, of $x^{p-1}e^{-x}$:

$$\Gamma(p)=\int_0^{\infty} e^{-x}x^{p-1}dx.$$

It is a function of parameter p defined by an improper integral. Indeed, the upper bound of the integral is infinite.

It can be shown that this integral exists for any value of p. It is denoted Γ(p).

A.13.1. *Calculation of Γ(1) and Γ(1/2)*

$$\Gamma(1)=\int_0^{\infty} x^0 e^{-x}dx=-\left[e^{-x}\right]_0^{\infty}=1$$

Γ(1/2) can also be calculated by making some variable changes. It can be found that:

$$\Gamma\left(\frac{1}{2}\right)=\sqrt{\pi}.$$

A.13.2. *Γ function properties*

It can also be easily shown that: $\forall p \in R \;:\; \Gamma(p+1)=p\Gamma(p)$.

For any natural number *p*, it can be deduced that:

$$\Gamma(p)=(p-1)(p-2)(p-3)...1.$$

Let: $\forall n \in N \;:\; \Gamma(n+1)=n!$.

The previous two properties allow us to calculate the values of the function for non-zero natural numbers. Table A.18 gives an idea of the evolution of this function. A rapid growth can be noted:

p	1	2	3	4	5	6	7	8	9	10
Γ (p)	1	1	2	6	24	120	720	5,040	40,320	362,880

Table A.18. *Evolution of function Γ for the first natural numbers*

Note that the Gamma function ceases to exist for negative integer p and that Γ(0) is infinite.

In an Excel spreadsheet, it is represented by the GAMMA(x) function.

A.14. Bessel functions

A.14.1. *Definition and properties*

Let us consider the following differential equation:

$$\frac{\partial^2 y}{\partial x^2}+\frac{1}{x}\frac{\partial y}{\partial x}+\left(1-\frac{v^2}{x^2}\right)y=0 \quad \text{with } v \in \Re$$

It is a linear differential equation, of order 2, with non-constant coefficients. It is called a Bessel equation.

The solution to this equation, denoted $J_v(x)$ is called a "Bessel function of the first kind and order v". It is given by:

$$J_v(x)=\left(\frac{x}{2}\right)^v \sum_{j=0}^{\infty}\frac{(-1)^j}{j!\Gamma(v+j+1)}\left(\frac{x}{2}\right)^{2j}$$

thus $\Gamma(x)$ is the Gamma function defined in section A.13 above.

For integer values of v, knowing that $\forall n \in N \;:\; \Gamma(n+1)=n!$, Bessel functions of the first kind are calculated using the following factorials:

$$J_n(x)=\left(\frac{x}{2}\right)^n \sum_{j=0}^{\infty}\frac{(-1)^j}{j!(n+j)!}\left(\frac{x}{2}\right)^{2j}$$

Recurrence relations

$$J_{n+1}(x) = 2\frac{n}{x}J_n(x) - J_{n-1}(x)$$

This relation is important because it allows us to calculate higher order Bessel functions, using the knowledge of $J_0(x)$ and $J_1(x)$.

Integral form

$$J_n(x) = \frac{1}{2\pi}\int_0^{2\pi} \cos(x \sin\psi - n\psi)d\psi$$

For $n = 0$:

$$J_0(x) = \frac{1}{2\pi}\int_0^{2\pi} \cos(x \sin\psi)d\psi = \frac{1}{\pi}\int_0^{\pi} \cos(x \sin\psi)d\psi = \frac{2}{\pi}\int_0^{\frac{\pi}{2}} \cos(x \sin\psi)d\psi$$

A.14.2. *Heat transfer Bessel equations*

Bessel functions can be found in the description of transient heat propagation in cylindrical coordinates. Indeed, this type of problem is determined using the following differential equation:

$$\frac{1}{\alpha}\frac{\partial\theta}{\partial t} = \frac{\partial^2\theta}{\partial r^2} + \frac{1}{r}\frac{\partial\theta}{\partial r}$$

in which we notice that the second number is a particular Bessel equation of the form:

$$\frac{\partial^2 y}{\partial x^2} + \frac{1}{x}\frac{\partial y}{\partial x} = 0.$$

The Bessel functions involved in solving the equation of transient heat transfer in cylindrical coordinates are $J_0(x)$ and $J_1(x)$. Indeed, for a cylindrical conductor under transient conditions, the temperature profile is given by (see Chapter 6, section 7.2):

$$\bar{T}(\bar{r}, Fo) = \sum_{j=1}^{\infty}\left[\frac{2}{\omega_j}\frac{J_1(\omega_j)}{J_0^2(\omega_j) + J_1^2(\omega_j)} e^{-\omega_j^2 Fo} J_0(\omega_j \bar{r})\right]$$

where ω_j are the solutions to the equation $\omega_j \dfrac{J_1(\omega_j)}{J_0(\omega_j)} = \mathrm{Bi}$; Bi being the Biot number corresponding to the situation considered.

A.14.3. *Calculation of $J_0(x)$ and $J_1(x)$ Bessel functions*

A.14.3.1. *Expressions*

Using the factorial form and knowing that 0! = 1, $J_0(x)$ and $J_1(x)$ functions can be calculated:

$$J_0(x) = \sum_{j=0}^{\infty} \frac{(-1)^j}{j!(j)!}\left(\frac{x}{2}\right)^{2i}.$$

Let: $J_0(x) = 1 - \dfrac{x^2}{2^2} + \dfrac{x^4}{2^4 2!^2} - \dfrac{x^6}{2^6 3!^2} + \dfrac{x^8}{2^8 4!^2} - \dfrac{x^{10}}{2^{10} 5!^2} + \ldots$

Similarly, we obtain: $J_1(x)$: $J_1(x) = \sum_{j=0}^{\infty} \dfrac{(-1)^j}{j!(j+1)!}\left(\dfrac{x}{2}\right)^{2j+1}.$

Let: $J_1(x) = \dfrac{x}{2} - \dfrac{x^3}{2^3 3} + \dfrac{x^5}{2^5 2!3!} - \dfrac{x^7}{2^7 3!4!} + \ldots$

A.14.4. *Practical calculations*

In practice, $J_0(x)$ and $J_1(x)$ are calculated using their respective factorial expressions, knowing that:

$$J_0(x) = 1 - \frac{x^2}{2^2} + \frac{x^4}{2^4 2!^2} - \frac{x^6}{2^6 3!^2} + \frac{x^8}{2^8 4!^2} - \frac{x^{10}}{2^{10} 5!^2} + \ldots$$

and:

$$J_0(x) = 1 - \frac{x^2}{2^2} + \frac{x^4}{2^4 2!^2} - \frac{x^6}{2^6 3!^2} + \frac{x^8}{2^8 4!^2} - \frac{x^{10}}{2^{10} 5!^2} + \ldots$$

The different summations are stopped when their contributions appear to be lower than the desired precision.

On the other hand, Bessel functions are computed directly using the function module in an Excel spreadsheet, where the Bessel function of the first kind, of order n and variable x, is referred to by: BESSELJ (x; n).

A.14.5. *Graphic representations of $J_0(x)$ and $J_1(x)$*

The shapes of Bessel functions $J_0(x)$ and $J_1(x)$ are presented in Figures A.2 and A.3, which have been calculated using the BESSELJ(x; n) function in the Excel spreadsheet.

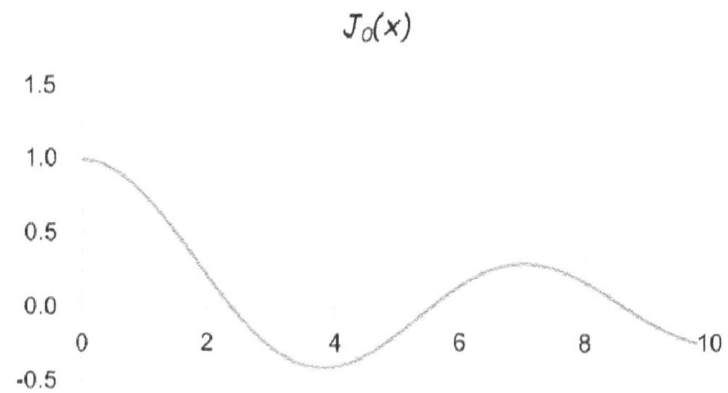

Figure A.2. *Bessel function of order 0*

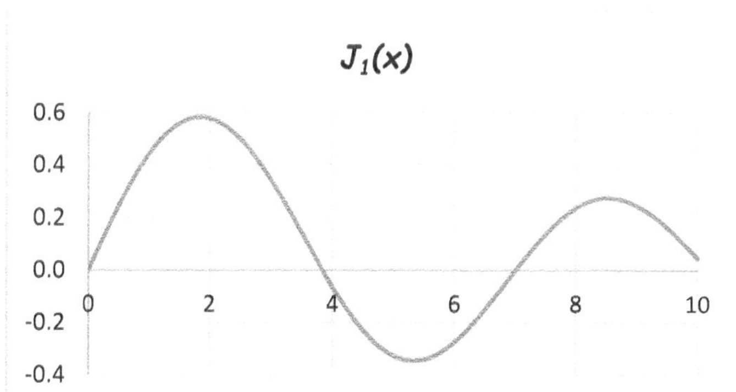

Figure A.3. *Bessel function of order 1*

A.15. Unit conversion tables

The different units are arranged in alphabetical order.

Initial unit	Final unit	Multiply by
Acres	ha	4.074E-01
	m^2	4.074E+03
Acre-feet	m^3	1.233E+03
Ah	Cb	3.600E+03
Angström	m	1.000E-10
Atmosphere	Bar	1.013E+00
	cm H_2O	1.000E+02
	g/m^2	1.033E+03
	lb/in^2 (psi)	1.470E+01
	mmHg	7.600E+02
	N/m^2	1.013E+05
	Pa	1.013E+05
Bags (cement)	kg (cement)	4.264E+01
Bar	Atm	9.869E-01
	cm H_2O	9.869E+01
	lb/in^2 (psi)	1.450E+01
	mm Hg	7.500E+02
	N/m^2	1.000E+05
	Pa	1.000E+05
Barrel (cement)	kg (cement)	1.705E+02
Barrel (oil)	liters	1.590E+02
Btu	cal	2.520E+02
	Joules	1.055E+03
	kWh	2.930E-04
Btu/h	W	2.931E+00
Btu/h-ft-°F	W/m-°C	1.731E+00
Btu/h-ft^2-°F	kcal/h-m^2-°C	4.880E+00
	cal/s-cm^2-°C	1.355E-04
	W/cm^2-°C	5.680E-06
	W/m^2-°C	5.678E+00
Btu/lb	J/kg	2.326E+03
Btu/lb°F	cal/g°C	1.000E+00
Btu/sec	W	1.054E+03

Initial unit	Final unit	Multiply by
Bushels (US dry)	m^3	3.524E-02
cal	Btu	3.968E-03
	kWh	1.163E-06
	J	4.187E+00
	liters-Atm	4.130E-02
Carat	kg	2.000E-04
	lb	4.409E-04
Centipoises (Cp)	g/s-cm or Poise	1.000E-02
	kg/h-m	3.600E+00
	kg/s-m	1.000E-03
	lb/s-ft	6.720E-04
	lb/h-ft	2.420E+00
	N-s/m, that is to say Pa s or Poiseuil	1.000E-03
cm	in	3.937E-01
	ft	3.281E-02
	yd	1.094E-02
cm H_2O	Atm	1.000E-02
	Bar	1.013E-02
	mm Hg	7.600E+00
	N/m^2	1.013E+03
	Pa	1.013E+03
cm^2	ft^2	1.076E-03
Curie	Disintegrations/mn	2.200E+12
Dynes	N	1.000E-05
Ergs	J	1.000E-07
eV	Btu	1.520E-22
Faraday (F)	Coulomb (Cb)	9.650E+04
ft	m	3.048E-01
	in	1.200E+01
ft/mn	cm/s	5.080E-01
ft^2	m^2	9.290E-02
ft^2/h	m^2/s	2.581E-05
ft^3	m^3	2.832E-02
Gallons	liters	3.785E+00
	m^3	3.785E+03
Hectares (ha)	Acres	2.471E+00
	m^2	1.00E+04

Initial unit	Final unit	Multiply by
in	cm	2.540E+00
	m	2.540E-02
in^2	cm^2	6.452E+00
in^3	m^3	1.639E-05
Joule	Btu	9.478E-04
	cal	2.388E-01
	Wh	2.778E-04
	kWh	2.778E-07
kcal	kWh	1.163E-03
kcal/hm °C	W/m °C	1.161E+00
kg	lb	2.205E+00
lb	kg	4.536E-01
lb/ft^2	Atm	4.725E-04
lb/ft^2	kg/m^2	4.882E+00
lb/ft^3	kg/m^3	1.602E+01
lb/in^2 (psi)	Atm	6.805E-02
	kg/cm^2	7.031E-02
	N/m^2	6.895E+03
lb/s-ft	$N\text{-}s/m^2$	1.448E+00
lb_f	N	4.448E+00
lb_f/ft^2	N/m^2	4.788E+01
Lumens	W	1.496E-03
m	ft	3.281E+00
	in	3.937E+01
	yd	1.094E+00
	mile	6.214E-04
	rods	1.988E-01
mile	m	1.609E+03
	rods	3.200E+02
mile/h	m/s	4.470E-01
	km/h	1.609E+00
mm Hg	Atmosphere	1.316E-03
	Bar	1.333E-03
	cm H_2O	1.316E-01
	N/m^2	1.333E+02
	Pa	1.333E+02

Initial unit	Final unit	Multiply by
Pa	N/m²	1.000E+00
	Atmosphere	9.869E-06
	Bar	1.000E-05
	cm H_2O	9.869E-04
	mm Hg	7.501E-03
Poise	Centipoises (Cp)	1.000E+02
	g/s-cm	1.000E+00
	lb/s-ft	6.720E-02
	lb/h-ft	2.419E+02
	kg/h-m	3.600E+02
	N-s/m, that is to say Pa s	1.000E-01
Poiseuil	Poise	1.000E+01
	Centipoises	1.000E+03
	N-s/m²	1.000E+00
Quarts (US, liquid)	liters	9.464E-01
rod (rd)	ft	1.650E+01
Slugs	kg	1.459E+01
Stokes	m^2/s	1.000E-04
TEP	Btu	3.981E+07
	Calories	1.003E+10
	kWh	1.167E+04
	Joules	4.200E+10
Torr	N/m^2	1.333E+02
W	Btu/h	3.413E+00
	Btu/s	1.411E-04
	Uranium nuclei fissions/sec	3.100E+10
W/m °C	cal/h-m-°C	8.612E+02
	kcal/h-m-°C	8.612E-01
	Btu/h-ft-°F	5.777E-01
W/m^2 °C	cal/h-m-°C	2.116E+06
	kcal/h-m-°C	2.116E+03
	Btu/h-ft-°F	1.761E+05

Table A.19. *Unit conversion table*

A.16. Fundamental constants

Symbol	Designation	Value	Unit
g_0	Acceleration of gravity	9.80665	m/s^2
R	Ideal gas constant	8.31439	J/mol-°C
R	Ideal gas constant	1.98719	cal/mol-°C
R	Ideal gas constant	0.0820567	liter-Atm/mol-°C
N	Avogadro's number	6.02283E+23	atoms/mol
h	Planck constant	6.62E-34	J-s
c_0	Speed of light in a vacuum	2.99776E+08	m/s2
F	Faraday constant	96501.2	Coulomb/g
e	Charge of an electron	1.60199E-19	Coulomb/g
eV	Electron-Volt/mol	9.65012E+04	J/mol
k	Boltzmann constant	1.38048E-23	J/ °C
σ	Stefan–Boltzmann constant	5.67E-08	$W/m^2\ °K^4$

Table A.20. *Fundamental constant values*

Bibliography

ALLEN T., Molten salt database, University of Wisconsin, available at: http://allen.neep.wisc.edu/shell/index.php/salts, 2010.

AMBROSEK J., Molten chloride salts for heat transfer in nuclear systems, PhD thesis, University of Wisconsin, Madison, 2011.

ARNOLD J.H., "Vapor viscosities and the Sutherland equation", *Journal of Chemical Physics*, no. 1, p. 170, 1933.

BAHLKE W.H., KAY W.B., "Specific Heats of Petroleum Vapors", *Industrial and Engineering Chemistry*, vol. 21, p. 942, 1929.

BAYS J.S., MCADAMS W.H., "Heat Transfer Coefficients in Falling Film Heater Streamline flow", *Industrial and Engineering Chemistry*, no. 29, p. 1240, 1937.

BENSON S.W., "Critical densities and related properties of liquids", *Journal of Physical and Colloid Chemistry*, no. 52, p. 1060, 1948.

BIRD R.B., STEWART W.E., LIGHTFOOT E.N., *Transport Phenomena*, Wiley, New York, 1975.

BLUNK R., Report 64-28, Department of Engineering, UCLA, Los Angeles, 1964.

BOUCHER D.F., ALVES G.E., "Dimensionless numbers", *Chemical Engineering Progress*, no. 55, pp. 55–64, 1960.

BOUSSINESQ J., *Théorie analytique de la chaleur*, Gauthier-Villars, Paris, 1901.

BRIDGMAN P.W., *Dimensional Analysis*, Yale University Press, New Haven, 1937.

BROKAW R.S., Measurement of viscosity at atmospheric pressure, Technical note D-4496, NASA, 1968.

BROMLEY L.A., WILKE C.R., "Viscosity behavior of gases", *Industrial and Engineering Chemistry*, no. 43, p. 1641, 1951.

BROOKS G., SU G.J., "Heat transfer in agitated kettles", *Chemical Engineering Progress*, no. 55, p. 54, 1959.

BROWN A.I., MARCO S.M., *Introduction to Heat Transfer*, McGraw-Hill, New York, 1958.

BROWN G.G., *Unit Operations*, Wiley, New York, 1950.

CARLSON G.J., OLNEY R.B., "Power and absorption in mixers: correlation with equipment dimensions and fluid properties", *Chemical Engineering Progress*, no. 43, pp. 473–480, 1947.

CARSLAW H.S., *Conduction of Heat in Solids*, Clarendon Press, Wotton-under-Edge, 1959.

CENGEL Y.A., *Heat Transfer: A Practical Approach*, McGraw-Hill, New York, 1998.

CHANG P., WILKE C.R., "Some measurements of diffusion in liquids", *Journal of Physical Chemistry*, no. 59, p. 592, 1955.

CHAPMAN S., DOOTSON F.W., "A note on thermal diffusion", *Philosophy Magazine*, no. 33, p. 248, 1917.

CHEN C.Y., HAWKINS G.A., SOLBERG H.L., "Heat transfer in annuli", *Transactions of the American Society of Mechanical Engineers*, no. 68, pp. 99–106, 1946.

CHEN N.H., "Correlation for latent heat of vaporization", *Journal of Chemical Engineering Data*, vol. 10, no. 2, pp. 207–210, 1965.

CHILTON T.H., DREW T.B., JEBENS R.H., "Heat transfer coefficients in agitated vessels", *Industrial and Engineering Chemistry*, no. 36, p. 510, 1944.

CLENDENIN J.D., Thermal and electrical properties of anthracite and bituminous coals, Technical note 160, Pennsylvania State College, Mineral Industries Experimental Station, 1949.

COWAN D.O., "Overview of Siloxane Polymers", *Advances in Chemistry Symposium Series*, no. 27, p. 224, 1960.

COWLING C., *The Mathematical Theory of Non-uniform Gases*, Cambridge University Press, New York, 1970.

CRAGOE C.S., Thermal properties of petroleum products, Miscellaneous publication no. 97, National Bureau of Standards, 1929.

DUSINBERRE G.M., *Heat Transfer Calculations by Finite Differences*, International Textbook Company, Scranton, 1961.

DWYER O.E., "On the transfer of heat to fluids flowing through pipes, annuli, and parallel plates", *Nuclear Science and Engineering*, no. 17, p. 366, 1963.

EDMISTER W.C., "Applications of thermodynamics to hydrocarbon processing", *Petroleum Refiner*, p. 213, 1948.

EDMISTER W.C., "Phase relations for petroleum fractions", *Petroleum Refiner*, pp. 95–102, 1949.

EDMISTER W.C., "Applied hydrocarbon thermodynamics: part 4: compressibility factors and equations of state", *Petroleum Refiner*, no. 37, pp. 173–179, 1958.

EUCKEN A., "Thermal conductivity of polyatomic gases", *Physik Z.*, no. 14, p. 324, 1913.

EWING C.T., WALKER B.E., GRAND J.A. *et al.*, "Experimental determination of the thermal conductivity of molten salts", *Chemical Engineering Progress Symposium Series*, vol. 53, no. 20, pp. 19–24, 1957.

FOREST PRODUCTS LABORATORY, Technical note no. 248, U.S. Department of Agriculture, Madison, 1952.

FORGACS C., KOLOVSKY L., RABINOWITZ J., "A hydrodynamic theory of desalination by electrodialysis", *Desalination*, vol. 5, no. 3, p. 349, 1968.

FRIEDLER S.E., RICKLES L.A., *Analytical Chemistry*, vol. 27A, p. 37, 1965.

GAMBILL W.R., "You can predict heat capacities", *Chemical Engineering*, p. 243, 1957.

GAMBILL W.R., "How to estimate mixtures viscosities", *Chemical Engineering*, no. 66, p. 151, 1959a.

GAMBILL W.R., "Estimate engineering properties: how P and T change liquid viscosity", *Chemical Engineering*, p. 123, 1959b.

GANDER W., HREBICEK J., *Solving Problems in Scientific Computing Using Maple and MATLAB*, Springer-Verlag, Berlin, 2004.

GEBHART B., *Heat Transfer*, McGraw-Hill, New York, 1961.

GIACALONE A., "The invariance of heat of vaporization with temperature and the theory of corresponding states", *Gazzetta Chimica Italiana*, no. 81, p. 180, 1951.

GILLILAND E.R., "Diffusion coefficients in gaseous systems", *Industrial and Engineering Chemistry*, no. 26, p. 681, 1934.

GIOVANNINI A., BÉDAT B., *Transfert de chaleur*, Cépaduès, Toulouse, 2012.

GOSSE J., *Propriétés de transport des gaz à pression modérée*, Techniques de l'Ingénieur, Saint-Denis, 1991.

GOWEN E., PETERSON S., "Equilibria between aromatic acids and weak base anion exchangers", *Industrial and Engineering Chemistry*, no. 45, p. 2584, 1953.

GOYAL B.K., DORAISWAMY L.K., *Hydrocarbon Processing and Petrol Refining*, vol. 45, p. 200, 1966.

GRELE M.D., GEDEON L., Forced-convection heat-transfer characteristics of molten fLiNaK flowing in an inconel X system, NACA RM E53L18 report, National Advisory Committee for Aeronautics, Cleveland, 1954.

GREW K.E., IBBS T.L., *Thermal Diffusion in Gases*, Cambridge University Press, Cambridge, 1952.

GRIMES W.R., Annual progress report of reactor chemistry division for period ending December 31, 1965, ORNL-3913 report, Oak Ridge National Laboratory, 1966.

GROSSE A.V., *Revue hautes températures et refroidissements*, vol. 3, p. 115, 1966.

HAMMOND B.R., STOKES R.H., "Diffusion in binary mixtures", *Transactions of the Faraday Society*, no. 52, p. 781, 1956.

HANSON D.N., DUFFIN J.H., SOMERVILLE G.F., *Computation of Multistage Separation Processes*, Reinhold, New York, 1962.

HEISLER M.P., "Temperature charts for internal heat generation", *Transactions of the American Institute of Mechanical Engineers*, no. 78, pp. 1187–1192, 1956.

HELFFERICH F., *Ion Exchange*, McGraw-Hill, New York, 1962.

HENLEY E.J., LI N.N., LONG R.B., "Membrane separation processes", *Journal of Industrial and Engineering Chemistry*, no. 57, p. 18, 1965.

HERZOG R., "Correlations of critical constants with parachors", *Industrial and Engineering Chemistry*, no. 36, p. 997, 1944.

HIRSCHFELDER J.O., BIRD R.B., SPOTZ E.L., *Transactions of the American Society of Mechanical Engineers*, no. 71, p. 921, 1949.

HIRSCHFELDER J.O., CURTISS C.F., BIRD R.B., *Molecular Theory of Gases and Liquids*, Wiley, New York, 1964.

HOLLAND C.D., *Multicomponent Distillation*, Prentice-Hall, Englewood Cliffs, 1963.

HSU C.J., "Heat transfer to liquid metals flowing past spheres and elliptical-rod bundles", *International Journal of Heat and Mass Transfer*, vol. 8, no. 2, p. 303, 1965.

HUANG C.J., JOHNSON A.I., "Estimation of heat capacity of organic liquids", *Canadian Journal of Technology*, no. 33, p. 421, 1955.

IGNATEV V.V., MERZLYAKOV A.V., SUBBOTIN V.G. *et al.*, "Experimental investigation of the physical properties of salt melts containing sodium and lithium fluorides and beryllium", *Atomnaya Énergiya*, vol. 101, no. 5, pp. 822–829, 2006.

INCROPERA F., DEWIT D., *Fundamentals of Heat and Mass Transfer*, Wiley, New York, 1996.

INGERSOLL D., FORSBERG C., MACDONALD P., Trade studies on the liquid-salt-cooled very high-temperature reactor, ORNL/TM-2006/140 report, Oak Ridge National Laboratory, 2007.

JAKOB M., *Heat Transfer*, vols 1 and 2, Wiley, New York, 1949 and 1957.

JAKOB M., HAWKINS G.A., *Elements of Heat Transfer*, Wiley, New York, 1957.

JANZ G.J., "Thermodynamic and transport properties for molten salts, correlation equations for critically evaluated density, surface tension, electrical conductance, and viscosity data", *Journal of Physical and Chemical Reference Data*, vol. 17, no. 2, p. 1, 1988.

JANZ G.J., TOMKINS R.P., Physical properties data compilations relevant to energy storage: IV molten salts: data on additional single and multi-component salt systems standard reference data system, NSRDS-NBS 61 report, part 4, National Bureau of Standards, 1981.

JANZ G.J., TOMKINS R.P., ALLEN C.B., "Molten Salts, Properties of Fluorides and Mixtures", *Journal of Physical and Chemical Reference Data*, no. 4, part 2, pp. 871–1178, 1975.

JONES R.C., FURRY W.H., "The separation of isotopes by thermal diffusion", *Reviews of Modern Physics*, no. 18, p. 151, 1946.

KALISH S., DWYER O.E., "Heat transfer to NaK flowing through unbaffled rod bundles", *International Journal of Heat and Mass Transfer*, no. 10, p. 1533, 1967.

KATO H., NISHIWAKI N., HIRATA M., "On the turbulent by free convection from a vertical plate", *International Journal of Heat and Mass Transfer*, no. 11, p. 1117, 1968.

KAY W., "Gases and vapours at high temperatures, density of hydrocarbons", *Industrial and Engineering Chemistry*, no. 28, p. 1014, 1936.

KAYS W.M., CRAWFORD M.E., *Convective Heat and Mass Transfer*, McGraw-Hill, New York, 1993.

KINGREA C.L., ROBBINS L.A., "Thermal conductivity of hydrocarbons", *Hydrocarbon Processing and Petroleum Refining*, vol. 41, no. 5, p. 133, 1962.

KIRILLOV P., *Thermophysical Properties of Materials for Nuclear Engineering, a Tutorial And Collection of Data*, AIEA, Vienna, 2008.

KNUDSEN J.G., KATZ D.L., *Fluid Dynamics and Heat Transfer*, McGraw-Hill, New York, 1958.

LACEY R.E., LANG E.W., "Development and evaluation of the electrosorption process for the desalination of water", *Desalination*, no. 2, p. 387, 1967.

LAGRÉE P.Y., Résolution de l'équation de la chaleur instationnaire, Course Jussieu, Paris, 2010.

LAGRÉE P.Y., Convection Libre, Ensta course, Jussieu, Paris, 2015.

LAGRÉE P.Y., Convection Forcée, Ensta course, Jussieu, Paris, 2016.

LANCET R.T., "The effect of surface roughness on the convection heat transfer coefficient for fully developed turbulent flow in ducts with uniform heat flux", *Trans. ASME, Journal of Heat Transfer*, vol. 81, series C, no. 2, p. 168, 1959.

LANDAU L., LIFSHITZ E., *Mécanique des fluides*, Ellipses, Paris, 1989.

LANDOLT–BÖRNSTEIN., *Eigenschaften der Materie in ihren Aggregatzuständen*, Springer-Verlag, Berlin, 1968.

LANE J.A., RIGGLE J.W., "Membrane separations", *Chemical Engineering Progress*, no. 24, p. 127, 1959.

LANGHAAR H.L., *Dimensional Analysis and Theory of Models*, Wiley, New York, 1951.

LE BAS G., *The Molecular Volumes of Liquid Chemical Compounds*, Longmans, London, 1915.

LEE Y., "Turbulent heat transfer from the core tube in thermal entrance regions of concentric annuli", *International Journal of Heat and Mass Transfer*, vol. 11, no. 3, p. 509, 1968.

LÉONTIEV A., *Théorie des échanges de chaleur et de matière*, Ellipses, Paris, 1985.

LEVY S., FULLER R.A., NIEMI R.O., "Heat transfer to water in thin rectangular channels", *Trans. ASME, Journal of Heat Transfer*, no. 81, p. 129, 1959.

LONDSALE H.K., *Desalination by Reverse Osmosis*, MIT Press, Cambridge, 1966.

LUCAS L.-D., *Densité des principaux métaux et métalloïdes*, Techniques de l'Ingénieur, vol. 1, no. M65, 1984.

LUNDBERG R.E., MCCUEN P.A., "Heat transfer in annular passages", *International Journal of Heat and Mass Transfer*, no. 6, pp. 483–495, 1963.

LYON R.N., *Liquid Metals Handbook*, Navexos, Washington D.C., 1952.

MARRERO M.A., "Gaseous diffusion coefficients", *Journal of Physical and Chemical Reference Data*, no. 1, 1972.

MARTINELLI R.C., BOELTER L.M.K., "Analytical prediction of superimposed free and forced convections in a vertical pipe", *University of California Publications in Engineering*, no. 5, p. 23, 1942.

MASON E.A., KIRKHAM T.A., "Longitudinal mixing in gas-sparged tubular vessels", *Chemical Engineering Progress Series*, vol. 55, no. 24, p. 173, 1959.

MCADAMS W.H., *Heat Transmission*, McGraw-Hill, New York, 1954.

MCCABE W., SMITH J., *Unit Operations of Chemical Engineering*, McGraw-Hill, New York, 1967.

MCRAE W.A., GLASS W., LEITZ F.B. *et al.*, "Recent developments in electrodialysis at elevated temperatures", *Desalination*, vol. 4, no. 2, pp. 236–247, 1968.

MEISSNER H.P., "Critical constants from parachor and molar refraction", *Chemical Engineering Progress*, vol. 45, no. 2, pp. 149–153, 1949.

MERRIEN J.-L., *Analyse avec MATLAB*, Dunod, Paris, 2007.

METAIS B., ECKERT E.R.G., "Forced, mixed, and free convection regimes", *Trans. ASME, Journal of Heat Transfer*, no. 86, p. 295, 1964.

MICHAELS A.S., BIXLER H.J., HODGES R.M. JR, "Kinetics of water and salt transport in cellulose acetate reverse osmosis desalination membranes", *Journal of Colloid Science*, no. 20, p. 1034, 1965.

MICKLEY H.S., SHERWOOD T.K., REED C.E., *Applied Mathematics in Chemical Engineering*, McGraw-Hill, New York, 1957.

MILLER R., "Physical properties of liquid metals", in LYON R.N. (ed.), *Liquid Metals Handbook*, Navexos, Washington D.C., 1954.

MONROE K.P., KENDALL J., "The viscosity of liquids: ii. the viscosity-composition curve for ideal liquid mixtures", *Journal of the American Chemical Society*, no. 39, p. 1787, 1917.

MOODY L.F., "Friction factors for pipe flow", *Transactions of the American Society of Mechanical Engineers*, no. 66, p. 671, 1944.

NATIONAL RESEARCH COUNCIL, *International Critical Tables of Numerical Data, Physics, Chemistry and Technology*, McGraw-Hill, New York, 1929.

NERNST W., "Zur Kinetik der in Lösung befindlichen Körper: Theorie der Diffusion", *Zeitschrift für Physikalische Chemie*, no. 2, p. 613, 1888.

NOKAY R., "Estimate of Petrochemical Properties", *Chemical Engineering Journal*, p. 146, 1959.

OLDSHUE J.Y., GRETTON A.T., "Helical coil heat transfer in mixing vessels", *Chemical Engineering Progress*, no. 50, p. 615, 1954.

OLIVER E.D., *Separation Processes Theory, Design and Evaluation*, Wiley, New York, 1966.

OSHE R., BABELOT J., MAGILL J.E., "Assessment of the melting, boiling and critical point data of the alkali", in OSHE R. (ed.), *Handbook of Thermodynamic and Transport Properties of Alkali Metals*, pp. 329–347, IUPAC Blackwell Scientific Publications, Boston, 1985.

OWEN H.E., *Physical Chemistry of Electrolytes*, Reinhold, New York, 1950.

PARTINGTON J.R., *An Advanced Treatise on Physical Chemistry*, vol. 2, Longmans, London, 1949.

PERRY R.H., CHILTON C.H., *Chemical Engineers Handbook*, McGraw-Hill, New York, 1973.

POPEL P., STANKUS S., MOZGOVOY A. *et al.*, "Physical properties of heavy liquid metal coolants in a wide temperature range", *EPJ Web of Conferences*, vol. 15, no. 01014, 2011.

PORTER A.W., *The Method of Dimensions*, Methuen, London, 1933.

POWELL R.W., "Correlation of metallic thermal and electrical conductivities for both solid and liquid phases", *International Journal of Heat Mass Transfer*, no. 8, p. 1033, 1965.

POWERS J.E., *New Chemical Engineering Separation Techniques*, Interscience, New York, 1962.

PRAUSNITZ J.M., "Volumetric properties of nonpolar gaseous mixtures", *AIChE Journal*, no. 4, pp. 430–435, 1958.

QUARTERONI A., SACCO R., SALERI F., *Méthodes numériques pour le calcul scientifique, programmes en Matlab*, Springer, New York, 2006.

RAYLEIGH L., "The Principles of Similitude", *Nature*, no. 95, p. 66, 1915.

REID R.C., SHERWOOD T.K., *The Properties of Gases and Liquids*, McGraw-Hill, New York, 1966.

REYNOLDS O., "An experimental investigation of the circumstances which determine whether the motion of water in parallel channels shall be direct or sinuous and of the law of resistance in parallel channels", *Philosophical Transactions of the Royal Society*, vol. 174, pp. 935–982, 1883.

RICKLES R.N., "Membrane Permeation", *Journal of Industrial and Engineering Chemistry*, no. 58, p. 19, 1966.

ROCHE M., "L'utilisation d'un mélange de sels fondus pour le stockage de la chaleur", *Revue de Physique Appliquée*, no. 15, pp. 895–902, 1980.

ROHSENOW W.M., HARTNETT J.P., *Handbook of Heat Transfer*, McGraw-Hill, New York, 1973.

SACADURA J.-F., *Initiation aux transferts thermiques*, Lavoisier, Paris, 1993.

SCHLICHTING H., *Boundary Layer Theory*, McGraw-Hill, New York, 1987.

SEBAN R.A., "Heat transfer to a fluid flowing turbulently between parallel walls and asymmetric wall temperatures", *Transactions of the American Institute of Mechanical Engineers*, no. 72, p. 789, 1950.

SEBAN R.A., SHIMAKAZI T.T., "Heat transfer to a fluid flowing turbulently in a smooth pipe with walls at constant temperature", *Transactions of the American Society of Mechanical Engineers*, no. 73, p. 803, 1951.

SEIDER E.N., TATE G.E., "Heat transfer and pressure drop of liquids in tubes", *Journal of Industrial and Engineering Chemistry*, no. 28, p. 1429, 1936.

SHAFFER L.H., MINTZ M.S., "Electrodialysis", in SPIEGLER K. (ed.), *Principles of Desalination*, Academic Press, New York, 1966.

SHERWOOD T.K., HOLLOWAY F.A.L., "Performance of packed towers: liquid film data for several packings", *Transactions of the American Institute of Chemical Engineers*, vol. 36, p. 39, 1940.

SHERWOOD T.K., PIGFORD R.L., *Absorption and Extraction*, McGraw-Hill, New York, 1952.

SHERWOOD T.K., REED C.E., *Applied Mathematics in Chemical Engineering*, McGraw-Hill, New York, 1957.

SHPIL'RAIN E., "Density and thermal expansion of alkali metals", in OHSE R. (ed.), *Handbook of Thermodynamic and Transport Properties of Alkali Metals*, Blackwell Scientific Publications, Boston, 1985.

SIENICKI J., MOISSEYTSEV A., "STAR lead-cooled, small modular fast reactor for development at remote sites", *Proceedings of International Conference on Advances in Nuclear Power Plants*, p. 5426, Seoul, 2005.

SKUPINSKI E.S., TORTEL J., VAUTREY L., "Détermination des coefficients de convection d'un alliage sodium-potassium dans un tube circulaire", *International Journal of Heat Transfer*, no. 8, p. 937, 1965.

SMIRNOV M.V., KHOKLOV V.A., FILATOV E.S., "Thermal conductivity of molten alkali halides and their mixtures", *Electrochimica Acta*, vol. 32, no. 7, p. 1019–1026, 1987.

SMITH B.D., *Design of Equilibrium Stage Processes*, McGraw-Hill, New York, 1963.

SMITH W.T. JR, GREENBAUM S., RUTLEDGE G.P., "Correlation of critical temperatures with thermal expansion coefficients of organic liquids", *Journal of Physical Chemistry*, no. 58, pp. 443–447, 1954.

SMITHELLS C., *Metals Reference Book*, Interscience Publications, New York, 1955.

SOBOLEV V., *Database of Thermophysical Properties of Liquid Metal Coolants for GEN-IV*, NC, Mol, 2011.

SORET C., "État d'équilibre des dissolutions dont deux parties sont portées à des températures différentes", *Archives des Sciences Physiques et Naturelles*, no. 2, p. 48, 1879.

SOUTIF M., *Vibration, propagation et diffusion*, Dunod, Paris, 1991.

SPIEGLER K.S., *Salt Water Purification*, Wiley, New York, 1962.

STOKES R.A., *Electrolyte Solutions*, Butterworth, London, 1955.

STOUGHTON R.W., LIETZKE R.H., "Calculation of some thermodynamic properties of sea salt solutions at elevated temperatures from data on NaCI solutions", *Journal of Chemical and Engineering Data*, no. 10, p. 254, 1965.

SÜLI E., MAYERS D.F., *An Introduction to Numerical Analysis*, Cambridge University Press, Cambridge, 2003.

TAYLOR G.I., "The viscosity of a fluid containing small drops of another fluid", *Royal Society of London, Series A*, no. 138, No. 834, p. 41, 1932.

THODOS G., GATES D.S., "The critical constants of the elements", *AIChE Journal*, vol. 6, no. 1, pp. 50–54, 1960.

THOMAS L.O., "The dependence of the viscosities of liquids on reduced temperature, and a relation between viscosity, density and chemical constitution", *Journal of the Chemical Society*, no. 2, pp. 573–579, 1946.

TREYBAL R.E., *Mass Transfer Operations*, McGraw-Hill, New York, 1968.

TSUNODA Y., KATO M., "Compact apparatus for sea water desalination by electrodialysis using ion-exchange membranes", *Desalination*, no. 3, p. 66, 1967.

TUWINER S.B., *Diffusion and Membrane Technology*, Reinhold, New York, 1962.

UHL V.W., "Mechanically aided heat transfer to viscous materials", *Chemical Engineering Progress Symposium Series*, no. 17, p. 51, 1959.

VAN WINKLE M., *Distillation*, McGraw-Hill, New York, 1967.

VORMEN D., "Cellulose Membranes", *Journal of Industrial and Engineering Chemistry*, no. 54, p. 20, 1962.

VRIESEMA I.B., Aspects of molten fluorides as heat transfer agents for power generation, WTHD report, no. 112, Delft University of Technology, South Holland, 1979.

WASHINGTON L., MARKS W.M., "Rectangular air passages", *Journal of Industrial and Engineering Chemistry*, no. 29, p. 337, 1937.

WATSON K.M., "Thermodynamics of the liquid state", *Journal of Industrial and Engineering Chemistry*, no. 35, p. 398, 1943.

WEBER H.F., "Investigations into the thermal conductivity of liquids", *Wiedemanns Annalen der Physik und Chimie*, no. 10, p. 103, 1880.

WILKE C.R., "A viscosity equation for gas mixtures", *The Journal of Chemical Physics*, vol. 18, no. 4, p. 517, 1950.

WILKE C.R., CHANG P., "Correlation of diffusion coefficients in dilute solutions", *American Institute of Chemical Engineers Journal*, no. 1, p. 264, 1955.

WILLIAMS D.F., "The influence of Lewis acid/base chemistry on the removal of gallium by volatility from weapons grade plutonium dissolved in molten chlorides", *Nuclear Technology*, no. 136, p. 367, 2001.

WILLIAMS D.F., TOTH L.M., CLARNO K.T., Assessment of candidate molten salt coolants for the advanced high-temperature reactor, ORNL/TM-2006/12 report, Oak Ridge National Laboratory, 2006.

WILSON J.R., *Demineralization by Electrodialysis*, Butterworth, London, 1960.

ZIPPERER L., "Calculation of the viscosity of technical gas mixtures from the viscosity of individual gases", *Gas und Wasserfach*, vol. 79, no. 49, p. 69, 1936.

Index

A

B, C

E

F

G, H, I

L

M

N, O

P

R

Other titles from

iSTE

in

Energy

2018

BENALLOU Abdelhanine
Energy and Mass Transfers: Balance Sheet Approach and Basic Concepts (Energy Engineering Set – Volume 1)

RUFFINE Livio, BROSETA Daniel, DESMEDT Arnaud
Gas Hydrates 2: Geoscience Issues and Potential Industrial Applications

VIALLET Virginie, FLEUTOT Benoit
Inorganic Massive Batteries
(Energy Storage – Batteries, Supercapacitors Set – Volume 4)

2017

BROSETA Daniel, RUFFINE Livio, DESMEDT Arnaud
Gas Hydrates 1: Fundamentals, Characterization and Modeling

LA SCALA Massimo
From Smart Grids to Smart Cities: New Challenges in Optimizing Energy Grids

MOLINA Géraldine, MUSY Marjorie, LEFRANC Margot
Building Professionals Facing the Energy Efficiency Challenge

SIMON Patrice, BROUSSE Thierry, FAVIER Frédéric
Supercapacitors Based on Carbon or Pseudocapacitive Materials
(Energy Storage – Batteries, Supercapacitors Set – Volume 3)

2016

ALLARD Bruno
Power Systems-on-Chip: Practical Aspects of Design

ANDRE Michel, SAMARAS Zissis
Energy and Environment

DUFOUR Anthony
Thermochemical Conversion of Biomass for the Production of Energy and Chemicals

2015

CROGUENNEC Laurence, MONCONDUIT Laure, DEDRYVÈRE Rémi
Electrodes for Li-ion Batteries
(Energy Storage – Batteries, Supercapacitors Set – Volume 2)

LEPRINCE-WANG Yamin
Piezoelectric ZnO Nanostructure for Energy Harvesting
(Nanotechnologies for Energy Recovery Set – Volume 1)

ROBYNS Benoît, FRANÇOIS Bruno, DELILLE Gauthier, SAUDEMONT Christophe
Energy Storage in Electric Power Grids

ROSSI Carole
Al-based Energetic Nanomaterials
(Nanotechnologies for Energy Recovery Set – Volume 2)

TARASCON Jean-Marie, SIMON Patrice
Electrochemical Energy Storage
(Energy Storage – Batteries, Supercapacitors Set – Volume 1)

2013

LALOUI Lyesse and DI DONNA Alice
Energy Geostructures: Innovation in Underground Engineering

2012

BECKERS Benoit
Solar Energy at Urban Scale

ROBYNS Benoît, DAVIGNY Arnaud, FRANÇOIS Bruno, HENNETON Antoine, SPROOTEN Jonathan
Electricity Production from Renewable Energies

2011

GAO Fei, BLUNIER Benjamin, MIRAOUI Abdellatif
Proton Exchange Membrane Fuel Cell Modeling

MULTON Bernard
Marine Renewable Energy Handbook

2010

BRUNET Yves
Energy Storage

2009

SABONNADIERE Jean-Claude
Low Emission Power Generation Technologies and Energy Management

SABONNADIERE Jean-Claude
Renewable Energy Technologies

Printed and bound by CPI Group (UK) Ltd, Croydon, CR0 4YY

15/09/2026

14962345-0005